GEOLOGY OF THE GREAT LAKES

GEOLOGY OF THE GREAT LAKES

BY JACK L. HOUGH

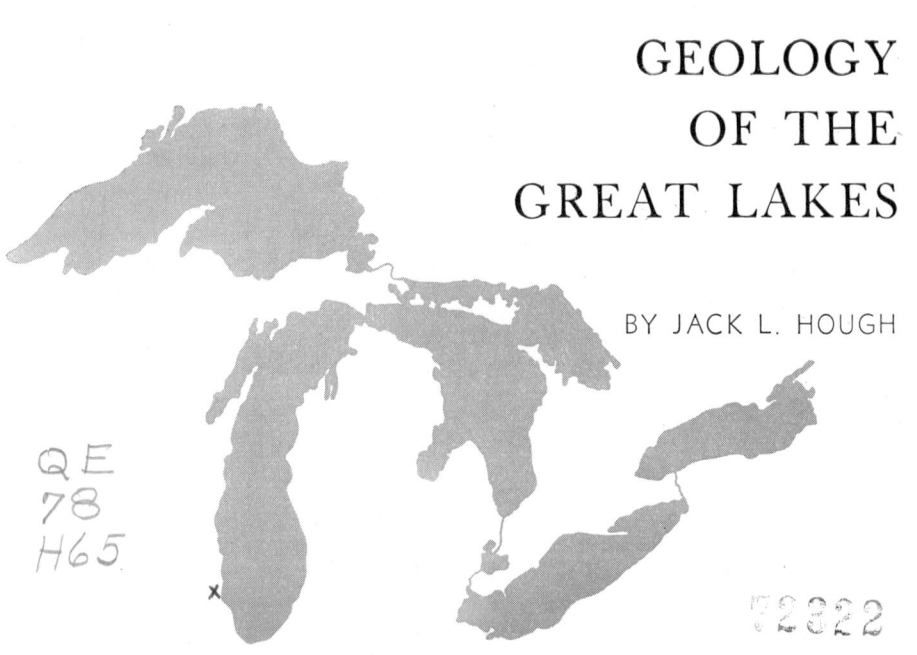

UNIVERSITY OF ILLINOIS PRESS, URBANA, 1958

© 1958 by the Board of Trustees of the University of Illinois.
Manufactured in the United States of America.
Library of Congress Catalog Card No. 58-6995.

INTRODUCTION

The Great Lakes of North America are a group composed of Lakes Superior, Michigan, Huron, Erie, and Ontario, all of which drain through the St. Lawrence River to the Atlantic Ocean (Fig. 1). They constitute a waterway which extends nearly halfway across the continent. Four of the lakes lie between the United States and Canada, while only one, Lake Michigan, lies wholly within the United States.

Lake Superior, at the head of the chain of lakes, stands 602 feet above sea level but its deepest point, at a depth of 1302 feet, lies 700 feet below sea level (Fig. 2). Three of the other lakes have depths extending far below sea level. Old shorelines lying above the present lakes are evidence of former higher levels, and they show that the surfaces of two of the lakes have been as much as 220 feet higher. Other evidence, including that revealed by studies of the lake-bottom deposits, shows that most of the lakes have had extremely low-level stages. Lake Superior has been at least 300 feet lower, Lake Michigan has been 350 feet lower, and Lake Huron has been approximately 400 feet lower. Lake Ontario, which now stands 246 feet above sea level, was at one time a brackish embayment of the sea.

The Great Lakes have attained their present form and connections as a result of a complicated series of events. Many of the basic attributes of the lakes, such as their locations, depths, and shapes, were indirectly influenced by events which occurred as much as a half-billion years ago, when the bedrock foundation

vi INTRODUCTION

Fig. 1. Geographic map of the Great Lakes region, showing drainage areas. (Reproduced from U.S. Lake Survey Chart 0.)

of the region was laid down. The bedrock terrain, with various degrees of resistance to erosion, was sculptured by weathering and stream erosion over a period of some 180 million years. During the last million years, continental ice sheets invaded the region several times and scoured and molded the landscape.

The earliest known predecessors of the modern Great Lakes are relatively recent arrivals on the scene. They came into existence probably not more than 20,000 years ago, when the wasting margin of the last continental ice sheet retreated into the lake basins. The earliest lakes were narrow, ice-margin bodies of water which expanded as the ice melted and which were compressed in area at various times when the ice sheets temporarily readvanced. The lake waters at first spilled southward over the divides of the various lake basins. During the northward retreat of the border of the continental ice sheet, the lake waters found new, lower outlets in the north, and the lakes periodically drained down to lower levels—only to be returned to higher levels when uplift

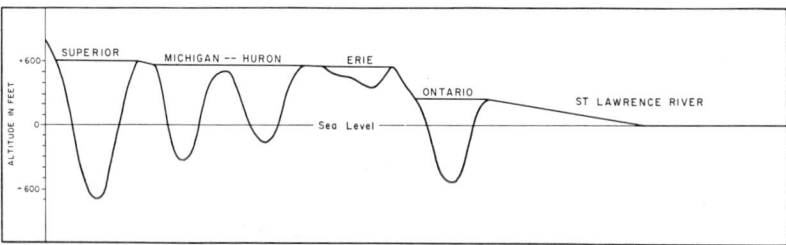

Fig. 2. Diagrammatic profile of the Great Lakes and St. Lawrence seaway showing relative depths of the lakes. (Lake Superior actually discharges directly into Lake Huron.)

of the land raised the northern outlets higher than the old southern outlets. This process of uplift continues today.

Man appeared on the scene sometime during the later part of the lake history, and some of his camp sites and tools have been found intimately associated with old beach deposits and other features of the lakes.

Part 1 of the book contains material which forms a frame of reference for the history of the lakes. It includes descriptions of the present lakes and of the processes operating in them, summaries of the events in the region which led up to the formation of the earliest lakes, and a review of the methods which are used to date the events of lake history. Because much of this material is familiar to the professional geologist, it is summarized here in a semipopular style, mainly for the non-geologist.

Part 2, the history of the lake stages, has been prepared primarily for the professional geologist and for specialists in related fields such as archeology. It contains some original material which has not been published elsewhere, and it includes a detailed review and re-evaluation of the extensive literature on the Great Lakes. The lay reader will have no difficulty in understanding this presentation. The Great Lakes history is summarized in a correlation chart with an absolute time scale in years, in Table 22 of Chapter 16, and the major events are shown on a series of lake stage maps, Figs. 53–75, which follow Chapter 16.

The history of the lakes is based on the observations of dozens of geologists, made over a period of more than a hundred years. Because the story is complicated and the record is fragmentary, some of the happenings can only be inferred. From time to time, new evidence has accumulated which has required a revision of the history. The last full treatment of the Great Lakes history was published by Leverett and Taylor in 1915, in U.S. Geological Survey Monograph 53, "The Pleistocene of Indiana and Michigan and the History of the Great Lakes." That monumental work contains a wealth of factual data, but it also contains many interpretations which have been proven erroneous in the light of newer investigations. It is the purpose of the present book to bring the geologic history of the Great Lakes up to date, and thus to facilitate further research on the subject. There is still much to be learned, and the history most certainly will be subject to revision in the future.

The author's own work on the Great Lakes has been supported and aided by several agencies and individuals. His early bottom-sampling work on Lake Michigan in 1932 and 1933 was carried out from the sailing vessel *Gubben Noak*, owned by Dr. Anton J. Carlson of the University of Chicago. Some of the apparatus used in this work was loaned by Dr. Parker D. Trask of the U.S. Geological Survey and by Dr. John VanOosten of the U.S. Bureau of Fisheries, and other equipment was purchased with funds granted by the American Academy of Arts and Sciences. Water chacteristics were studied from various U.S. Navy submarines which were launched and tested on Lake Michigan during World War II. The Woods Hole Oceanographic Institution provided the services of Mr. A. R. Miller, and apparatus, for bottom sampling done in Lake Huron from the U.S. Lake Survey vessel *Williams* in the summer of 1947. An extensive survey of Lake Michigan and of various areas on shore in Michigan, Wisconsin, and Ontario, was supported by the U.S. Navy, Office of Naval Research, under contract No. N 6 ori-07133, Project NR-018-122, administered by the University of Illinois. The U.S. Coast Guard provided the buoy-tender *Woodbine* for field work in the summer of 1950, and the U.S. Fish and Wildlife Service provided their research vessel *Cisco* for field work in the summer of 1951. Several graduate students in the Department of Geology, University of Illinois, have shared parts of the author's work, and the contributions of the following merit especial mention: Maxwell Silverman, Donald Baldwin, Raymond F. McAllister, and Donald Snodgrass. Grants from the University of Illinois Research Board supported several field investigations during the years 1953–57.

The Great Lakes Research Institute of the University of Michigan has sponsored the author's work in the summer of 1957 and 1958. The detailed results of this recent work are not reported in the present book, but the preliminary impressions gained, and the discussion of Great Lakes problems with Dr. David C. Chandler and Dr. John C. Ayers of the institute have been very helpful in the later stages of preparation of the manuscript. The entire manuscript has been read by Dr. James H. Zumberge, Department of Geology, University of Michigan, who made several valuable suggestions.

Adequate acknowledgment of the sources of information and interpretations contained in this book is impracticable, if not impossible. The Leverett and Taylor monograph of 1915 lists 421 references. In general, in the present book, the Leverett and Taylor reference will be used for any information quoted or recorded by those writers. For information published since 1915, reference is made in the text to the author's name and date of publication, and the complete reference will be found in the bibliography at the end of the book.

CONTENTS

PART I—THE GREAT LAKES REGION: THE PRESENT LAKES AND PRE-LAKE HISTORY. METHODS OF DATING EVENTS OF LAKE HISTORY.

1. THE MODERN GREAT LAKES	3
General Description	3
The Depths of the Lakes	7
Topography of the Lake Bottoms	13
Geologic Setting	13
General	13
Lake Superior	16
Lake Michigan	20
Lake Huron	25
Lake Erie	27
Lake Ontario	30
2. THE LAKE WATER	31
Wave Action	31
Tides	35
Surface Currents	35
Seiches and Other Short-Period Fluctuations in Lake Level	44
Seasonal and Longer-Period Fluctuations in Lake Level	46
Ice in the Lakes	49
Thermal Stratification and Deep Circulation	50
Precipitation and Evaporation	58
Chemistry of the Lake Water	59
3. SEDIMENTS OF THE GREAT LAKES	65
Gravel	66
Sand	67

xi

xii CONTENTS

Silt 69
Clay 70

4. PREGLACIAL HISTORY OF THE REGION 76
History of the Bedrock Formations 76
   Introduction 76
   The Geologic Time Scale 76
   The Precambrian Era 77
   The Paleozoic Era 81
   Post-Paleozoic Time 85
Preglacial Erosion and Drainage Systems 86
   Erosion Surfaces 86
   Drainage Systems 87

5. GLACIAL HISTORY OF THE REGION 90
Résumé of Pleistocene Glaciation 90
   Introduction 90
   Nebraskan Age 91
   Kansan Age 93
   Illinoian Age 93
   Wisconsin Age 93
   Substages of the Wisconsin 94
Centers of Glaciation 109
Glacial Scour of the Lake Basins 111
Interglacial Lakes 113

6. DATING THE EVENTS OF LAKE HISTORY 116
Relative Ages 116
Estimates and Measurements of Age 117
   Rate of Soil Formation on Glacial Deposits . . . 117
   Age of Niagara Gorge 119
   Varve Correlation 122
   Uranium-Ionium-Radium Method of Age Determination . 124
   Radiocarbon Method of Age Determination . . . 127

PART II—HISTORY OF LAKE STAGES

7. INTRODUCTION TO HISTORY OF THE LAKES 133
Evidences of Former Lake Stages 133
Warping of the Shoreline Features 135
Interpretation of the Evidence 137
Outline of Discussion of the Lake History 138

8. EARLY LAKE STAGES IN THE ERIE AND HURON BASINS 139
Introduction 139
Highest Lake Maumee 140
Lowest Lake Maumee 144
Middle Lake Maumee 145
Lake Saginaw 145
Lake Arkona 146
Low-Water Stage During the Cary–Port Huron Interval . . . 147
Second Lake Arkona 148
Lake Whittlesey and the Port Huron Glacial Substage . . . 148
Highest and Middle Warren Stages 149
Low-Water Stages of the Two Creeks Interval 150
Lowest Lake Warren and the Valders Glacial Substage . . . 151
The Grassmere, Lundy, and Early Algonquin Stages . . . 152
Further Development of the Early Algonquin Stage and Early Lake Erie 161

9. EARLY LAKE STAGES IN THE MICHIGAN BASIN 163
Introduction 163
Early Lake Chicago 164
The Tinley Glacial Advance 165
The First Glenwood Stage 165
Intra-Glenwood Low-Water Stage 168
The Second Glenwood Stage 169
The Two Creeks Low-Water Stage 170
The Third Glenwood Stage 171
The Bowmanville Low-Water Stage 172
Termination of the Glenwood Stage 174
The Calumet Stage 181
The Toleston Stage 182

10. EARLY LAKE STAGES IN THE SUPERIOR BASIN 184
Introduction 184
Cary–Port Huron–Interval Inferred Lake 185
Lake Keweenaw of the Port Huron–Valders Interval . . . 187
Post-Valders Ice-Margin Lakes 188
   Glacial Lake Nemadji 188
   Glacial Lake Brule 188
   Glacial Lake Ashland 189
   Glacial Lake Ontonagon 189
Lake Duluth 189
Transition from Lake Duluth to a Post-Algonquin Stage . . 192

CONTENTS

11. LAKE STAGES IN THE ONTARIO BASIN 194
Stages Described by Fairchild 194
Late Cary Events 195
A Possible Low-Water Stage During the Cary–Port Huron Interval 195
Cary–Port Huron–Interval Events 195
Port Huron–Substage Events 199
Two Creeks–Interval Events 199
Valders-Substage Events 199
Transition from Lowest Lake Warren to Lake Iroquois 200
Early Lake Iroquois 201
Lake Frontenac and the Termination of Lake Iroquois 202
St. Lawrence Marine Embayment 203
Early Lake Ontario 204
Correlation of Events in the Ontario Basin with Algonquin Stages
  in the Huron Basin 205

12. LAKE ALGONQUIN 207
Introduction 207
Early Lake Algonquin 209
The Kirkfield Stage 211
The Kirkfield–Main Algonquin Transition 212
The Main Algonquin Stage 214
  Evidence for the Main Algonquin Stage 214
  The Maximum Extent of the Main Algonquin Stage . . . 216
  Warping of the Algonquin Beach 223

13. POST-ALGONQUIN LOW STAGES 225
Close of the Main Algonquin Stage 225
The "Upper Group" of Beaches 228
The Wyebridge, Penetang, Cedar Point, and Payette Stages of the
  Huron Basin 229
The Sheguiandah and Korah Stages of the Huron Basin . . . 234
The Korah–Fort Brady Relationships 236
The Chippewa and Stanley Low-Water Stages 236
  Introduction 236
  North Bay Outlet 237
  The Submerged Mackinac River Valley 237
  Lake Chippewa in the Michigan Basin 238
  Lake Stanley in the Huron Basin 242
  Outlets 242
Post-Algonquin, Pre-Nipissing Stages of the Superior Basin . . 243

14. THE NIPISSING GREAT LAKES	248
Introduction	248
The Stanley-Nipissing Transition	249
Characteristics of the Nipissing Shoreline	249
North-South Correlations of the Nipissing Beach	251
Radiocarbon Dates Affecting Correlation of Nipissing Features	253
Extent	254
Outlets	259
Cause of the Apparently Static Level of the Nipissing Stage	260
15. THE TRANSITION FROM THE NIPISSING STAGE TO THE PRESENT GREAT LAKES	263
Close of the Nipissing Stage	263
The Algoma Stage	265
The Transition from the Algoma Stage to the Present Great Lakes	267
16. RADIOCARBON CHRONOLOGY OF GREAT LAKES HISTORY	269
Dates of the Two Creeks Interval and the Valders Glacial Substage	269
Dates of Lake Stages	271
The St. Lawrence Valley	278
The North Bay Outlet and the Cochrane Problem	279
Lake Barlow-Ojibway	279
Lake Agassiz	281
LAKE STAGE MAPS	284
BIBLIOGRAPHY	297
INDEX	311

List of Illustrations

1. Geographic map of the Great Lakes region, showing drainage areas vi
2. Diagrammatic profile of the Great Lakes and St. Lawrence seaway vii
3. Bottom topography of Lake Superior 8
4. Bottom topography of Lake Michigan 9
5. Bottom topography of Lake Huron 10
6. Bottom topography of Lake Erie 11
7. Bottom topography of Lake Ontario 12
8. Geologic map of the Great Lakes region 14
9. Geologic cross section of Lake Superior 17

10. Geologic cross section of the Michigan basin, showing Lakes Michigan and Huron 17
11. Geologic map of the western end of Lake Erie 28
12. Geologic cross section of the western end of Lake Erie . . 28
13. Geologic cross section of the Ontario and Erie basins . . . 29
14. A narrow sand beach at the foot of a cliff cut in clay; Ohio shore of Lake Erie 33
15. Surface currents of Lake Superior 37
16. Surface currents of Lake Michigan 38
17. Surface currents of Lake Huron 39
18. Surface currents of Lake Erie 40
19. Surface currents of Lake Ontario 41
20A. Chart of monthly mean water levels in the Great Lakes: 1860 to 1909 47
20B. Chart of monthly mean water levels in the Great Lakes: 1910 to 1958 48
21. Ice ridges and pack ice on the Indiana shore of Lake Michigan 49
22. The annual temperature cycle in lakes of the temperate zone 51
23. Terminology of the lake water temperature profile 53
24. A gravel beach on the eastern shore of Lake Superior . . . 66
25. A wide sand beach on the eastern shore of Lake Michigan . . 67
26. Log of core sample from the deepest point in Lake Michigan . 72
27. Diagrammatic summary of the classification of Precambrian rocks 79
28. Map showing regional structures which controlled the distribution of Paleozoic seas 83
29. Preglacial drainage systems 87
30. Maximum extent of the four principal glacial stages in the central and northeastern United States 92
31. The Two Creeks forest bed and associated deposits near Manitowoc, Wisconsin 102
32. A dated climatic record in a southeastern Pacific Ocean core sample, compared with the North American Pleistocene chronology 126
33. Climatic record of the last 70,000 years in three ocean-bottom cores 128
34. Hinge lines of various uplifts in the Great Lakes region . . 136
35. Lake stages of the Erie basin 141
36. Lake stages of the Huron basin 143
37. Lake stages of the Michigan basin 166
38. The Jordan valley in the Grand Traverse Bay region, Michigan 180

LIST OF ILLUSTRATIONS xvii

39. Lake stages of the Superior basin 186
40. Lake stages of the Ontario basin 198
41. The northern shore of the main stage of Lake Algonquin in the Michigan basin 220
42. Algonquin beach isobases 224
43. Relationships of the sub-Algonquin beaches, as interpreted by Leverett and Taylor 226
44. Relationships of the sub-Algonquin beaches, as interpreted by Stanley and by Hough 227
45. Profile of the eastern slope of Manitoulin Island, Ontario . . 233
46. Profile of the slope west of Sault Ste. Marie, Ontario . . . 235
47. The submerged Mackinac River channel 238
48. Unconformity in Lake Michigan bottom sediments, indicating the Chippewa low-water stage 239
49. The Lake Chippewa stage of the Michigan basin 241
50. The Nipissing beach on the west side of Torch Lake, Antrim County, Michigan 250
51. Nipissing beach isobases 256
52. Nipissing-stage variations in level of the lake surface and of the land 261
53. The Great Lakes region immediately prior to the first known lakes; the late Valparaiso and Fort Wayne phases of the Cary glacial substage 284
54. Early Lake Chicago and Highest Lake Maumee; post-Valparaiso glacial retreat 284
55. Highest Lake Maumee; Tinley and Defiance glacial advances 285
56. Glenwood I stage of Lake Chicago and Lowest Lake Maumee of the Erie basin; post-Tinley and post-Defiance glacial retreat 285
57. Glenwood I stage of Lake Chicago and Middle Lake Maumee of the Erie basin; Lake Border glacial advance 286
58. Glenwood I stage of Lake Chicago and Highest Lake Arkona of the Erie and Huron basins; early post–Lake Border glacial retreat 286
59. Low-water stages; Cary–Port Huron interval 287
60. Glenwood II stage of Lake Chicago and Lake Whittlesey of the Erie basin; Port Huron (Mankato) glacial advance 287
61. Glenwood II stage of Lake Chicago and Highest Warren stage of the Erie and Huron basins; early post–Port Huron glacial retreat 288
62. Two Creeks low-water stages and Lake Keweenaw of the Superior basin; Two Creeks interval 288

xviii   LIST OF ILLUSTRATIONS

63. Glenwood III stage of Lake Chicago and Lake Wayne of the Huron, Erie, and Ontario basins; glacial advance, before the Valders maximum 289
64. Glenwood III stage of Lake Chicago and Lowest Lake Warren of the Huron, Erie, and Ontario basins; Valders maximum glacial advance 289
65A. Glenwood III stage of Lake Chicago, Lake Grassmere of the Erie and Huron basins, and early Lake Duluth of the Superior basin; early post-Valders glacial retreat 290
65B. Alternative interpretation; Toleston stage of Lake Chicago, Lake Grassmere of the Erie and Huron basins, and early Lake Duluth of the Superior basin; early post-Valders glacial retreat 290
66A. Calumet stage of Lake Chicago, Lundy stage of the Erie and Huron basins, and early Lake Duluth of the Superior basin; continued post-Valders glacial retreat 291
66B. Alternative interpretation; Toleston stage of Lake Chicago, Lundy stage of the Erie and Huron basins, and early Lake Duluth of the Superior basin; continued post-Valders glacial retreat 291
67A. Toleston stage of Lake Chicago, Early Lake Algonquin of the Erie and Huron basins, and Lake Duluth of the Superior basin; further post-Valders glacial retreat 292
67B. Alternative interpretation; Toleston stage of Lake Chicago, Early Lake Algonquin of the Huron basin, early Lake Erie, Lake Iroquois of the Ontario basin, and Lake Duluth of the Superior basin 292
68. Early Lake Algonquin of the Michigan and Huron basins, early Lake Erie, Lake Iroquois, and Lake Duluth 293
69. Kirkfield stage of the Michigan and Huron basins, early Lake Erie, Lake Iroquois, and Lake Duluth 293
70. Main Algonquin stage of the Michigan and Huron basins, early Lake Erie, Lake Iroquois, and Lake Duluth 294
71. Post-Algonquin "upper group" lake stages of the Michigan and Huron basins, first marine embayment of the Ontario basin, and a sub-Duluth stage of the Superior basin . . . 294
72. Lake Payette, one of several post-Algonquin low-water stages of the Michigan and Huron basins 295
73. Lakes Chippewa and Stanley, the lowest stages of the Michigan and Huron basins 295
74. Lake Nipissing, confluent in the three upper Great Lakes basins 296
75. The modern Great Lakes 296

PART ONE

THE GREAT LAKES REGION

The Present Lakes
And Pre-Lake History

Methods Of Dating Events
Of Lake History

chapter one

THE MODERN
GREAT LAKES

GENERAL DESCRIPTION

The five Great Lakes are among the first fourteen of the world's big lakes, listed in the order of surface area (see Table 1). Lake Superior is the largest body of fresh water on earth, and among inland waters it is exceeded in size only by the salty Caspian Sea in Asia. The Great Lakes region is shown in Fig. 1, and the principal dimensions of the lakes are listed in Table 2.

At the head of the system, Lake Superior stands at an average altitude of 602 feet above sea level. Its overflow of 73,700 cubic feet per second goes to Lake Huron via the St. Marys River, a 63-mile-long channel. Almost all of the 22-foot drop in water-surface altitude occurs in a distance of one mile, at the St. Marys Falls, which lies between the twin cities of Sault Ste. Marie, Michigan, and Sault Ste. Marie, Ontario. Canals and locks constructed on both the United States and Canadian sides of the falls provide for navigation between Lakes Superior and Huron. Navigation is closed for four months by ice, but during the remainder of the year the St. Marys waterway (the "Soo") carries a greater tonnage of shipping than the Panama Canal and the Suez Canal combined.

Lakes Huron and Michigan are connected by the wide Straits of Mackinac, and their surfaces are at the same altitude, 580 feet above sea level. Though the direction of flow through the straits alternates from east to west, depending on barometric pressures

Table 1. *The first fourteen of the world's lakes, listed in order of surface area.*[a]

Lake	Area (square miles)	Continent
1. Caspian Sea	169,300	Asia-Europe
2. Superior	31,820	North America
3. Aral Sea	26,233	Asia
4. Victoria Nyanza	26,200	Africa
5. Huron	23,010	North America
6. Michigan	22,400	North America
7. Baikal	13,300	Asia
8. Tanganyika	12,700	Africa
9. Great Bear Lake	11,490	North America
10. Great Slave Lake	11,170	North America
11. Nyasa	11,000	Africa
12. Erie	9,930	North America
13. Winnipeg	9,390	North America
14. Ontario	7,520	North America

[a] From a list by Lane (1948, p. ix).

and winds, the net flow is eastward. The entire discharge of the Michigan basin goes to the Huron basin, with the exception of a small amount of water diverted at Chicago.

Lake Michigan is separated from the Illinois River watershed by a low divide. In 1848 the Illinois and Michigan canal was constructed from the Chicago River across the divide to the Illinois River at a point just below La Salle, thus permitting through navigation. In 1900, the Sanitary District of Chicago completed a drainage canal from the Chicago River to the Des Plaines River at Joliet. This reversed the flow of the Chicago River and diverted Lake Michigan water to the Illinois River system. The quantity so diverted was limited to an average of 8500 cubic feet per second in 1925. Following a period of unusually low lake level, legal action resulted in a Supreme Court decision in 1929 to restrain the Chicago authorities from diverting water. After further consideration of the issue, a Federal permit was issued in 1930 for the diversion of an average of 1500 cubic feet per second, exclusive of pumpage for municipal water supply. It may be noted, incidentally, that a discharge of 1500 c.f.s. would

Table 2. *Physical characteristics of the Great Lakes.*[a]

Lake	Surface altitude (feet above sea level)[b]	Maximum depth (feet)	Length (miles)	Maximum breadth (miles)	Surface area of lake (square miles)	Area of drainage basin excluding lake (square miles)	Rainfall, average annual	Discharge (cubic feet per second)
Superior	602	1,302	350	160	31,820	49,180	29	73,700
Michigan	580	923	307	118	22,400	45,460	31	—
Huron	580	750	206	101	23,010	49,610	31	177,500[c]
Erie	572	210	241	57	9,930	22,560	34	196,300[d]
Ontario	246	778	193	53	7,520	27,280	34	234,500

[a] Data from Corps of Engineers, U.S. Army (1958).
[b] Mean altitude above mean tide at New York, to nearest foot.
[c] Exclusive of diversion at Chicago of 1500 c.f.s. to Mississippi River system.
[d] Exclusive of diversions by Welland Canal, Black Rock Canal, and New York State Barge Canal. These diverted waters are discharged to Lake Ontario at various points.

be sufficient to lower the water surface in the Michigan and Huron basins by only 0.037 foot, if all natural inflow into those basins were stopped for one year, and that the former higher discharge rate of 8500 c.f.s. would lower the water surface only 0.21 foot, if all natural inflow were stopped for one year. It appears that the observed low-water periods were caused by climatic variations rather than by diversion of water.

The channels between Lake Michigan and the Illinois River were constructed simply by further deepening the existing channelways which had been used by discharge of earlier, higher stages of the lake.

The outflow of Lake Huron, 177,500 cubic feet per second, goes down the St. Clair and Detroit rivers to Lake Erie. The water surface is lowered six feet in the 33-mile-long channel of the St. Clair River to Lake St. Clair, and another two feet in the 22 miles of the Detroit River channel connecting Lake St. Clair to Lake Erie. The Huron to Erie waterway has been improved by dredging, but it is essentially a natural channel.

Lake Erie, at an altitude of 572 feet above sea level, owes its elevation to the Niagara escarpment. The Niagara River, with a discharge of 196,300 cubic feet per second, descends 60 feet from Lake Erie to the brink of Niagara Falls, drops 167 feet in the falls, and then descends nearly 100 feet farther in the course of the Niagara gorge. Emerging from the gorge, the river traverses a lake plain and enters Lake Ontario at an altitude of 246 feet. Niagara Falls is working headward by erosion at a rate of between four and five feet per year; within about 25,000 years it may be expected that the gorge will extend all the way back to Lake Erie, and will completely drain that shallow lake.

Navigation between Erie and Ontario was made possible by construction of the Welland Canal and locks in Ontario, between Port Colborne on the Erie shore and Port Weller on the Ontario shore.

Lake Erie is also connected to the Hudson River, by canal. The New York State Barge Canal, or Erie Canal, extends from Tonawanda on the Niagara River to Troy on the Hudson and has a length of 340 miles. Thirty-five locks in this distance provide for adjustments in level. Other canals in the system provide connections with Lake Ontario and with Lake Champlain. Construction of the Erie Canal was feasible because it was laid out in part along old natural dischargeways which had carried water from the Erie basin to the Hudson during some of the earlier lake stages.

Lake Ontario, the smallest of the five Great Lakes in area, ranks third in maximum depth. Like Erie and Superior, its long axis is roughly from east to west (those of Huron and Michigan are more nearly north and south). The waters of Ontario pour northeastward into the St. Lawrence River, with an average discharge of 234,500 cubic feet per second.

The St. Lawrence River flows 270 miles from Lake Ontario to Quebec, and 370 miles from Quebec to Anticosti Island in the Gulf of St. Lawrence. From there to Belle Isle and the open Atlantic Ocean is another 440 miles. The distance, by water, from Duluth at the head of Lake Superior to the open Atlantic Ocean is more than 2000 miles.

Detailed descriptions of the Great Lakes relating to physical conditions pertinent to navigation are contained in the "Great Lakes Pilot" published by the U.S. Lake Survey.[1]

THE DEPTHS OF THE LAKES

Lake Erie, the shallowest of the five Great Lakes, has an average depth of only 58 feet. Its water content is but 109 cubic miles, or a little more than one-thirtieth of Lake Superior's volume. A relatively small area near the eastern end of Lake Erie contains depths greater than 100 feet, with the deepest point, 210 feet, lying nine miles east and a little south of the tip of Long Point (see Fig. 6). The bottom of the lake at its deepest point is 362 feet above sea level and 116 feet above the surface of Lake Ontario.

All of the other Great Lakes contain depths extending well below sea level (see Figs. 3, 4, 5, 7). The deepest point in Lake Superior, 1302 feet, is 700 feet below sea level; in Lake Huron the deepest point, 750 feet, is 170 feet below sea level; in Lake Michigan the maximum depth, 923 feet, is 343 feet below

---

[1] 630 Federal Building, Detroit 26.

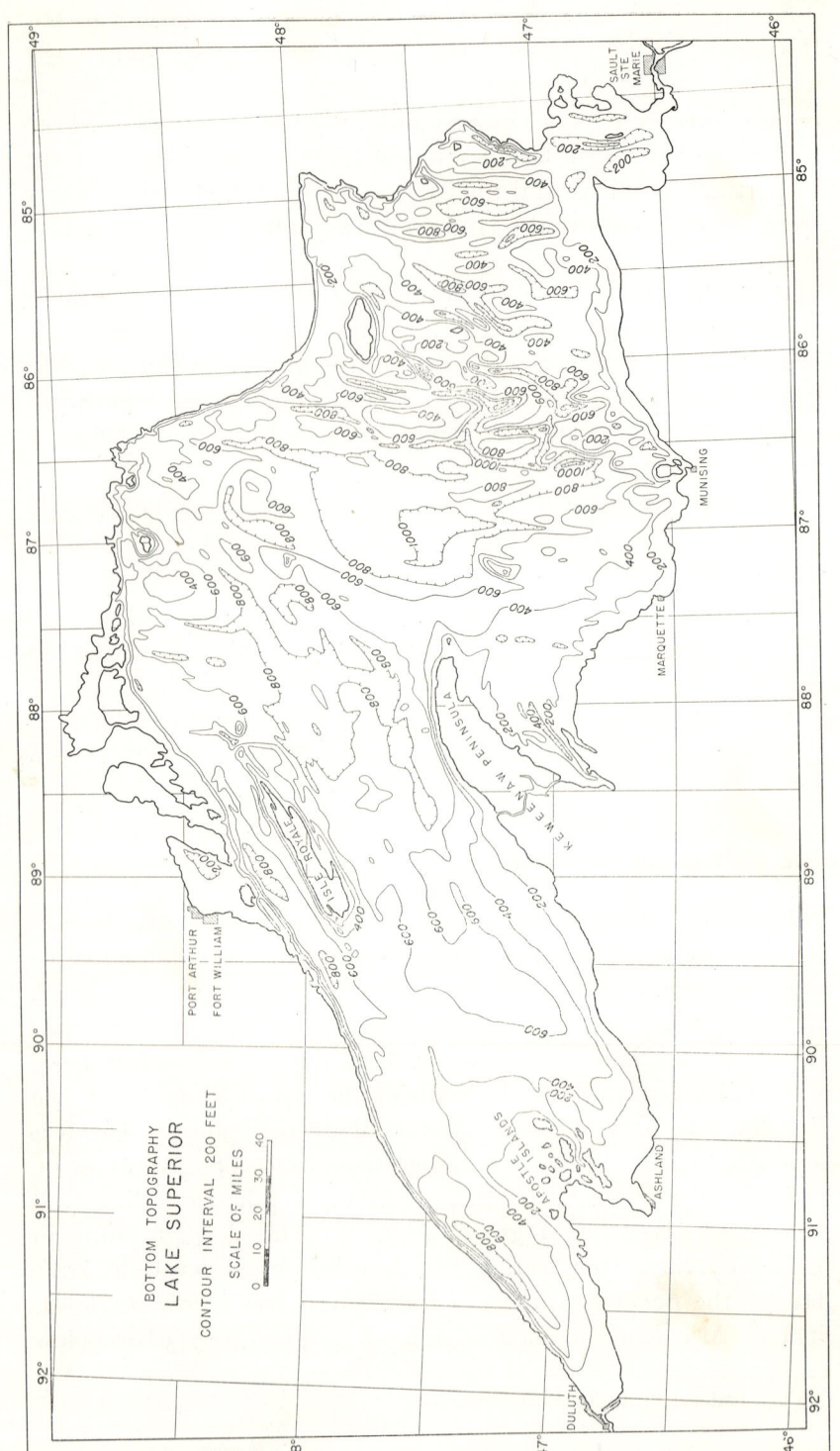

Fig. 3. The bottom topography of Lake Superior.

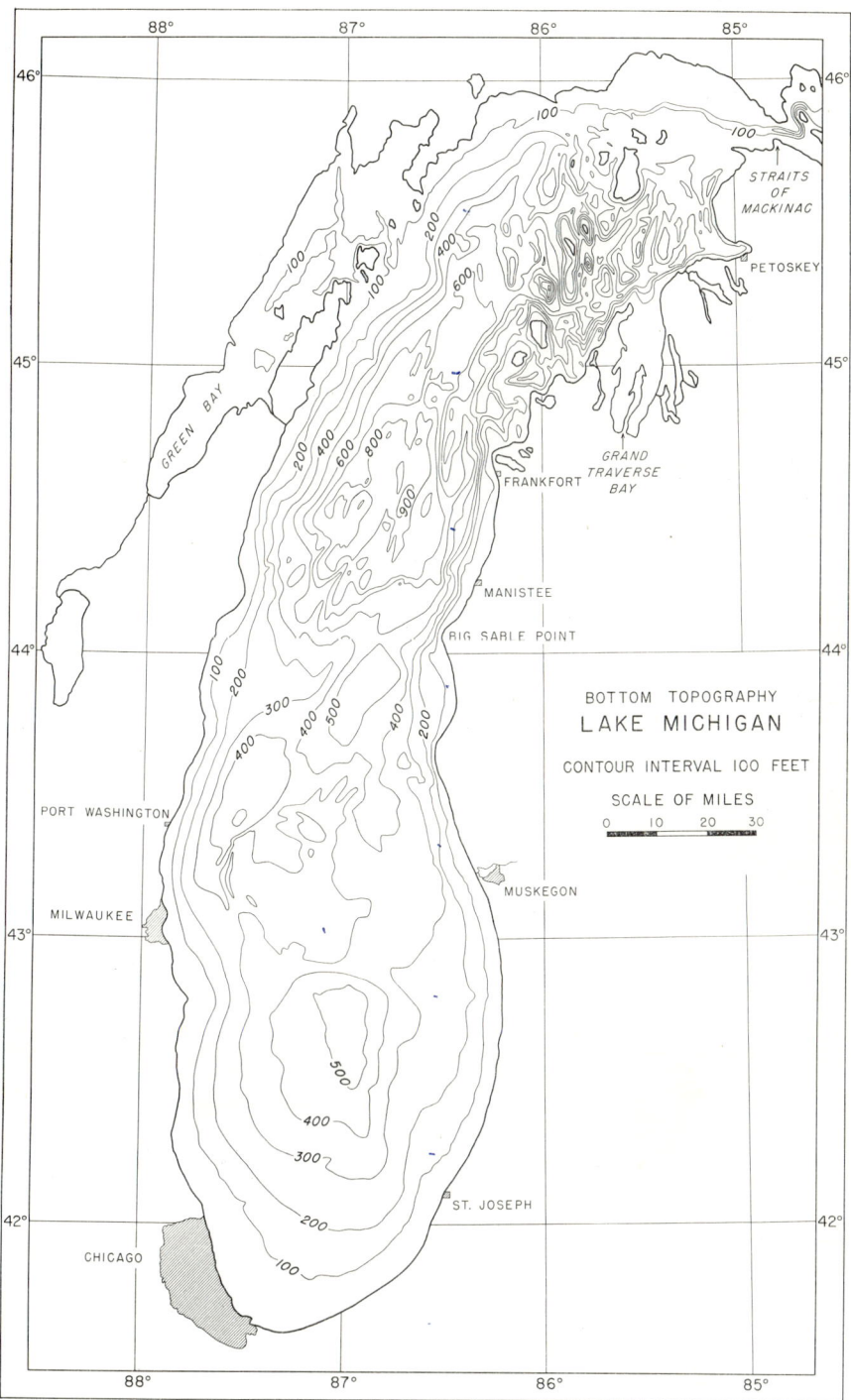

Fig. 4. The bottom topography of Lake Michigan.

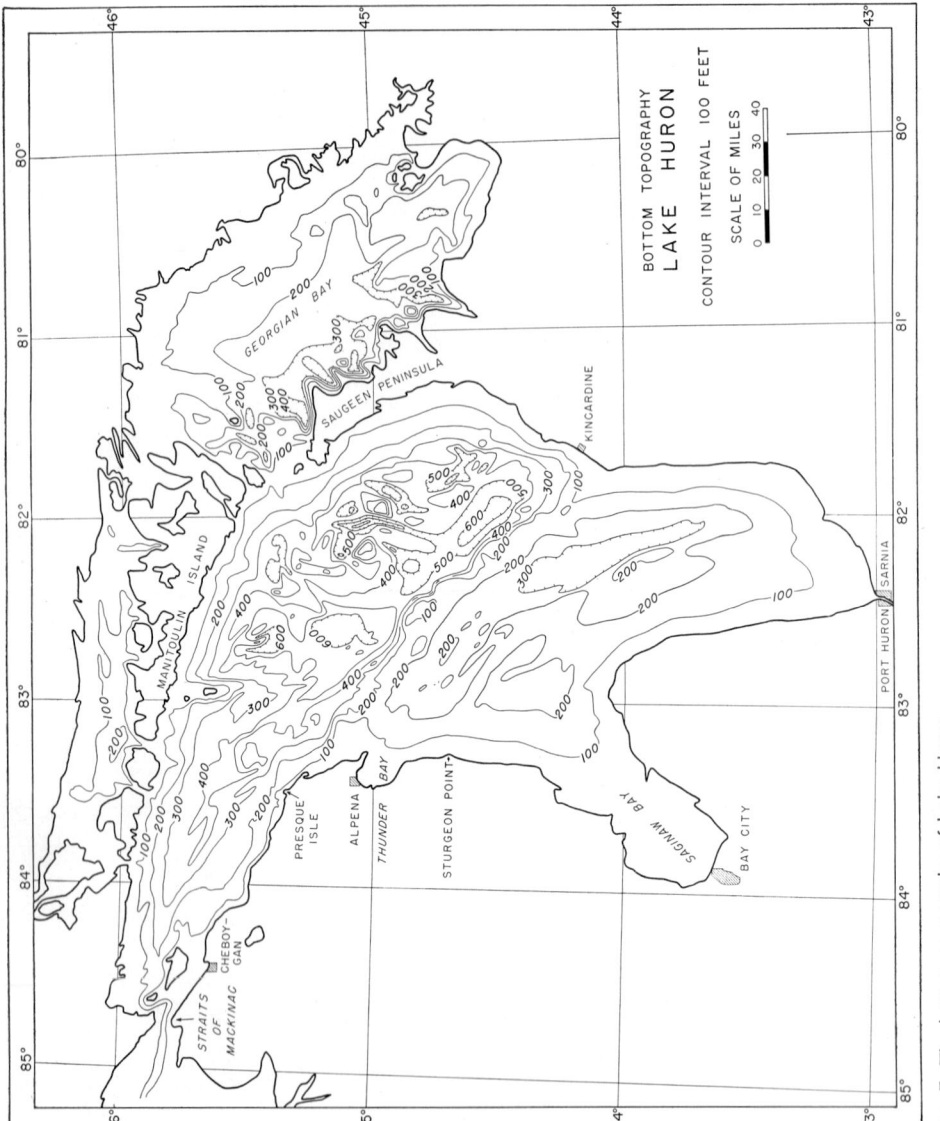

Fig. 5. The bottom topography of Lake Huron.

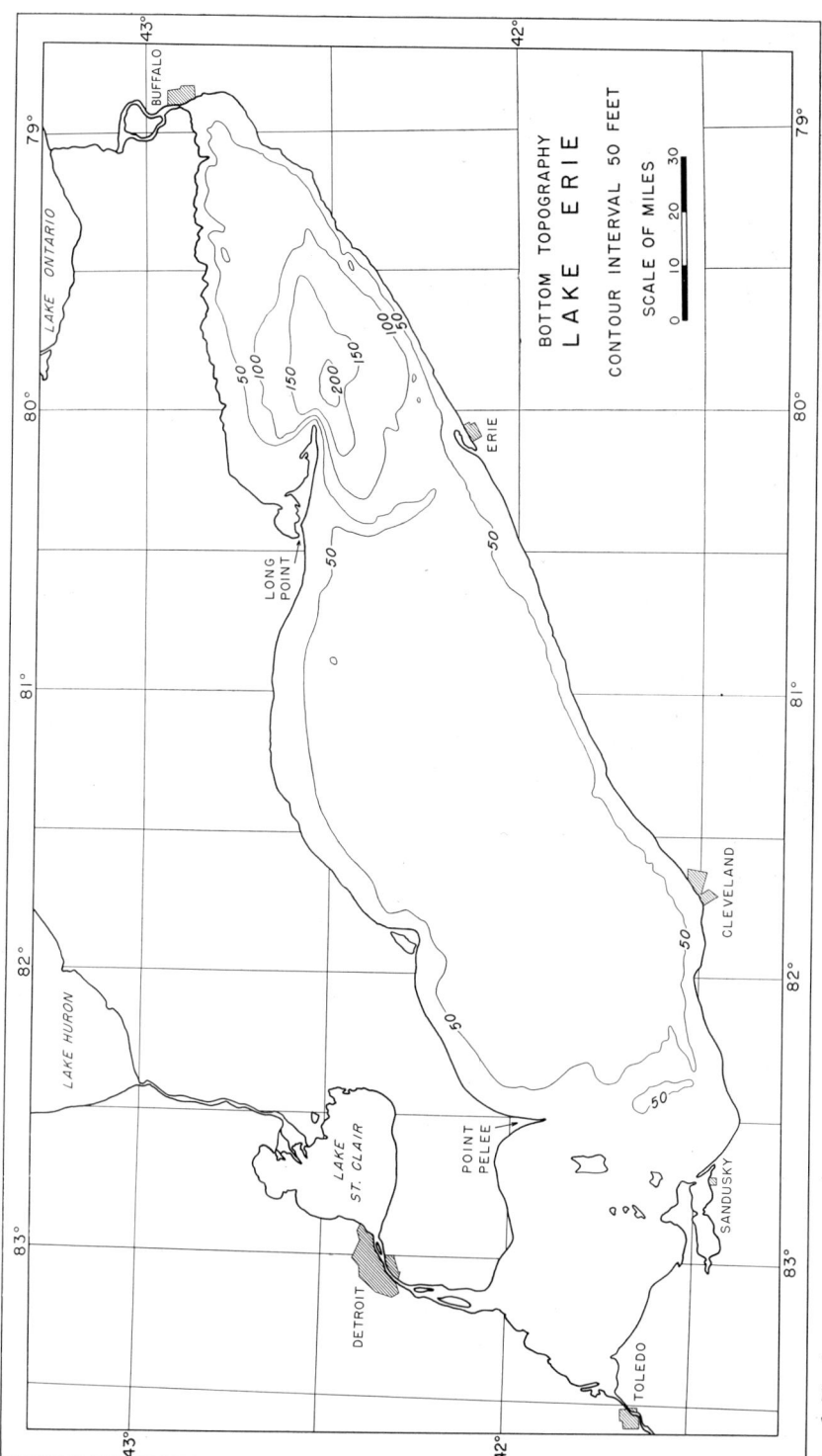

Fig. 6. The bottom topography of Lake Erie.

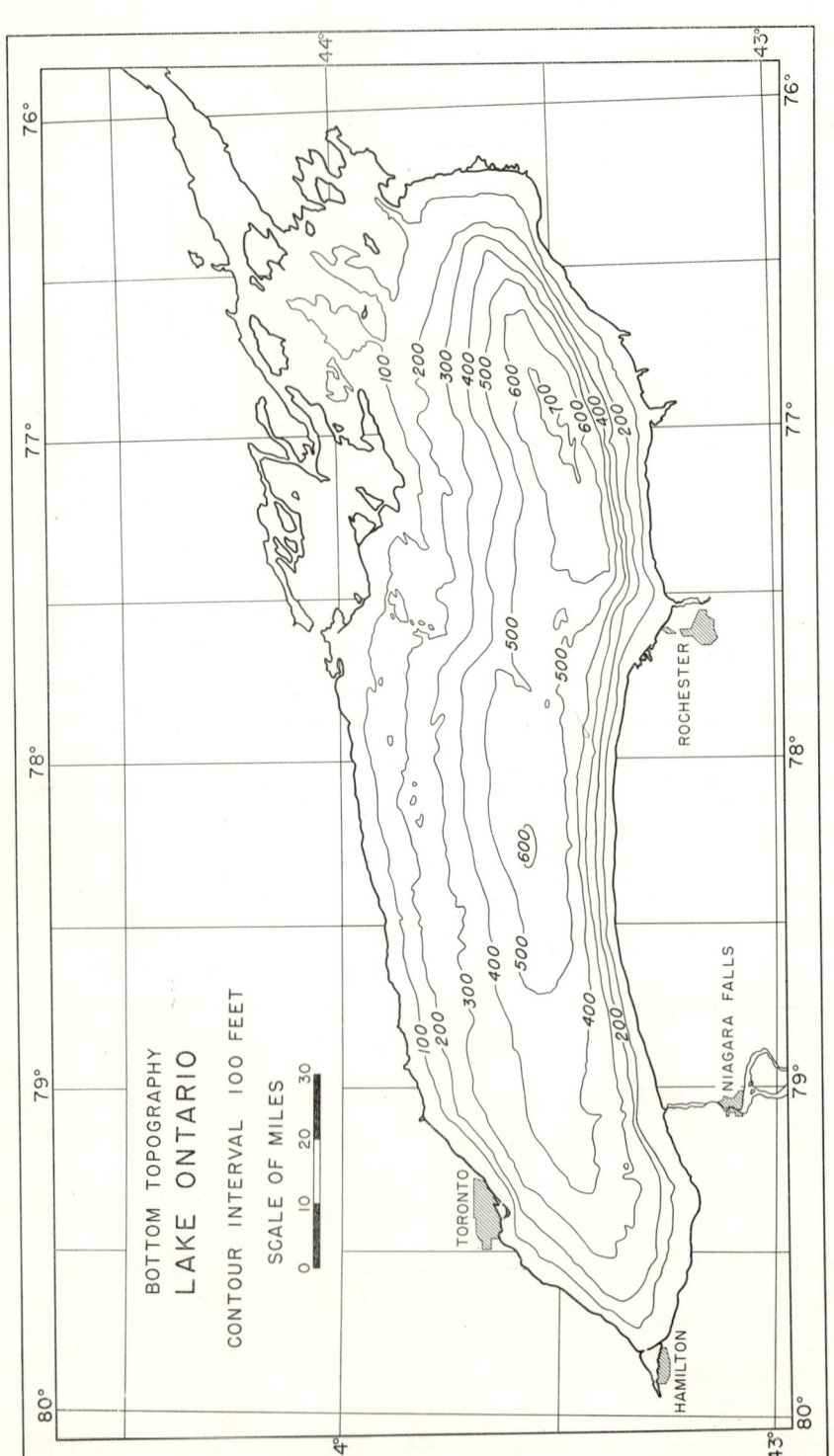

Fig. 7. The bottom topography of Lake Ontario.

sea level; and in Ontario the 778-foot depth is 532 feet below sea level. The origin of these deep basins is a question which will be discussed in a subsequent chapter.

TOPOGRAPHY OF THE LAKE BOTTOMS

The United States Lake Survey and the Canadian Hydrographic Service have made detailed surveys of the bottom topography of all of the Great Lakes, not only to locate hazardous shoals but to provide information for use in determining the position of a ship by sounding.[2] As a result of these surveys, the shapes of the lake bottoms even in deep water are known in considerable detail. The depth contours shown in Figs. 3, 4, 5, 6, and 7 were drawn from the soundings shown on the largest-scale published charts of the two government surveys, supplemented by unpublished data provided by the U.S. Lake Survey. It is interesting to note that three of the lakes, Superior, Michigan, and Huron, have rather complicated bottom shapes and contain prominent islands, while two of the lakes, Erie and Ontario, are relatively smooth troughs containing only a few small islands. These different characteristics are directly related to the geology of the region and they have important effects on the circulation of the lake waters.

The topography of the lakes is discussed in relation to the bedrock, to glacial scour and deposition, and to postglacial sedimentation, in subsequent parts of the book.

GEOLOGIC SETTING

General

The bedrock formations of the Great Lakes region are shown in Fig. 8. A large area of very old rocks, which generally are hard and dense, forms the Canadian shield which lies north of the lakes and extends southward into central Wisconsin. These rocks

---

[2] For a complete list of available charts, see Catalog of charts of the Great Lakes and connecting waters: U.S. Lake Survey, 630 Federal Building, Detroit 26, and Catalogue of nautical charts and sailing directions for the Great Lakes: Canadian Hydrographic Service, Dept. of Mines and Tech. Surveys, Ottawa, Canada.

14 GEOLOGY OF THE GREAT LAKES

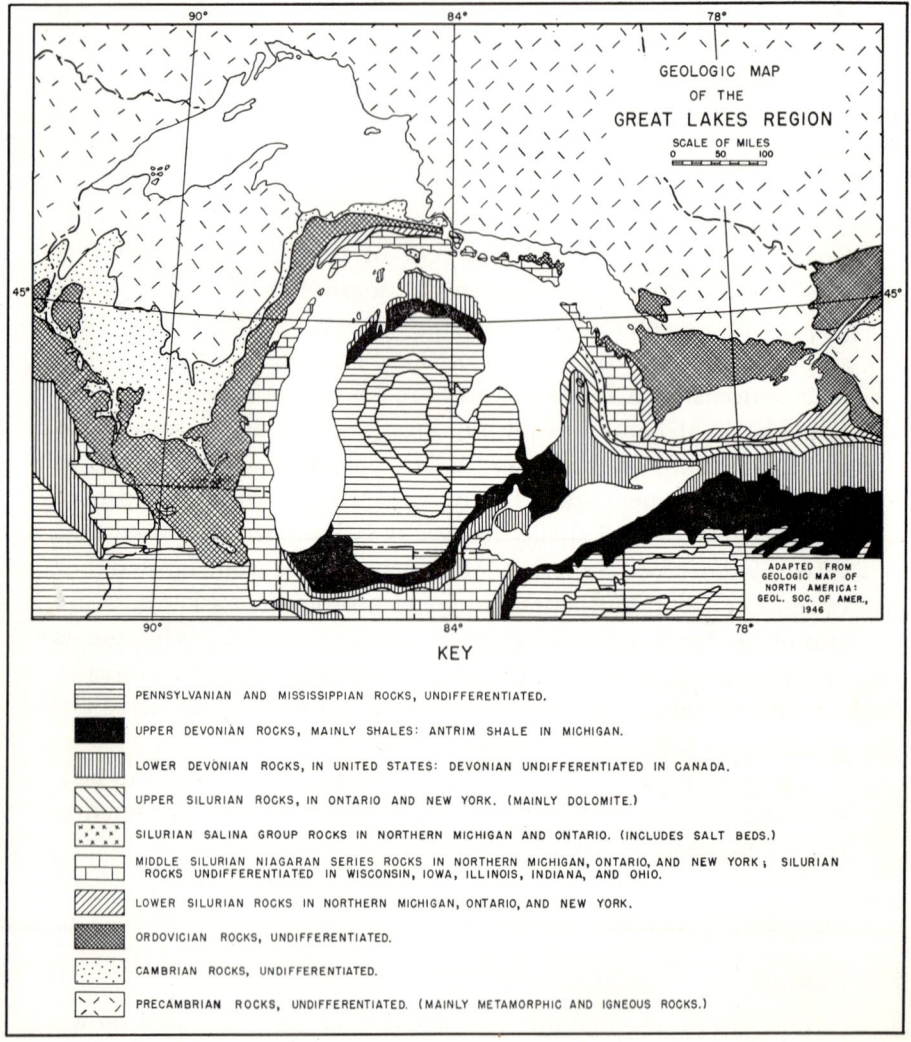

Fig. 8. Geologic map of the Great Lakes region.

were formed more than a half-billion years ago. Some of them were laid down in extensive seas of either salty or fresh water, but they have been metamorphosed into slates, quartzites, phyllites, gneisses, or other types, depending on their composition and degree of metamorphism. Large areas of the shield contain granite and other igneous rocks, cooled from the molten state. Earth forces have folded these rocks into mountain ranges, and the ranges have been worn down, leaving a surface of moderate relief.

The details of this Precambrian history have little bearing on the history of the Great Lakes, except for the fact that the present distribution of the rock types, and to some extent their structure, form a framework which has been sculptured to contain Lake Superior.

Lapping onto the shield are Paleozoic sedimentary rocks, approximately 185 to 520 million years old (Ladd, 1957, p. vii), which were deposited in marine water at various times when the sea flooded the continent. During the Paleozoic era there was a tendency for certain parts of the region to sink downward; this was most marked in the Michigan structural basin, centered in the Lower Peninsula of Michigan (Fig. 10), and in the Appalachian geosyncline, which extends northward into eastern Ohio, across Pennsylvania, and into southern New York. In these basins the Paleozoic rocks are many thousands of feet thick. Elsewhere in the region the Paleozoic rocks are relatively thin, and they wedge out on the flanks of the Canadian shield. The rock types formed in this part of the history are mainly limestones, dolomites, shales, and sandstones. These have been consolidated but otherwise they are not strongly altered.

As with the Precambrian rocks, the details of the Paleozoic history are of small importance in the history of the Great Lakes. The sea water drained from the continent at the end of the Paleozoic era, and there is no evidence that any appreciable portion of the region has been under the sea since that time. The present Great Lakes bear no relationship to the boundaries of the Paleozoic seas. The Paleozoic rock formations have, however, been tilted slightly, and they have been excavated by different erosional processes in a manner to produce the major topographic features which later determined the locations of the Great Lakes.

A brief examination of the geologic map (Fig. 8) shows that there is a definite relationship between the rock types and the locations of the lake basins. The most obvious feature is the control exerted by the Niagaran Dolomite of Silurian age (see Table 5, Chap. 4, for a list of the geologic ages). This rock forms the western shore of Lake Michigan and the Door Peninsula, which separates

Green Bay from Lake Michigan; it occurs along the northern shore of Lake Michigan and extends eastward, forming the islands which separate Lake Huron from the North Channel and Georgian Bay. Curving southeastward, it forms Saugeen Peninsula between Lake Huron and southern Georgian Bay. This belt of rock, to this point, has swung around the Michigan structural basin (Fig. 10). Farther southeast, the Niagaran Dolomite swings eastward south of Lake Ontario, where it forms the escarpment over which the Niagara River drops, at Niagara Falls, and on to the east it parallels the Ontario basin as it swings across the northern end of the Appalachian geosyncline.

Devonian shales, shown in black in Fig. 8, are relatively weak in their resistance to erosion, and they have been excavated by various erosional processes to form some of the basins in the Great Lakes region. Parts of the Michigan, Huron, and Erie basins lie in the outcrop belt of these shales.

A belt of weak Ordovician shales underlies Green Bay on the west side of Lake Michigan and extends through the North Channel and the deeper parts of Georgian Bay, in the northern and northeastern parts of Lake Huron. The deeper parts of Lake Ontario lie in the outcrop belt of the same formation.

The foregoing brief review has served to establish the general principle of correlation of rock types with the larger topographic features of the Great Lakes region. In general, it is seen that dolomites and limestones tend to form the higher parts of the region, while the shales have been removed more extensively to form the lower areas, in the four lower lake basins.

A more detailed analysis of the correlation between bedrock formations and topography of the lake bottoms for the individual lakes is given in the following paragraphs. In these discussions and in the accompanying figures the generally thin mantle of unconsolidated material is ignored.

Lake Superior

Lake Superior lies almost wholly within the Precambrian Canadian shield. The Superior basin is exceptional, in being nearly surrounded by highlands. Going outward from the lake in any direction except the southeast, one soon comes to an escarpment

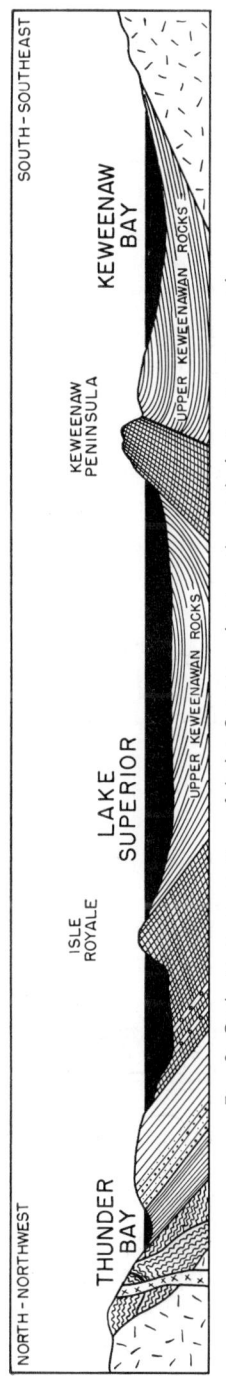

Fig. 9. Geologic cross section of Lake Superior, showing the rock basin structure and relations of the lake basins to zones of weak Keweenawan rocks. (After Van Hise and Leith, 1911, sec. A–B, pl. 1.)

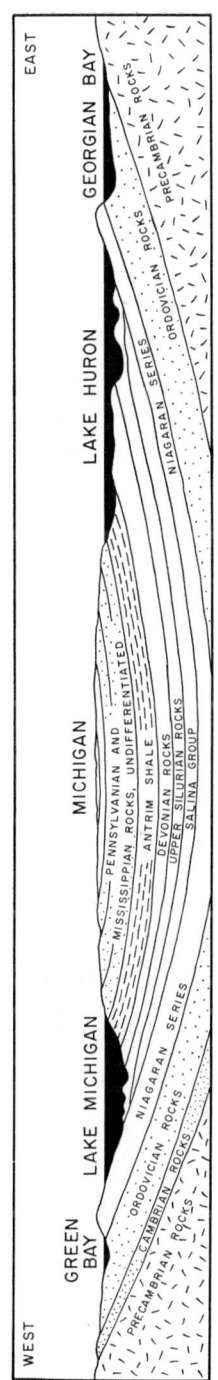

Fig. 10. Geologic cross section of the Michigan sedimentary rock basin, showing Lakes Michigan and Huron lying in zones occupied by relatively weak rocks of the Antrim Shale and Salina Group, Green Bay and Georgian Bay lying in zones of weak Ordovician rocks, and the escarpment of the Niagaran Dolomite which separates the lakes from the bays. (Based on Geologic map of the United States, U.S. Geol. Survey, 1932, and Geologic map of Michigan, Michigan Geol. Survey, 1936.)

above which is a distinct upland overlooking the lake basin. At many places the escarpment, which is 400 to 800 feet high, descends into water 500 to 900 feet deep. In the western half of the basin there is a distinct relationship between rock structure and the orientation of the basin and its bottom topography (Fig. 3). The strike of the rocks on the Keweenaw Peninsula, on Isle Royale, and on the northwestern shore of the lake is generally parallel to the east-northeast, south-southwest trend of this part of the basin. A cross section at a right angle to this trend is shown in Fig. 9. On this section it may be seen that Keweenaw Bay and the main basin of Lake Superior, between Isle Royale and Keweenaw Peninsula, are located in areas underlain by Upper Keweenawan (late Precambrian) sedimentary rocks. These consist of conglomerates, sandstones, arkoses, and shales, and in their slight degree of metamorphism they are more like the Cambrian than the underlying, older, Precambrian series. (In fact, the higher parts of the Upper Keweenawan have been classified as Lower Cambrian by some writers; however, the classification of Leith, Lund, and Leith [1935, p. 8] is followed in the present discussion.)

In the main basin of Fig. 9 there appears to be only a slight unconformity between the Upper Keweenawan sediments and the older Keweenawan lava flows and associated rocks which form the backbone of Keweenaw Peninsula and the greater part of Isle Royale. The main basin is essentially a simple synclinal structure and the softer sedimentary rocks in its center have been excavated to a considerable depth. The Keweenaw Bay trough is bounded along its northwestern side by a great thrust fault which extends along the center of the peninsula, bringing the older lava flows up and over the edges of the sedimentary beds. The Upper Keweenawan sediments lap onto Precambrian granite to the southeast. The present trough of Keweenaw Bay has been formed by excavation of the relatively soft sedimentary rocks.

The northwestern shore of Lake Superior in the vicinity of Duluth (at the western end of the lake) apparently is bounded by a major fault which may extend northeastward along the shore for many miles. This fault is not shown in the section of Fig. 9, but Van Hise and Leith (1911, pp. 112–16) have suggested that the trough between the northwestern shore and Isle Royale may be a down-faulted block between the Duluth fault and another possible fault along the northwestern shore of Isle Royale.

Thwaites (1935, pp. 226–28), in a study of the geomorphology of Lake Superior, has suggested other possible faults at various places on the shores and the lake bottom, but a discussion of these is beyond the scope of the present brief summary.

For the eastern part of the Superior basin there is no obvious explanation of the topography. Here, the predominant topographic trends are north-south (Fig. 3). The strike of the older Precambrian rocks east of Lake Superior is east-west, conforming with the general trends in the Canadian shield in this region (Moore, 1927, p. 76). There are, however, a number of diabase dikes and faults east of Lake Superior which trend north-south, indicating a later structural development with this orientation. Furthermore, the late Precambrian Keweenawan lava flows and conglomerates of the Mamainse Point area, on the eastern shore of the lake, strike nearly north-south and dip toward the lake basin (Thomson, 1954, p. 16). It is possible that the eastern part of the Superior basin is a distinct structural province with north-south trends predominating, but there is little indication of this on shore.

The origin of the Lake Superior basin has been summarized by Schwartz (1949) as follows:

. . . Lake Superior probably owes its origin to a combination of conditions. The first important event was the formation of the great syncline following the extensive igneous extrusions and intrusions of the Keweenawan. This syncline no doubt was expressed at the surface by a basin that was filled by later and softer rocks than the older rocks around the edges. Faulting at a still later time modified the structure of portions of the syncline. Some of the faulting has been considered of late Keweenawan age; part of the movement is Paleozoic or later. It is generally assumed that a large river valley occupied the present site of the lake and guided the early glacial erosion. The immediate cause of the present topographic basin was deepening by successive lobes of glacial ice that occupied the bottom of the syncline and eroded out the soft sediments but modified only in a moderate degree the resistant pre-Cambrian rocks on the sides [p. 83].

The importance of scour of the lake bottom by glacial ice is discussed in Chapter 3.

Lake Michigan

Lake Michigan lies wholly within the Paleozoic rock province of the region. Bedrock exposures are not common around the lake, because glacial deposits mantle the rock almost everywhere. Rock is exposed in a sufficient number of places, however, to give a good indication of its distribution and structure. A detailed study of the shape of the lake and its bottom topography (Fig. 4) indicates that the Paleozoic formations form the major topographic features.

Lake Michigan is bounded on the west and north by the Silurian Niagaran Dolomite cuesta, and the formation dips under the lake toward the center of the Michigan structural basin centered under the Lower Peninsula of Michigan (Fig. 10). The relatively smooth slope of the lake bottom from the shore down to the depths, on the west and northwest side of the lake, is essentially a dip-slope surface on the dolomite beds. The Niagaran Dolomite extends across the north shore to within 20 miles of the Straits of Mackinac, but the lake is relatively shallow off this shore.

The Bois Blanc Formation (lower Middle Devonian) cuesta can be traced from the south shore of the Straits of Mackinac westward through Waugoshance Point (Wilderness State Park area), and Grays Reef, Hog Island, parts of Garden and Beaver islands, Trout Island, High Island, and Gull Island, to the submerged elevation southwest of the last island. Beyond that area the cuesta cannot be traced. The Bois Blanc is composed of cherty dolomite, dolomitic limestone, and limestone (Landes, Ehlers, and Stanley, 1945, pp. 80–109).

The Traverse Group (Middle Devonian) cuesta forms the shore of Lake Michigan from Bayview in Little Traverse Bay westward and southwestward through Petoskey, Charlevoix, and the headlands of Grand Traverse Bay; the group underlies the shore as far southwest as Frankfort, Michigan. From there, it can be projected with some confidence southwestward across Lake Michigan where there is a distinct ridge 35 miles long in midlake, lying on a straight line drawn from Frankfort, Michigan, to Port Washington, Wisconsin. A small patch of Middle Devonian rocks is present in Wisconsin just north of Port Washington, and there is another just north of Milwaukee. The Traverse Group is composed mainly of limestones.

The Antrim Shale (Upper Devonian) is a weak rock which forms lowlands. It occurs in a zone across the northern part of the Lower Peninsula of Michigan which includes the deep southern arms of Grand Traverse Bay and the area south of Frankfort, where several hundred feet of glacial drift overly the bedrock. It runs offshore into Lake Michigan between Frankfort and Manistee, and it may be projected southwestward through a broad hollow which lies along a line drawn from just south of Frankfort to Milwaukee. Depths of 450 to 570 feet are common in this hollow. The Antrim Shale then very probably extends through a relatively deep channel in the lake bottom, having a minimum depth of 290 feet, which passes between the western shore and a mid-lake shallow area. The axis of this channel lies about 17 miles east of Milwaukee. From this area the Antrim Shale zone apparently widens into the broad southern basin on the lake, where a maximum depth of 564 feet is attained, and it continues southward to the known outcrop belt of the shale on the Indiana and southwestern Michigan shore.

There is an unconformity in the sequence of rocks underlying Lake Michigan. At the north end of the lake the Niagaran Dolomite is separated from the Antrim Shale by a number of formations, including the Bois Blanc and the Traverse Group. At the south end of the lake the Antrim Shale lies directly on the Niagaran Dolomite. The pinchout of the intervening beds apparently occurs under the lake in the area northeast of Milwaukee.

The Marshall Sandstone (Middle Mississippian) is the first relatively resistant rock above the Antrim Shale. This sandstone forms a bedrock high feature beneath the glacial drift of the Lower Peninsula of Michigan (Thwaites, 1949, p. 249). It appears likely that the Marshall Sandstone forms the principal irregularity of the southern basin of the lake shown by a definite westward bulge in the contour lines along the eastern side of the basin from St. Joseph, Michigan, northwestward. Further, it is possible that the Marshall Sandstone forms the distinct ridge at the western edge of the mid-lake shallow area, passing about 20 miles northeast of Milwaukee. The author has attempted to sample the bottom at the crest of this ridge; trials with both a heavy coring tube and a Petersen clamshell dredge yielded no samples. It is concluded that the bottom is bedrock at this point. Two miles to the east, glacial till was found on the bottom (for details see Hough, 1952, pp. 165–66, fig. 3). The Marshall Sandstone may be projected northeastward along the northern edge of the shallow area, which is also the southeastern edge of the Antrim Shale hollow. This edge of the mid-lake shallow area has a very steep to vertical face, as shown by fathogram traces made along three lines crossing it (two of these are illustrated by Hough, 1952, figs. 3 and 8). The northeastern part of the shallow area tapers into a ridge trending northeastward toward Big Sable Point, where the Marshall Sandstone zone returns to the Michigan shore.

The extreme westward excursions of the Michigan basin rocks in the latitude of the shallow area east of Milwaukee are proved by the occurrence of Middle Devonian rocks at Port Washington and Milwaukee. This westward extension seems to reflect a considerably gentler dip of the formations, and it indicates the possibility of local reversal of dip. A reversal of dip would make the area of interest as a possible oil province. Exploration for oil under the lake is not likely to be undertaken, however, because the comparatively low productivity of the Michigan basin rocks would not be expected to pay for costly offshore drilling operations.

The interpretations given in the foregoing paragraphs differ from those previously published (Thwaites, 1949, and Emery, 1951) in some important details. With the small amount of information available, no one can be sure of the precise location of the various formations on the lake bottom. The author has sounded and sampled the bottom of Lake Michigan rather extensively, and found that most of the ridges in deep water are covered by glacial till or lake sediments. No samples of bedrock were obtained. The hollows in the bottom are floored with lake clay (Hough, 1952 and 1955).

Within the deep northern basin of the lake, northwest of the ridge which has been correlated with the Traverse cuesta, there are other ridges trending northeast-southwest. These undoubtedly are formed by strata occurring between the Traverse Group and the Niagaran Dolomite. The ridges terminate within the deep basin, however, and they cannot be correlated with any particular formations.

The deepest point in the lake, 923 feet, occurs in the northern basin, northwest of the Traverse cuesta. The absence of definite topographic trends in this deep area seems to require some special explanation. As Thwaites (1949, p. 247) has pointed out, the northern basin of Lake Michigan is evidently underlain by the Salina Evaporite Series of Upper Silurian age; this is mainly thin layers of dolomite separated by salt and gypsum. Solution of the salt, at some time long before the Great Lakes came into existence, would have caused slumping of the overlying formations and a rather haphazard structural adjustment. This supposition for the deep basin area is based on known features of the Mackinac Straits area which are confidently assigned to the process of slumping and faulting due to solution of salt. These will be described in a subsequent paragraph.

East of the northern basin and south of the Bois Blanc cuesta is a portion of Lake Michigan which may appropriately be called a ridge-and-valley province. This area contains a number of islands and its bottom topography is characterized by a number of deep troughs, 250 to 500 feet deep, separated by ridges with only 25 to 50 feet of water over them. The valleys are connected in what appears to be a drowned drainage system: the master valley, or deepest trough, extends along the south shore from a point west of Petoskey, Michigan, west-southwestward along the northern face of the Traverse cuesta, across the mouth of Grand Traverse Bay, and thence westward to the deep northern basin. The depth of water in this valley increases fairly regularly westward. Most of the valleys and ridges in the lake bottom in this area have a generally north-south trend. Grand Traverse Bay may be considered a part of the area, as it is oriented north-south and contains two elongate southern arms, with depths of 300 to 600 feet, separated by a narrow peninsula. East of Grand Traverse Bay there are several deep lakes separated by north-south-trending ridges.

The ridge-and-valley province of Lake Michigan is underlain

by rocks of the Detroit River Group, the Dundee Limestone, and the Rogers City Formation. Onshore to the east, these have been mapped as striking nearly east-west. This regional structure does not coincide with the major north-south trends of the lake-bottom topography. There is, however, an unusual structural phenomenon in the area which probably has affected the topography; this is the Mackinac breccia. The brecciated or shattered rocks occur in many parts of the straits region, and they involve all of the formations between the Niagaran Dolomite and the Dundee Limestone. Just above the Niagaran, in the subsurface in the Lower Peninsula of Michigan, there is a thick sequence of evaporites, the Salina Formation, over half of which is salt. The average aggregate thickness of salt under the state of Michigan is approximately 1000 feet. On the north side of the straits area, where this formation crops out, the salt is absent, but other members known in the subsurface are present. These nonsoluble salina beds and all of the overlying formations up to the Dundee have been broken in a manner suggesting collapse. The breakage ranges from brecciation on a minute scale to the formation of large fault blocks measuring hundreds of feet in diameter. The blocks are tilted at angles of from 6° to 25° and their orientation is entirely random.

The origin of these structures has been explained by Landes, Ehlers, and Stanley (1945, pp. 143–45) in the following manner: after deposition of the Salina to Detroit River sequence of beds, the region emerged from the sea, and percolating underground waters leached great quantities of salt from around the rim of the Michigan basin. Extensive caves may have been produced locally, into which the overlying rocks suddenly collapsed. Elsewhere, the overlying rocks slowly settled downward as the salt was removed. During and after the collapsing, erosion of the surface produced a relatively smooth plain. In a subsequent period of submergence of the region the Dundee Limestone was deposited on the eroded surface, and this was later covered by the younger Paleozoic formations. The final events of this history include the relatively recent erosion which brought about the exhumation of the collapsed rocks and the partial isolation of indurated breccia masses by differential erosion.

It appears possible that the rough topography of the ridge-and-valley province between the Bois Blanc and the Traverse cuestas has been produced by differential erosion of the Mackinac breccia. A similar explanation may be suitable for the deepest part of the Michigan basin, which was described in a foregoing paragraph, and for the deepest part of the Huron basin.

North of the ridge-and-valley province of Lake Michigan, or north of the Bois Blanc cuesta, the lake bottom has a different character. It is essentially a plain with an average depth of about 75 feet, on which there are several reefs with depths of less than 15 feet. Incised in the plain is a narrow valley with a depth ranging from 150 feet on the west to 250 feet on the east (Fig. 47, Chap. 13). This submerged valley extends 70 miles, from the north end of the northern deep basin of Lake Michigan, on the west, eastward to the Straits of Mackinac. From there the valley extends eastward into Lake Huron, where it swings around the north side of Mackinac Island. The role of this valley as a river which drained the Lake Michigan basin eastward to the Huron basin, during a low-water stage, will be described in a subsequent chapter.

Lake Huron

The northern shore of Lake Huron, along the North Channel and northeastern shore of Georgian Bay, is on the edge of the Canadian shield. The lake basin otherwise is within the Paleozoic rock province of the region. As in Lake Michigan, some of the Paleozoic formations may be correlated with major topographic features of the lake bottom (Figs. 5 and 10).

The Niagaran Dolomite (Silurian) forms the north shore of Lake Huron for a distance of 40 miles east of the Straits of Mackinac, and it forms the southern and southwestern shores of the chain of islands and of the Saugeen Peninsula, which separate the main body of the lake from its North Channel and Georgian Bay. On the northeast side of the Niagaran cuesta there is a belt of shales, including the Cabot Head Shale (Silurian) and the Queenston Shale (Ordovician), which have been excavated to form the deep holes in Georgian Bay lying along its southwestern side. Farther northwest, on Manitoulin Island, these shales crop out on the northeastern slopes of the island where they are capped by the more resistant Niagaran Dolomite.

The Bois Blanc Formation forms the southern part of Bois Blanc Island (north of Cheboygan, Michigan), and it lies along the shore of Lake Huron from the Straits of Mackinac eastward to a point east of Cheboygan, a distance of 30 miles. This rock does not appear at the surface anywhere else in the Huron basin, but a study of the bottom topography of the lake suggests that it forms a shelf lying off the shore and extending from the straits southeastward 100 miles to the vicinity of North Point, east of Thunder Bay. Beginning in this locality there is a ridge which extends southeastward across the lake bottom to the southeastern shore at Clark Point, ten miles southwest of Kincardine, Ontario (Fig. 5). This ridge is the most striking feature of the bottom of Lake Huron: its northeastern face is very steep and descends to depths of more than 600 feet, while its crest generally is between 100 and 200 feet deep. The water is only 36 feet deep at one point (Six Fathom Bank) on the ridge in mid-lake, 50 miles east-southeast of Alpena. The southwestern side of the ridge slopes more gently, to depths of 200 to more than 300 feet in the southwestern basin of the lake. It appears to be possible that the Bois Blanc Formation extends along the northeastern edge of this ridge.

Other resistant formations of the Rogers City and the Traverse Group are recognized from Thunder Bay northward to Presque Isle, and they form shores and headlands in this area. The outcrop belt of these rocks may be projected southeastward through the prominent mid-lake ridge to the Ontario shore.

The Antrim Shale outcrop belt runs offshore into Lake Huron between the center of Thunder Bay and Sturgeon Point, 22 miles to the south. The same formation is mapped at the south end of the lake; the two areas may be connected through the southwestern basin of the lake, where depths in excess of 200 feet are common.

The main or northeastern basin of Lake Huron lies between the Niagaran cuesta and the principal ridge. The deepest point in the lake, 750 feet, occurs in the main basin 23 miles southwest of the northern tip of the Saugeen Peninsula. The bottom in all of the deep-water portion of the basin is irregular. In several places the depths range from less than 200 to more than 600 feet in a distance of a few miles. There is no recognizable trend to the topography. The Salina Evaporite Formation undoubtedly extends through the main basin, because it is mapped onshore at both the northwestern and southeastern ends of the basin. An explanation of the irregular topography may well be the same as that invoked for the Mackinac breccia of the straits area: the solution of salt in the Salina Formation (during Paleozoic time) with consequent collapse and faulting of the overlying formations.

Saginaw Bay is a shallow extension of the lake across the boundaries of Mississippian and early Pennsylvanian rocks.

Lake Erie

The western end of Lake Erie is extremely shallow, with most of the lake floor lying between 25 and 35 feet deep (Fig. 6). Point Pelee on the north shore and a number of islands and shoals partly close off the western portion of the lake. The bedrock structure of this area, shown in Fig. 11, has a definite relationship to the topography. Pelee Island and Kelleys Island lie in the zone of the resistant Columbus Limestone, and the Bass Islands and Sister Islands lie in the zone of the resistant Upper Bass Island Dolomite. These relationships are illustrated in Figs. 11 and 12. A more detailed description of this area is given in a publication by Carman (1946, pp. 279-83).

The central and eastern parts of Lake Erie are located along the strike of a simple structure in which the beds are tilted to the south, as is shown in Fig. 13. The basin of Lake Erie east of Sandusky has been excavated in soft Devonian shales and it lies in part on the surface of the underlying, more resistant Devonian limestones. In the narrow eastern part of the basin, where the angle of dip of the rocks is steeper, the Devonian shales were eroded more deeply to form the deepest basin in the lake (Fig. 6). Along the southern border of the Erie basin eastward from Cleveland there is an escarpment composed mainly of Mississippian sandstone and shales rising 200 to 300 feet above the floor of the lake basin. This is the northwestern edge of the Appalachian plateau (Carman, 1946, p. 279).

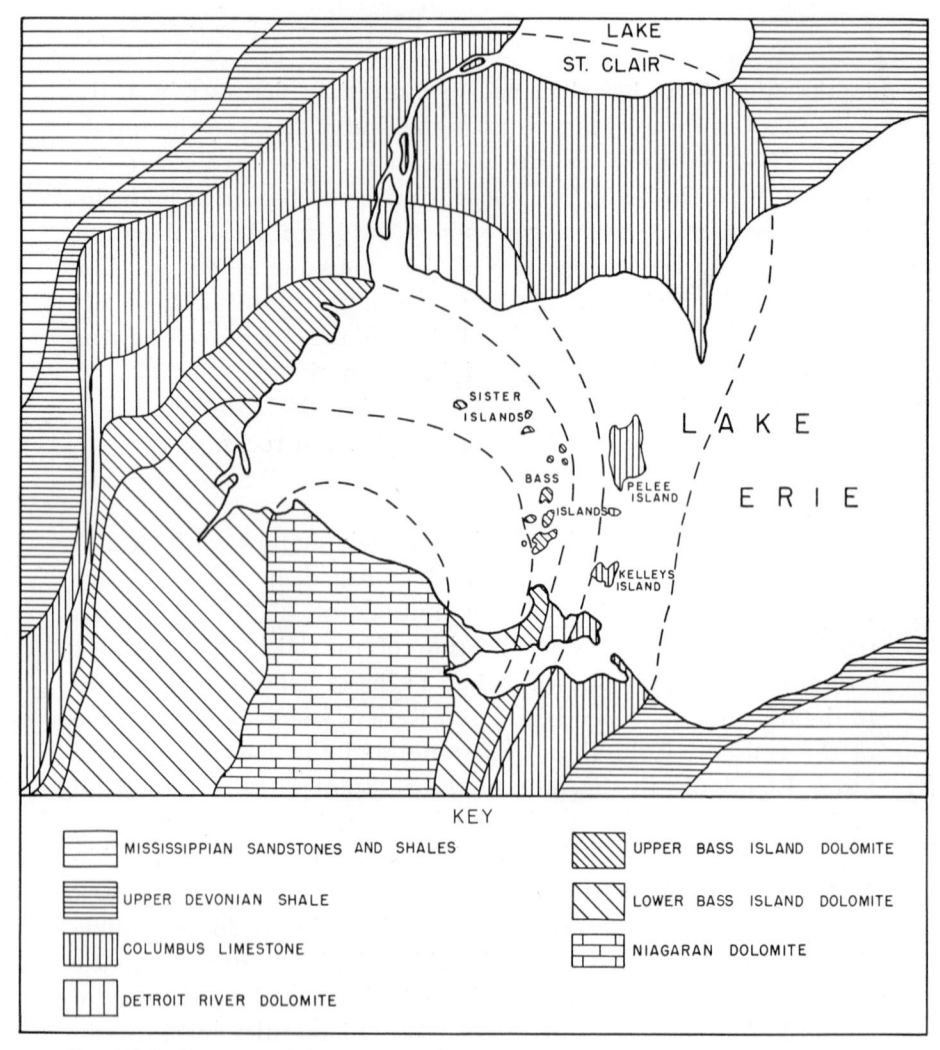

Fig. 11. Geologic map of the western end of Lake Erie, showing a portion of the Findlay arch (see Fig. 28). The islands occur in zones of more resistant rocks, the structure of which is shown in Fig. 12. (After Carman, 1946, fig. 35.)

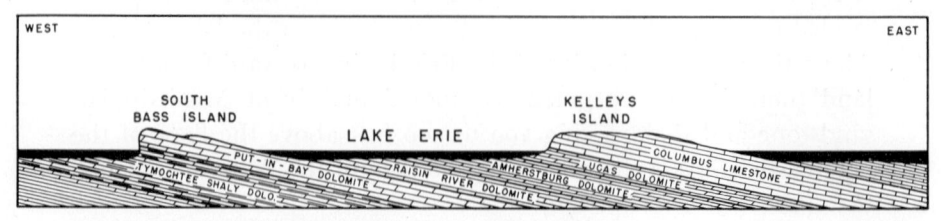

Fig. 12. Geologic cross section through South Bass and Kelleys islands in western Lake Erie, showing the rock structure. (After Carman, 1946, fig. 37.)

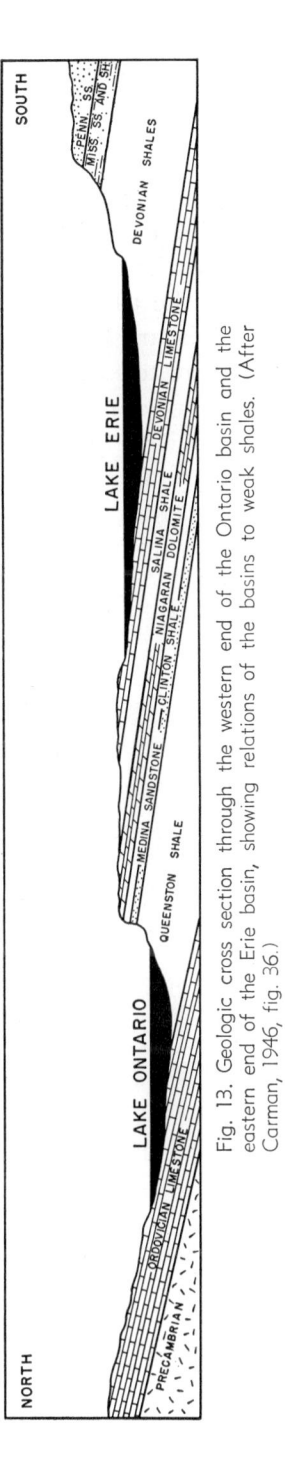

Fig. 13. Geologic cross section through the western end of the Ontario basin and the eastern end of the Erie basin, showing relations of the basins to weak shales. (After Carman, 1946, fig. 36.)

Lake Ontario

The Ontario basin, like the greater part of the Erie basin, is oriented parallel to the strike of beds which dip gently to the south (Fig. 13). The southern rim of the Ontario basin is formed by the cuesta or outcrop of the tilted Niagaran Dolomite. This rock also forms the sill of Niagara Falls. The greater part of the Ontario basin has been excavated in the soft Queenston Shale of Ordovician age, and the northern half of the lake bed is underlain by a more resistant Ordovician limestone. As is shown on the topographic map of the bottom of Lake Ontario (Fig. 7), the deeper parts are located south of the center and the lake has a relatively steep slope where it rises from the depths to the south shore.

chapter two

THE LAKE WATER

The strength of wave action and its effects on the bottom and shores are important considerations in unraveling the history of the Great Lakes. A migration of the zone of wave action through a considerable vertical range has been observed over a period of a few years. This range must be kept in mind while studying the raised and abandoned beaches of earlier stages in the lake history. The circulation of the lake waters, both horizontally and vertically, has an important effect on the character and distribution of the lake sediments and the life in the lakes. These and other factors in the present-day environment are reviewed here in order to give a background for the interpretations to be presented in subsequent chapters.

WAVE ACTION

The size of waves formed by wind action depends on the velocity of the wind, how long it blows, the distance over which it acts (the "fetch"), and the depth of water. In the Great Lakes, all of these factors may be of such magnitude that the resulting waves should compare in size with those observed on many seacoasts. The author has observed waves which he estimated to be 15 feet high and well over 100 feet long at a point in Lake Michigan 30 miles from the nearest shore and 85 miles from the windward shore, during an October gale. Because he was aware of the tendency of casual observers to overestimate wave heights, and because he was aboard a 64-foot schooner which served as a scale for estimating lengths, he believes that this estimate is conservative. Most of the other personnel aboard the vessel estimated the waves to be 30 feet high. Cornish (1910, p. 33) quotes captains of vessels navigating Lake Superior who say that during unusual storms waves have been encountered in deep water of a height of from 20 to 25 feet and a length of 275 to 325 feet. An aerial photograph of the wreck of the steamship *Henry W. Cort* [1] enabled a fairly accurate measurement of wave length to be made. The vessel was driven against the north breakwater at Muskegon, Michigan, during a storm on November 30, 1934. The photograph shows waves from the open lake beating against the vessel. Since the length of the ship was known (315.9 feet, registered with the U.S. Steamboat Inspection Service, Chicago), caliper measurement of the ship in the picture gave a scale for measurement of the wave lengths. By this method an average wave length of 160 feet was obtained (Hough, 1935, pp. 62–63). The extreme examples cited are certainly rare occurrences. A recording device on a breakwater at Milwaukee, Wisconsin, failed to record waves higher than six feet during the first two years it was in operation.[2]

Waves in shallow water tend to be steeper than those in deep water, and Lake Erie is known for "choppy" seas in time of storm.

The effects of wave action are shown on all of the exposed shores of the Great Lakes, where the broad beaches or steep sea cliffs compare favorably with those on the oceanic margins. Some of the older Great Lakes shores, now abandoned and lying above the present lakes, show a comparable strength of development. The question will be raised, in a subsequent chapter, of how much time is required for the development of shore features of this magnitude. Observations bearing on this question were made in an area near Woods Hole, Massachusetts, after the hurricane of 1938.[3] The amount of shore erosion which occurred during the hurricane, in a period of a few hours, exceeded that which had occurred during the previous 50 years. The geological doctrine of

---

[1] Published in the Chicago *Tribune,* Dec. 2, 1934, pt. 1, p. 3.
[2] Caples, Col. W. G., U.S. Engineers Office, Chicago, personal communication.
[3] Stetson, H. C., Woods Hole Oceanographic Inst., Woods Hole, Massachusetts, personal communication.

uniformitarianism, stressing the slow, orderly working of familiar processes, has perhaps blinded us to the importance of the unusual and catastrophic event.

The shores of the Great Lakes have changed very little from one year to the next, during periods of normal or low-water stages, even though violent storms occurred. During high-water stages, notably in 1951 and 1952, shore erosion proceeded at an alarming rate. It may very well be that the greater part of the development of a beach profile generally occurs quite early in the history of a given lake stage, particularly if the water reached that stage by rising from a lower elevation. The stability of a shore profile is, of course, affected by the materials present. Glacial till, composed of boulders to clay-sized material, provides coarse material to the beach and this takes the brunt of the wave attack. If only clay is present, it is carried away in suspension and the waves continue to cut into the land (Fig. 14).

The depth to which wave action is effective on the bottom is a

Fig. 14. A narrow sand beach at the foot of a cliff cut in materials consisting mainly of clay. The clay is transported to deep water, leaving very little coarse material on the beach to protect the shore from wave action. Ohio shore of Lake Erie. (Photograph by P. R. Shaffer.)

controversial subject and one of some importance in evaluating the evidence for changes in lake level. For example, gravel is common on the bottom of Lake Michigan in an area extending several miles offshore from Chicago, occurring in depths as great as 132 feet. Were these gravel deposits formed in shallow water during some low stage of the lake, or are they the result of processes acting in the present environment?

Textbooks of geology written during the past 50 years have generally stated that wave action is effective on the bottom to a depth equal to the wave length, or equal to one-half the wave length. The depths on the continental shelves of the ocean basins have been described as being adjusted to "wave base"; and "wave base" in the oceans has been assumed to be about 500 feet, because that is the approximate depth on the outer edges of some continental shelves. No one has proved that the shelves are in adjustment with the present depth of the ocean; therefore we are not justified in deducing the depth of wave base from their depths. Dietz and Menard (1951, pp. 1994–2016) have considered this question and have concluded that the continental shelves are not in adjustment to present sea level, and that the maximum depth of *vigorous wave abrasion* is generally about 25 feet. Their most significant point, however, is that currents capable of stirring sediment exist far below the level at which the motion of wind waves has become negligible. Ripple-marked sediment has been found to occur to depths as great as 4500 feet in the oceans. No level of negligible motion appears to exist in the ocean; therefore there is no depth that can be assigned to wave base.

Returning to the question of the origin of the gravel deposits on the bottom off Chicago, a sample collected from a depth of 50 feet contained pebbles which were subangular, and some bore glacial striae. Because surface markings are destroyed easily and pebble-sized fragments are rounded readily (Pettijohn, 1949, pp. 410–11) these were not beach pebbles, either from the present shore or of some earlier shoreline now submerged. The pebbles were found to occur overlying a pebbly, sandy clay, undoubtedly glacial till, and they were interpreted as a lag concentrate of the coarser constituents of the till, resulting from erosion of the bottom under present conditions (Hough, 1932, pp. 131–32, and 1935, p. 74). The pebbles and core samples described in the foregoing statement are illustrated in a publication by Bretz (1955, figs. 41 and 42).

Kindle (1925, p. 30) gives a striking piece of evidence for vigorous action on the bottom of Lake Ontario. The tug *Walker* sank in 65 feet of water near Nicholson's Island. Eleven years later divers examined the wreck, which was positively identified, and they found that it had worked down five feet into a firm bottom, that the hull of the vessel was battered to pieces, and that 75 tons of coal which had been aboard were washed away.

The mid-lake shallow area of Lake Michigan (Fig. 4), which lies between Milwaukee, Wisconsin, and Muskegon, Michigan, and is separated from both shores of the lake by relatively deep channels, has a minimum depth of 132 feet. Nowhere in the area, in depths down to 200 feet, has a cover of recent clay or sand been found. The bottom is composed of glacial till, or glacial till with a thin veneer of gravel. This area has been above water during a low stage of the lake (see the Chippewa stage, described in Chap. 13), but the point of interest here is that the bottom is somehow kept free of finer-grained sediments at the present time.

Whether wave action or currents of some kind were involved in all of the foregoing examples of bottom scour is, perhaps, unimportant; the significant point is that a bottom showing scour or non-deposition may exist at a considerable depth.

TIDES

True tides, caused by the attraction of the moon and the sun, are negligible in the Great Lakes. The true tide at Chicago and Milwaukee produces a total range of oscillation of 0.14 foot or less (Harris, 1907, pp. 483–86). At other stations on the Great Lakes the tide is, as a rule, smaller than at Chicago and is nowhere much larger. For all practical purposes, the lunar tides in the lakes may be ignored (Hayford, 1922, p. 113).

SURFACE CURRENTS

In the absence of any appreciable tidal action, the surface currents of the lakes are produced mainly by wind action and differences in barometric pressure over different parts of the lake basins. Brief violent windstorms may create surface waves which cause strong currents locally which exist for a very short time. Strong winds blowing for many hours or a few days will produce a transfer of water toward the leeward shore and a temporary circulation which is affected by the shape and topography of the lake basins. The velocities of currents involved in this type of movement are appreciable but generally are not great enough to be a hazard to navigation. Close to shore in shallow water the alongshore drift produced by waves approaching at an oblique angle may reach velocities of one or two miles per hour. All of these water movements are of a temporary nature. In addition, there appear to be certain patterns of permanent, or at least seasonal, circulation in all of the lakes, involving a slow drift of the water.

A study of the surface currents of the Great Lakes was made by Harrington (1895) by means of drift bottles, which were released at various points in the lakes and recovered at their points of stranding on the shores. The bottles contained post cards on which the location and date of recovery could be noted. The study was made during the warmer half of the year for each of three years in 1892 through 1894. For the details of the individual bottle tracks and the method of interpretation the reader must consult the original reference, but Harrington's conclusions are reproduced in somewhat simplified form in Figs. 15 through 19, which give the inferred surface-current systems of the five Great Lakes. These represent the currents in the summer season only.

An evaluation of the inferences regarding the Michigan basin has been made possible by further studies. Harrington's conclusions, as illustrated in Fig. 16, may be summarized as follows: there is a southward drift along the whole of the western side of the lake, which continues around the south end and turns northward on the eastern side, becoming more pronounced. Around the Beaver Island group in the northern end of the lake there is a counterclockwise swirl, and in the major southern basin there is a similar swirl. Between these two areas there is a tendency for the surface water to move eastward along lines which are curved with their convex sides to the south.

Townsend (1916, pp. 297–303) criticized these views and pointed out that since only the point of release and the point of

SURFACE CURRENTS
LAKE SUPERIOR

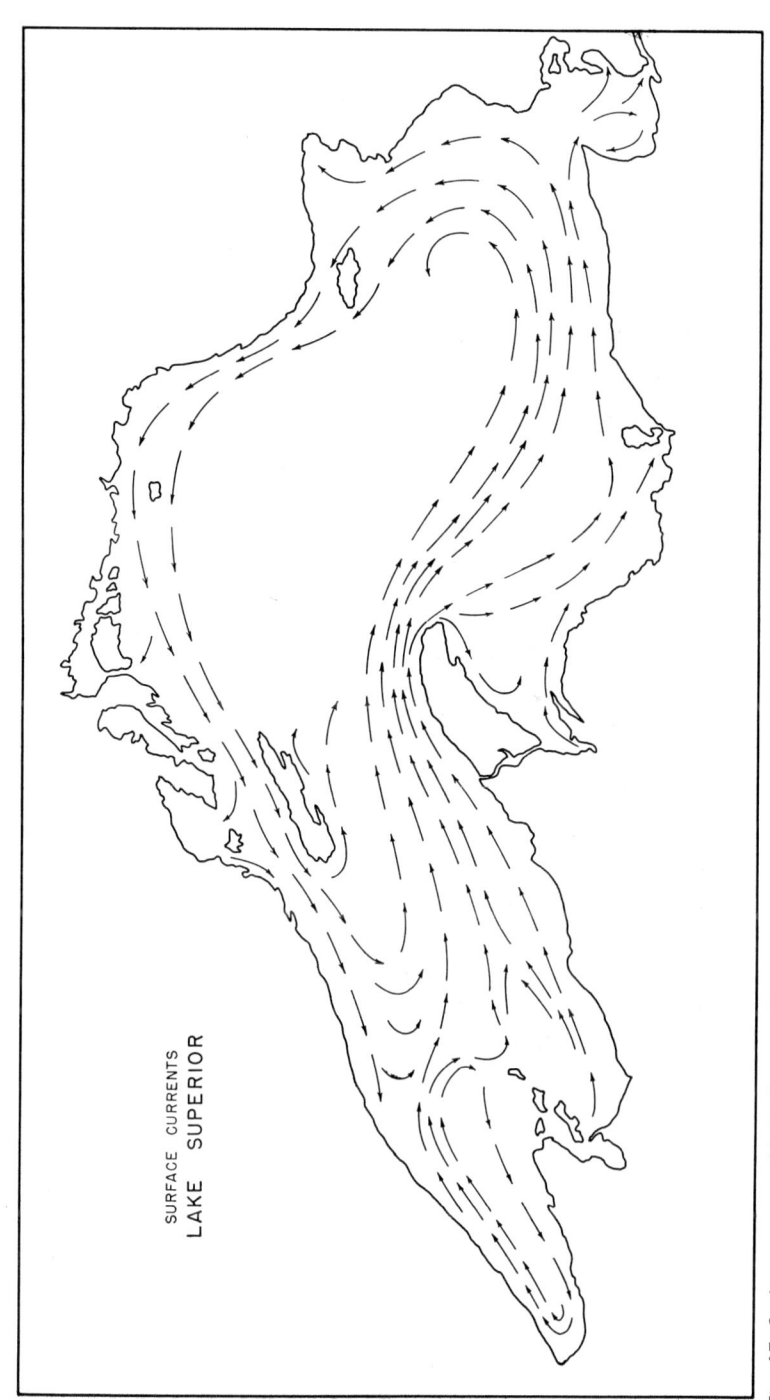

Fig. 15. Surface currents of Lake Superior. (After Harrington, 1895.)

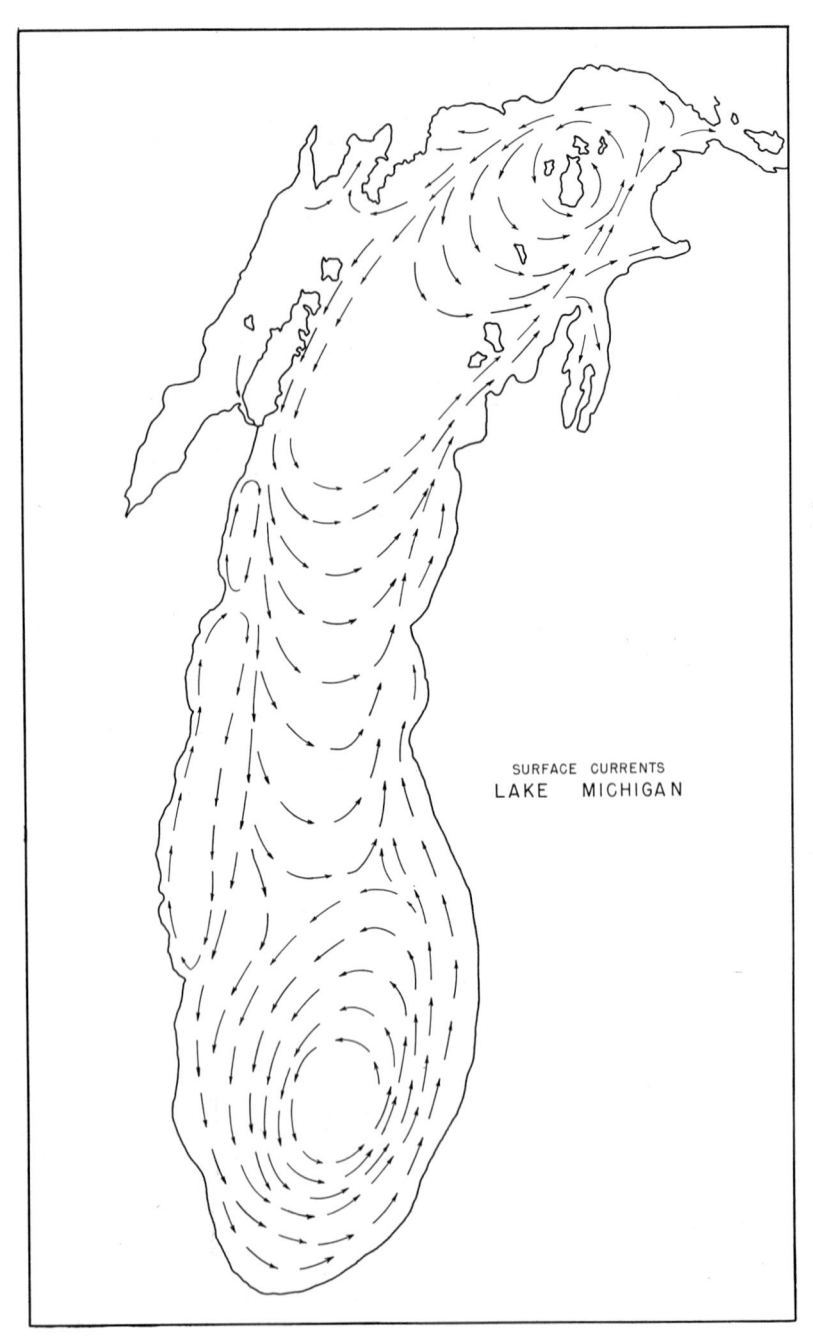

Fig. 16. Surface currents of Lake Michigan. (After Harrington, 1895.)

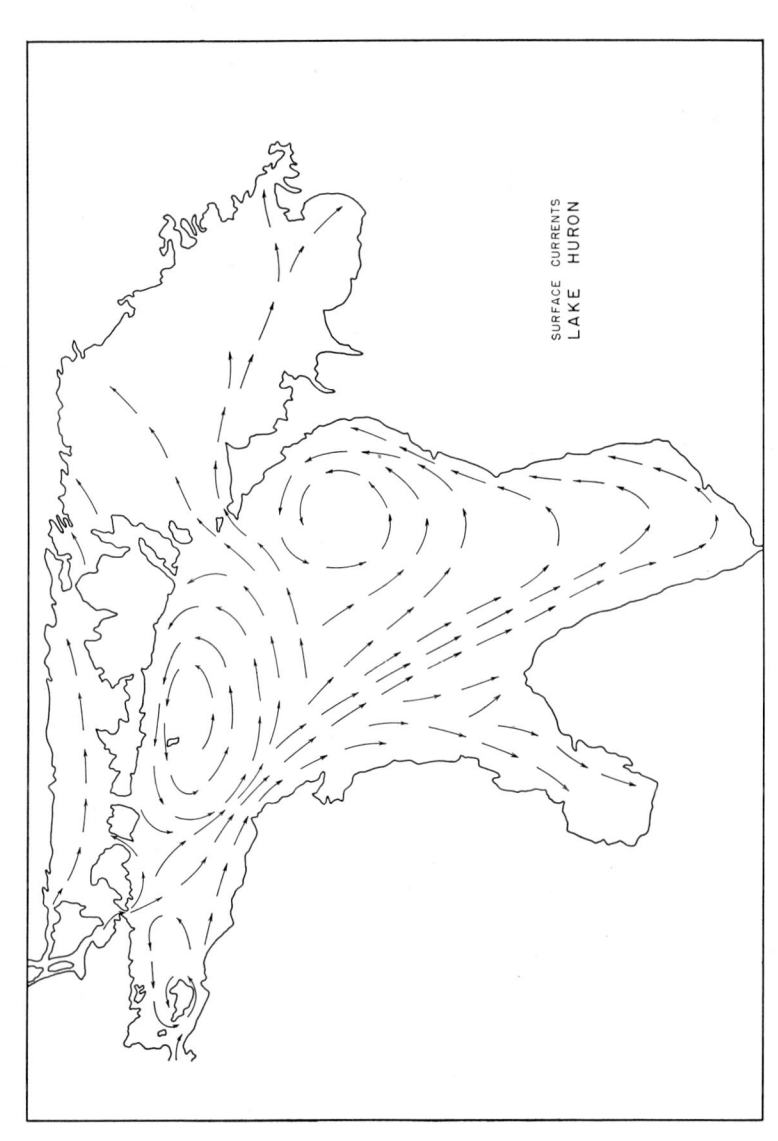

Fig. 17. Surface currents of Lake Huron. (After Harrington, 1895.)

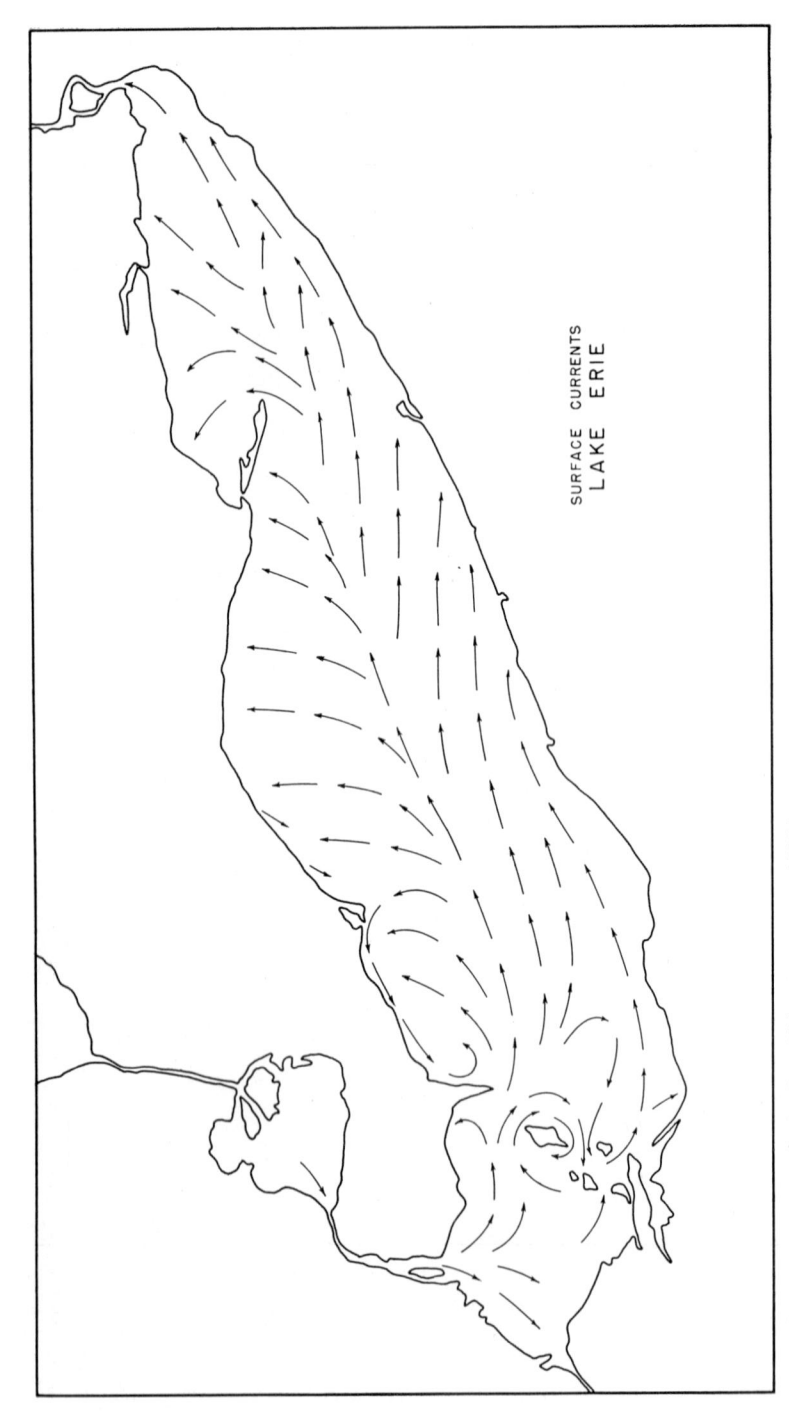

Fig. 18. Surface currents of Lake Erie. (After Harrington, 1895.)

Fig. 19. Surface currents of Lake Ontario. (After Harrington, 1895.)

recovery of a bottle are known, it is not possible to know the path taken. Townsend believed that the conception of a swirling movement in the southern basin is without sufficient evidence.

The U.S. Bureau of Fisheries made an investigation of the surface currents of Lake Michigan, also using drift-bottle apparatus. In a preliminary report the following conclusions were given: along the east shore of Lake Michigan the current tends to move northward at an average rate of several miles a day. Bottles set adrift on the west shore of the lake show a tendency to move over toward the east or Michigan shore and then move northward. In the southern part of the lake, in the area south of Waukegan and St. Joseph, the drift bottles showed a very irregular movement, indicating somewhat of a swirling action in that area. Bottles released there were usually out a very long time, which seems to indicate that they followed a very wide and indeterminate path before reaching their destination. The point of recovery was often less than 50 miles from the point of release although the bottles were out 30 days or more.

There is little doubt but that the action of the prevailing westerly winds in Lake Michigan has much to do with the rate and direction of the surface currents. The prevailing westerly winds coupled with the general direction of flow toward the outlet into Lake Huron may be predicted as the cause of the northerly current along the east shore and the tendency to move from the west to the east shore (Deason, 1932, p. 12). These conclusions follow, in general, those of Harrington in the original drift-bottle survey, but they do not specify the two definite counterclockwise swirls that are shown in Fig. 16.

Additional studies of a different nature cast further light on the problem of determining the surface currents of the lake. A survey of the temperature distribution in Lake Michigan was made by Church (1942), and he was able to draw conclusions regarding the horizontal flow of the water which supported Harrington's interpretations. The results of Church's work are reviewed in some detail in a later section entitled "Thermal Stratification and Deep Circulation."

More recently, a program of investigation of the limnology of the upper Great Lakes has been initiated by the Great Lakes Research Institute [4] in cooperation with several other agencies. Preliminary studies of the circulation of Lake Michigan, based on a dynamic-height method (Ayers, Anderson, Chandler, and Lauff, 1956), have indicated that on June 28, 1955, the surface-current pattern was similar to that shown in Fig. 16 in its broad aspects but that in addition, two clockwise swirls were present, one being located close to the eastern shore in the southern basin and the other occurring in a small area at the southwestern corner of the lake. The usual north-flowing current near the eastern shore, which extends to the outlet at the Straits of Mackinac (Fig. 16), may be reversed by northeasterly winds blowing over the straits and the northwestern end of Lake Huron. The reversal may extend several miles into Lake Michigan.[5]

A more comprehensive study of the waters of Lake Huron, made in the summer of 1954, has been published (Ayers, Anderson, Chandler, and Lauff, 1956). As a part of this study, the surface currents of Lake Huron have been obtained by application of the dynamic-height method, and found to be in general agreement with drift-bottle surveys which were made concurrently. The results of three synoptic studies, made on June 29, July 27, and August 25, indicate that the surface-current pattern shown in Fig. 17 (after Harrington) probably is a fair representation of the average circulation of the lake during the summer months. The studies show, however, that the surface currents are variable and that the generalized or average pattern is not always present. Specifically, "the fundamental circulation pattern in the upper and central portions of the lake appeared to be counter-clockwise. In the lower end of the lake outflow to the St. Clair River appeared to consist of a meandering surface current, near or east of the middle of the lake, which approached the entrance of the river from the northeast. The surface circulation of the lake appeared to reflect the direction and velocity of the winds of the preceding twelve days. . . . The circulation of Lake Huron has certain characteristics which are at least pseudo-oceanic and others which are definitely lacustrine" (Ayers, Anderson, Chandler, and Lauff, 1956, pp. 99–100).

---

[4] Univ. of Michigan, Ann Arbor.
[5] Ayers, J. C., personal communication, July, 1957.

The pseudo-oceanic characteristics include the visible effects of Coriolis force (the tendency of the strongest currents to swing to their own right), the apparent tendency of wind-driven surface water to move to the right of the wind direction, and the distribution of upwellings and sinkings on both upwind and downwind shores according to the relationship of current streamlines and the shore. . . . Although the circulation exhibits several "oceanic" characteristics, the surface current pattern is in the reverse direction from the inertia circle typical in oceans. The flow-through of Lake Michigan and Lake Superior waters is an external force which appears to be sufficient to overpower and reverse the tendency toward the clockwise inertia circle circulation which would result if Coriolis force were the dominant one. The surface circulation which exists appears to be the result of an equilibrium attained between the flow-through, the wind-driven transport of surface water, and the field of mass [Ayers, Anderson, Chandler, and Lauff, 1956, pp. 97–98].

SEICHES AND OTHER SHORT-PERIOD FLUCTUATIONS IN LAKE LEVEL

The most spectacular changes in level of the lakes occur rather suddenly, and these periods of higher or lower level are of short duration—from a few minutes to a few days, but generally measured in hours. They result from an imbalance or tilting of the lake surfaces induced primarily by winds and differential barometric pressures. The maximum temporary rises of any appreciable areal extent, which have been recorded, range from 8.4 feet on Lake Erie to 2.5 feet on Lake Huron (McDonald, 1954, p. 251).

Any of these short-period oscillations which are produced by some meteorological force, and whose period is longer than that of surface waves, are frequently referred to as a "seiche." In scientific discussion, the term "seiche" is restricted to an oscillation in the waters of a lake under the influence of inertia. It is a free oscillation as distinguished from a forced oscillation; it is a wave motion involving both horizontal transfer of water back and forth and a vertical oscillation of the water surface. The initial impulse which starts a seiche in the Great Lakes is probably much more frequently due to the wind than to barometric gradients (Hayford,

1922, p. 125, and Welch, 1935, p. 39). This has been observed repeatedly in Lake Erie, where a strong wind blowing for a period of several hours along the axis of the lake will drive the surface waters toward the leeward end of the lake, raising the water surface there as much as 8.4 feet, and lowering the level at the windward end of the lake. When the wind shifts or stops blowing, the lake surface begins to swing, with alternating rise and fall at each of the two ends of the lake, the swing diminishing rapidly in amplitude. The period of oscillation in Lake Erie has been observed to be from 14 to 16 hours (Welch, 1935, p. 40).

The reasons for the occurrence of the extreme changes in water level in Lake Erie apparently are that the lake is oriented nearly parallel with prevailing winds, that the lake is shallow, so that no deep return flow can occur while the surface is under wind stress, and that the basin has a fairly regular and simple shape, which permits the water to oscillate as a simple cell in its seiche mechanism. Short-period fluctuations in level have been observed in the other Great Lakes basins. Lakes Superior, Huron, and Michigan are sufficiently irregular to prevent their waters from oscillating in simple, single cells during seiche activity.

Local changes in level of considerable magnitude have occurred as a result of sudden violent windstorms. On the morning of June 26, 1954, an abrupt increase in the level of Lake Michigan occurred along the water front in the vicinity of Chicago. At least seven lives were lost in the Chicago area as a result of the first unexpected increase, which reached eight feet at Montrose Harbor and ten feet at North Avenue. The wave approached Chicago from the east to southeast. It was first observed at Michigan City, Indiana, with a height of about six feet at 8:10 A.M., and it reached the Chicago shore at about 9:30 A.M. A few hours earlier a severe squall line with winds up to 60 miles per hour had arrived from the northwest and crossed southern Lake Michigan. The squall and its associated pressure jump arrived at Michigan City simultaneously with a wave from the northwest. The wave was reflected from the Indiana shore and refracted in its travel westward so that it was concentrated on the Chicago lake front. A detailed analysis of this occurrence has been made by Ewing, Press, and Donn (1954, pp. 684–86).

SEASONAL AND LONGER-PERIOD FLUCTUATIONS IN LAKE LEVEL

The principal natural factors which affect the longer-period fluctuations of the levels of each of the Great Lakes are precipitation, evaporation, and flow in the rivers connecting the lakes. Precipitation falling directly on the lake surfaces immediately raises the levels. Precipitation falling on land surfaces of the drainage basins has a delayed and variable effect on lake levels due to variations in runoff. The average annual rainfall in the region ranges from 29 inches in the Superior basin to 34 inches in the Ontario basin.

Evaporation from the lake surfaces lowers their levels by the number of inches of water evaporated, but unfortunately no measurements of evaporation have been made. The depth of water removed each year by evaporation has been estimated to range from approximately 1.5 feet on Lake Superior to as much as 3 feet on Lake Erie (McDonald, 1954, p. 251).

The levels of the lakes follow a seasonal pattern with highs occurring in the summer and lows in the winter or early spring months. Within almost any year, the seasonal variations average from 1.1 feet on Lakes Huron and Michigan to 1.8 feet on Lake Ontario. During the period of record since 1860 the lakes have departed widely, however, from their average seasonal behavior both as to the magnitude of the changes and to the timing of the high and low levels. A chart of the monthly mean water levels, published by the U.S. Lake Survey, is reproduced in modified form in Figs. 20A and 20B.

In addition to the seasonal changes, the mean levels of all of the lakes fluctuate from year to year. The over-all long-range fluctuations over a period of nearly 100 years, based on the differences between the high and low monthly mean average levels, range from a little over 4 feet on Lake Superior to more than 6.5 feet on Lake Ontario. As a generalization, it may be stated that the present Great Lakes maintain levels for several months at a time which vary through a range of approximately five feet about their mean elevations, and that occasional storms produce temporary additional changes which may extend the vertical range of the shore zone considerably. This range in elevation will be of inter-

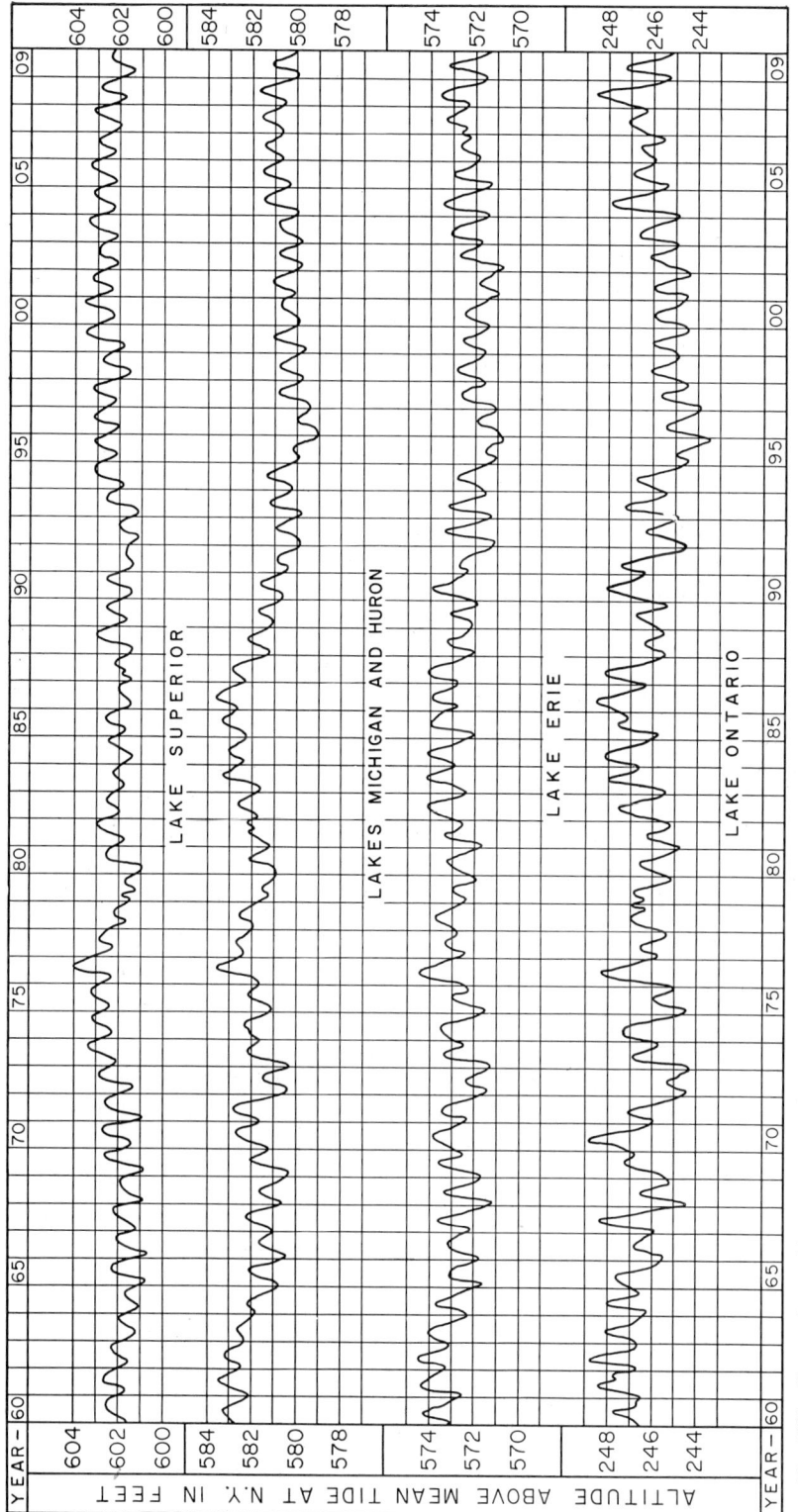

Fig. 20A. Chart of monthly mean water levels in the Great Lakes: 1860 to 1909. (From U.S. Lake Survey.)

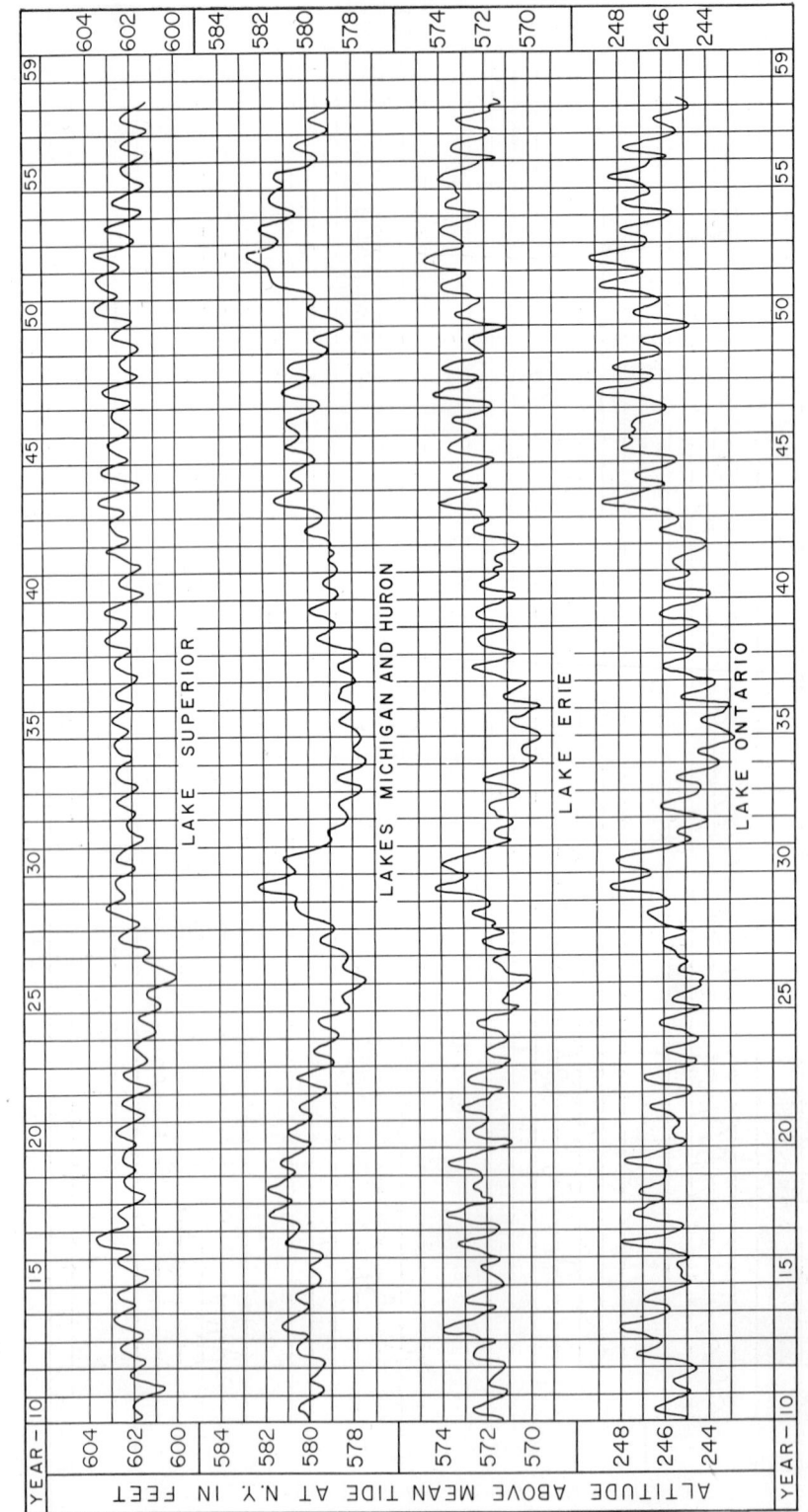

Fig. 20B. Chart of monthly mean water levels in the Great Lakes: 1910 to 1958. (From U.S. Lake Survey.)

est in connection with studies of some of the older shore features, formed during earlier stages of the lakes.

ICE IN THE LAKES

Freezing of the lake water closes the lake ports and most of the connecting waters to navigation in mid-December of the average year. Navigation remains closed until mid-April over the greater part of the lakes, though the opening date may be as early as late February in some localities (Corps of Engineers, U.S. Army, 1958, pp. 39, 103, 237, 300, 379). Navigation is maintained throughout the winter season in Lake Michigan by the railway car ferries which are designed to break ice.

A solid sheet of ice generally forms in the more protected embayments and channels, but in the main lakes only a relatively small percentage of the surface is covered with ice, even during the coldest winter periods, and this ice occurs in the form of floes which drift with the wind and the surface currents. Fast ice forms along the main shores in shallow water and ice floes may be blown against a shore and immobilized there by freezing to the fast ice. In this way a belt of shore ice is formed which may extend offshore for half a mile or more (Fig. 21).

Fig. 21. Ice ridges and pack ice on the Indiana shore of Lake Michigan, accumulated during an unusually severe winter. (Photographed about 1919 by L. W. Hough.)

Zumberge and Wilson (1954, pp. 201–5) have described another way in which a belt of shore ice may be extended offshore. A *shore ice foot* is first formed at the beach and extended a short distance into shallow water by the addition of ice formed from the spray of breaking waves. Eventually, the water surface in contact with the base of the ice foot may become frozen into a smooth sheet for some distance offshore, during a time of extremely cold weather and little wave action. Later, strong wave action may break the thin ice along its outer edge and pile the broken blocks along the edge; accretion of frozen spray then cements the ridge, forming a new ice foot called the *offshore ice foot*. This protects the entire shoreward belt of ice from further wave action. Wave erosion of the shores of the lakes is halted during the several months in which the fast or shore ice occupies the shore zone.

THERMAL STRATIFICATION AND DEEP CIRCULATION

The surface waters of the Great Lakes have a temperature range of from 32° F. in the winter to more than 70° F. in the summer. It is well known that the deeper lakes in the temperate zone go through an annual temperature cycle as follows. In the spring there is a period during which the water has the same temperature from top to bottom, approximately 39.2° F., which is the temperature of maximum density of fresh water (see Fig. 22A). At this time the density of the water is the same everywhere, and the water will circulate or mix from top to bottom under the influence of wind stress and the resulting currents. This is a phenomenon known as the *spring overturn*. During the warmer months of the year the surface waters gain heat, mainly by absorption of solar radiation and by conduction from the atmosphere. As the temperature rises the water becomes lighter and can no longer sink; a definite layer of warm surface water is formed, which floats on the denser water below (Fig. 22B). Only the warm surface layer can be circulated by the wind, and the colder, heavier, bottom water is left to stagnate throughout the summer. The chemical and other changes which result from the stagnation will be described later. During the fall there is a net loss of heat from the

THE LAKE WATER 51

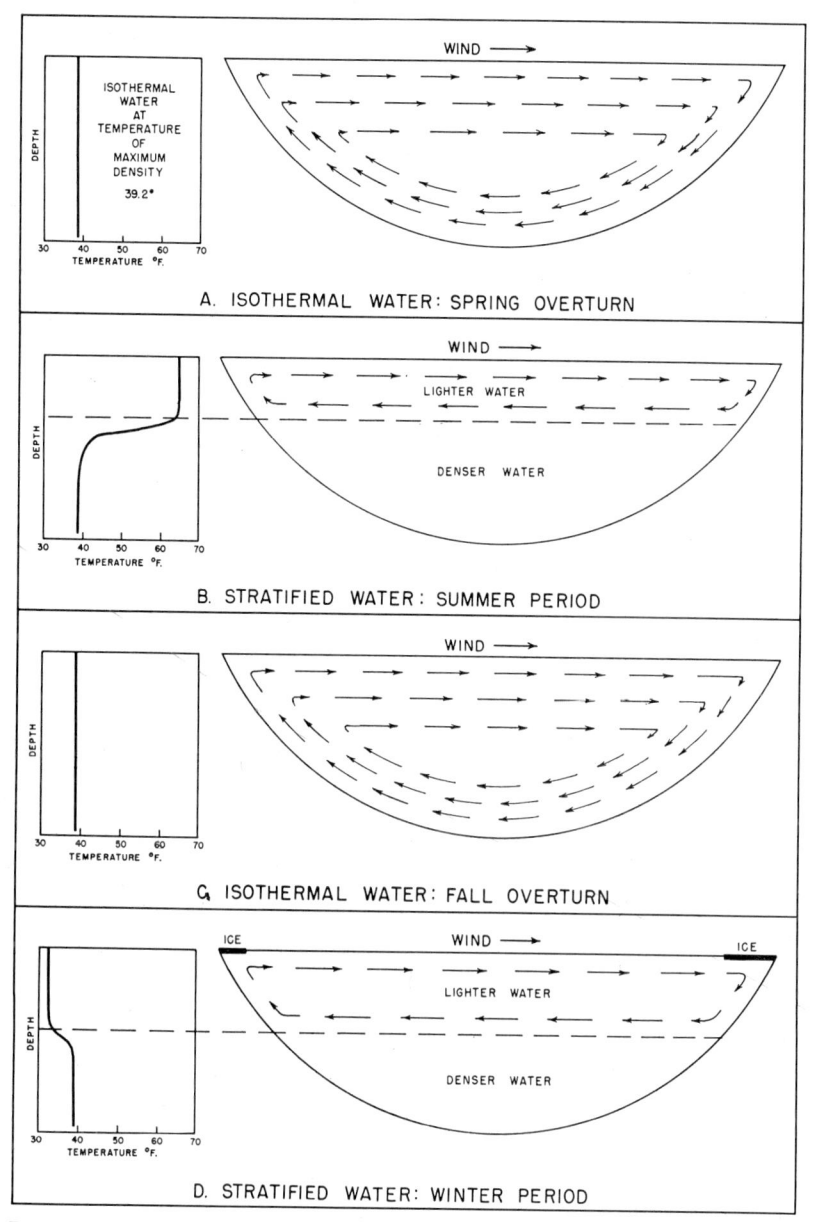

Fig. 22. The annual temperature cycle in typical deep lakes of the temperate zone. In the deep part of a lake basin the water temperature remains close to 39° F. throughout the year.

water to the atmosphere and the temperature of the surface water gradually lowers until it is equal to that of the bottom water, or about 39.2° F. An isothermal condition again exists (Fig. 22C), and complete circulation occurs in the *fall overturn*. With further loss of heat, the temperature of the surface water drops below 39.2° F. Because that is the temperature at which the water attains its maximum density, a further lowering of the temperature causes the water to expand and become lighter. As the surface water cools toward the freezing point, 32° F., it becomes increasingly lighter, and once again a distinct zone of lighter water is formed which floats on the denser water below (Fig. 22D). If the lake does not entirely freeze over, the wind will circulate the surface water, but the circulation will be limited to the upper zone of lighter water while the warmer, denser, bottom water is left to stagnate throughout the winter. Warming of the surface water in the spring returns the lake to the isothermal condition with which the cycle started.

The cycle as described above is an ideal one. Because of a number of factors, such as rate of heat exchange between the water and the atmosphere and the turbulence of the water during critical parts of the cycle, certain variations may occur (Welch, 1935, pp. 45–53, and Ruttner, 1953, pp. 32–36).

A curve representing the distribution of temperature in a lake in the typical summer condition is shown in Fig. 23, with the terminology of its parts indicated. The upper layer, composed of warmer and lighter water, is called the *epilimnion;* the zone of rapid decrease of temperature with increasing depth is called the *thermocline;* and the lowest layer of water, which is cold and dense, is called the *hypolimnion*. At the bottom, in a typical deep lake, the temperature approximates 39.2° F., the point of maximum density of fresh water. Actually, pressure lowers the temperature for maximum density, by approximately 0.045° F. per atmosphere, and this effect is noticeable in very deep lakes (Wright, 1931, p. 413).

The annual temperature cycle of Lake Michigan has been studied in detail by Church (1942 and 1945) over a period of two years. He found that the lake was vertically homogeneous, or isothermal, during the spring months. The temperature of the entire column of water was slightly above that of maximum

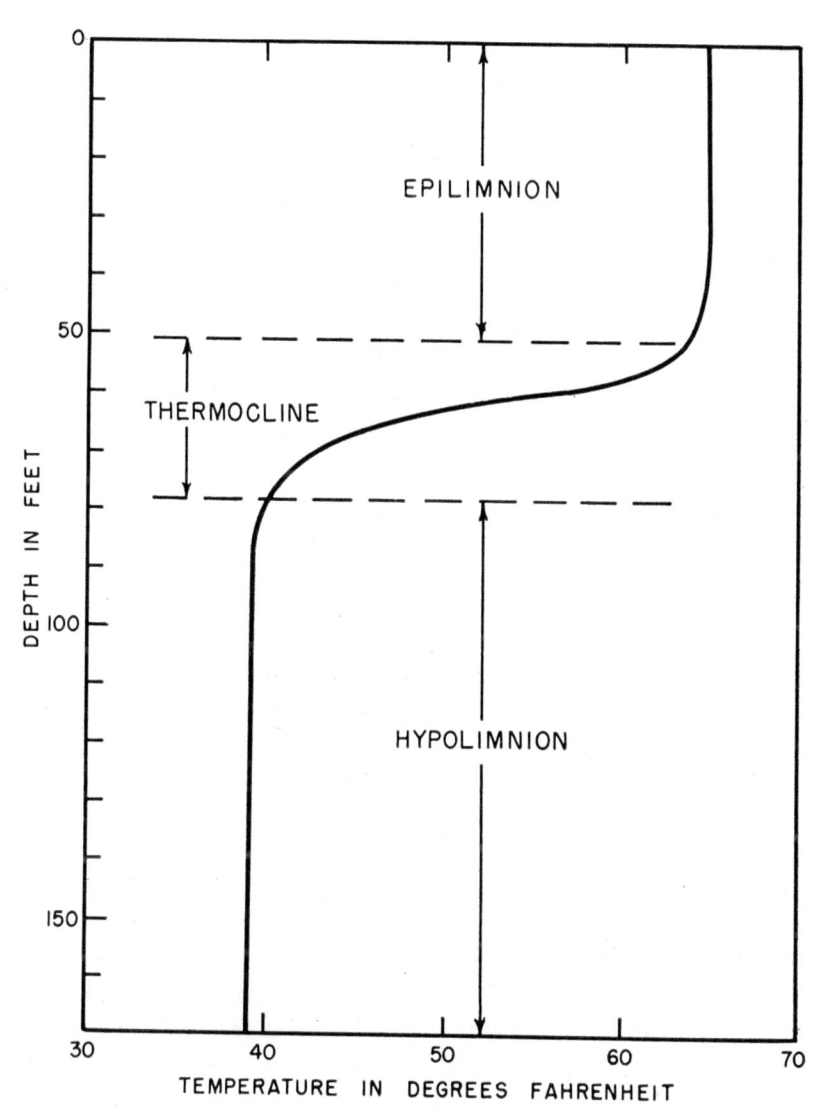

Fig. 23. Terminology of the lake water temperature profile. A typical summer distribution of temperature, with warm isothermal water in the upper zone (the epilimnion), and cold isothermal water in the lower zone (the hypolimnion).

density during the last week of May. The summer phase then was initiated by the formation of a shallow, protective thermocline at the surface on a relatively windless day, and after that the surface temperature increased rapidly. By mid-July the surface-layer temperature was between 65° and 72° F.; there was thus a difference of about 28°F. between the epilimnion and the hypolimnion. The temperature of the epilimnion did not change materially between mid-July and September, but stirring by the wind gradually deepened this homogeneous layer until it attained a thickness of from 32 to 50 feet. The bottom-water temperature remained nearly constant at approximately 40° F. all through the summer. The temperature of this layer is that which the entire column had reached at the time the surface protective layer began forming.

During the fall cooling period the water maintained a well-defined upper isothermal layer, a thermocline, and a cold bottom layer, until the surface had cooled to about 43.2° F. Wind stirring was then effective enough to destroy the thermocline and mix surface and bottom water, producing an isothermal condition at a temperature above the temperature of maximum density. Vertically isothermal water then prevailed for a period of about three months, from early December to late February, until the lake had cooled to approximately 35.6° F. While this isothermal condition existed the temperature of the entire column of water slowly dropped through a range of nearly 8° F. Further cooling then produced sufficiently light surface water, with a temperature at or just above 32° F., and prevented stirring to the bottom because a reverse thermocline had developed. This initiated the winter stationary period.

The spring warming period, which began on about March 21, had two phases. In the first, a slow increase in the temperature of the colder and lighter surface water took place until the water was isothermal at about 35.6° F. The spring overturn then began and the entire column of water was warmed until it reached a temperature of approximately 40° F. If there had been no wind stirring of the water, further warming of the surface above 39.2° F. would have produced a lighter surface layer immediately, and a bottom layer with maximum density would have been isolated. Actually, wind stirring kept the water in an isothermal condition until the temperature was slightly above the temperature of maximum density. Then a lighter surface layer was formed, completing the annual cycle which has been described.

Within this annual cycle, which was observed in Lake Michigan during the years 1941 to 1943, there were two periods during which complete circulation or overturn of the water occurred, and two periods during which there was stratification of the water which prevented overturn. The summer stationary period lasted from June through November, with a very stable stratification of the water. The fall overturn was maintained for three months, lasting well into the winter season. The winter stationary period was of relatively short duration, approximately one month, with only a low degree of stability. The spring overturn existed for two months.

During the year 1945, in the course of submarine test diving and underwater sound training, the author had occasion to observe a reverse thermocline in Lake Michigan in January and February and to verify the expected spring isothermal condition and the summer stable period. In August of that year the depth of the epilimnion in mid-lake over the deep northern basin was 60 feet.

Horizontal circulation in the lake may be deduced from the thermal patterns. During both the summer and winter periods of thermal stratification the boundaries between the different water masses are not horizontal. Church (1945, p. 1) states that during the summer stationary period the warm, light, surface layer became deeper on the east side of the lake, and the lower part of the thermocline or transition zone dipped downward on both the east and west sides; in other words, the cold, dense, bottom water mass was convex upward in the middle of the lake, the light upper layer was thickest on the east side, and the transitional zone of the thermocline was compressed and bent downward on the east side. This arrangement reflects a counterclockwise circulation of the water. Church (1942) further states, "The autumnal thermal pattern between Milwaukee and Muskegon demands a northbound current on the east side and a southbound current on the west side. These currents diminish to the vanishing point as the lake passes through a thermal nodal point near 4.0° C. The isothermal orientation of the one synoptic chart, made at the onset of vernal warming, further supports the cyclonic pattern of circulation. However, the chart shows the existence of two closed cyclonic circulation cells, which are centered over the two deep basins of the lake" (p. 1). This conclusion confirms the pattern of surface currents deduced by Harrington (Fig. 16). Variations of this general thermal pattern were found under different wind velocities and directions. For example, upwelling of deeper water was relatively common along the west shore whenever a wind having a westerly component was blowing.

During three synoptic studies which were made in Lake Huron on June 29, July 27, and August 25, 1954 (Ayers, Anderson, Chandler, and Lauff, 1956), it was found that a thermocline was absent in the wide central portion of the lake in late June. By late July a pronounced thermocline had developed over all but the ends of the lake, and the thermocline was present everywhere except in the extreme southern end in late August. Two parent water masses, Lake Michigan and Lake Superior waters, and a mixture of the two, Lake Huron water, were present. Lake Superior water could be traced for only a short distance. Lake Michigan and Lake Huron waters were both widespread on the surface. At greater depths Lake Huron water dominated the east side of the lake and was present at some places on the west side, while Lake Michigan water occurred only on the west side of the lake.

Sinking and upwelling occurred in Lake Huron during the summer of 1954, but the development of the thermocline generally limited vertical movement of the water to the thermocline depth, or less. The thermocline generally lay between 50 and 100 feet. For further details the reader is referred to Ayers, Anderson, Chandler, and Lauff (1956).

Temperatures in Lake Erie were observed from June to September in 1928, as a part of a cooperative survey carried on by various government agencies (Fish, 1929, pp. 7–220). The observations covered only the summer period, and they were made in the eastern quarter of the lake which includes the single deep basin with a depth of 210 feet. Summer stratification of the water was well advanced when the survey was begun. Throughout the summer months the mean temperature of the surface water in the open lake was approximately 70° F., and this temperature generally prevailed down to depths of 35 to 50 feet. A rather sharp thermocline was present, and the temperature of the bottom water in the deep hole was between 40.0° and 40.5° F., just above the point of maximum density (Parmenter, 1929, pp. 28–45). Elsewhere in Lake Erie the shallow water is nearly isothermal throughout the year. In the western part of the lake the daily water temperatures did not vary more than 2° C. from the surface to the bottom in 1942, except on a few occasions in spring and early summer (Chandler and Weeks, 1945, p. 437).

From the foregoing summary of the annual temperature cycle of Lake Michigan and the observed water stratification during the summer in Lake Huron and in the deep basin in Lake Erie, it is seen that these lakes conform in a general way to the pattern which has been observed in most other fresh-water lakes of the temperate zone. It is undoubtedly safe to assume that the other two Great Lakes have similar annual cycles. Lakes Superior and Ontario are comparable to Lakes Michigan and Huron in having deep basins. The only data available to confirm this assumption of similar thermal patterns are a few random observations. Some of these, assembled by Wright (1931, p. 413) are as follows: in Lake Superior in August, 1871, most of the bottom-water temperatures were 38.8° to 39.0° F. In Georgian Bay, Lake Huron, in the summer of 1889, temperatures of 38.7° to 39.0° F. were reported for depths below 350 feet. Smith (1874, p. 692) reported for Lake Superior that in the summer the temperature everywhere below 180 to 240 feet was very uniform, varying only slightly from 39° F., while at the surface it varied from 50° to 55° F. Several temperature curves for the Silver Bay area in northwestern Lake Superior, obtained in July, 1954, show a strong thermocline existing between depths averaging about 30 and 75 feet for the upper and lower limits (Swain and Prokopovich, 1957, p. 531). Kindle (1925, p. 24) reported temperatures in Lake Ontario, observed in June, 1916, which show a distinct stratification of warm surface water over cold bottom water.

The annual temperature cycle of the typical temperate-zone lake has an important effect on the chemistry of the lake waters, on the plants and animals living in them, and on the character of the bottom sediments. These effects will be described in the following pages.

PRECIPITATION AND EVAPORATION

The water of the Great Lakes is derived almost entirely from precipitation, so far as is known. There may be small additions and small losses by underground seepage beneath the watershed boundaries, but these are considered negligible. The quantities of water falling on the Great Lakes region have been calculated from the data of Table 2, and are given in Table 3 in cubic feet

Table 3. *The water budget of the Great Lakes: precipitation, discharge, and losses attributed mainly to evaporation.*

Lake	Area of lake plus drainage basin (square miles)	Precipitation, average annual (feet)	Precipitation (cubic feet per second)	Discharge (cubic feet per second)	Losses (cubic feet per second)	Per cent loss
Superior	80,000	2.42	170,950	73,700 [a]	97,250	57
Michigan and Huron	115,170	2.58	231,680	105,300 [b]	126,380	45
Erie	32,490	2.83	81,210	18,800 [c]	62,410	77 [f]
Ontario	34,800	2.83	86,960	38,200 [d]	48,760	56 [g]
Total	262,460	—	570,800	236,000	334,800	59

[a] Discharge of St. Marys River.
[b] Discharge at Port Huron plus discharge at Chicago minus inflow from Lake Superior.
[c] Discharge of Niagara River minus inflow from Lake Huron; does not include an unknown amount of diversion by Welland Canal, Black Rock Canal, and New York State Barge Canal. These diversions reach Lake Ontario by other channels.
[d] Discharge of Lake Ontario to St. Lawrence River minus inflow from Niagara River. Inflow from canals [e] not taken into account.
[e] Discharge of Lake Ontario plus diversion at Chicago.
[f] High because of unaccounted-for diversions.[e]
[g] Low because of unaccounted-for inflow.[d]

per second (mean precipitation) for each of the lakes and its drainage area. The mean discharge attributable to precipitation in each basin (actual discharge minus inflow from other basins) is given, insofar as it is known. Because the discharge through the Straits of Mackinac has not been measured, the data for the Michigan and Huron basins are combined. The difference between precipitation and discharge is given in cubic feet per second and as a percentage loss of the total precipitation. The losses are due to evaporation from the surfaces of lakes, rivers, and swamps, and from the land surface; to transpiration by plants; and to various presumably minor factors such as seepage, hydration of minerals, and municipal and industrial uses. It is believed that most of the water used by man is ultimately returned to the lakes, except for an appreciable amount lost by evaporation. The per cent loss shown in the last column of Table 3 is attributed mainly to evaporation. It does not represent the evaporation from a free water surface; it is the net result, mainly, of evaporation and transpiration from the entire area considered. The losses, whatever their causes may be, amount to about 60 per cent of the precipitation, or to about 19 inches of the annual precipitation.

CHEMISTRY OF THE LAKE WATER

One of the first questions raised by early investigators of the lakes was whether the water in the deep basins might be salty—either because of solution of underlying Paleozoic salt beds or because of a possible encroachment by the sea during some early lake stage. Stimpson, in 1870, writing on the deep-water fauna of Lake Michigan, said, "At present the bottom water, judging from a specimen we obtained from a depth of fifty fathoms approximately, is entirely fresh" (p. 405). Smith (1874), in a paper on the fauna of Lake Superior, stated, "Water was taken from the bottom at many points, and was everywhere perfectly fresh. That from 169 fathoms [1014 feet] gave no precipitate with nitrate of silver" (p. 692). Schermerhorn (1887) reported, "A chemical analysis of water taken from the deepest part of Lake Superior failed under the application of delicate tests to indicate the presence of salt" (p. 281). Since these reports were published, it appears that the freshness of the deep waters of all of the lakes has been taken as an established fact. From what is known of the circulation of the lake waters, it is a reasonable assumption that the waters are mixed sufficiently to have essentially the same content of dissolved solids at all depths. The thermal profiles of Lake Michigan indicate that only fresh water is present there.

The water of the Great Lakes is characterized as "fresh," but even fresh water contains some material in solution. Selected analyses of surface-water samples from the various lakes are given in Table 4, and are discussed in the following paragraphs.

Table 4. *Analyses of surface water from the Great Lakes.*[a]

Lake	Percentage composition of dissolved solids									Total dissolved solids (parts per million)
	Ca	Mg	Na + K	CO_3	SO_4	Cl	NO_3	SiO_2	Fe_2O_3	
Superior[b]	22.42	5.35	5.52	47.42	3.62	1.89	.86	12.76	.16	60
Michigan[c]	22.21	7.01	4.02	49.45	6.15	2.31	.26	8.54	.05	118
Huron[d]	22.33	6.52	4.10	47.26	5.77	2.42	.38	11.16	.06	108
Erie[e]	23.45	5.75	4.92	44.70	9.83	6.58	.23	4.46	.08	133
Ontario[f]	23.66	5.49	4.81	45.70	9.15	5.87	.23	5.03	.06	134

[a] From Clarke (1924, p. 74).
[b] Mean of 11 samples taken monthly between Sept. 22, 1906, and Aug. 22, 1907, at Sault Ste. Marie.
[c] Mean of 11 samples taken between Sept. 20, 1906, and Aug. 20, 1907, at St. Ignace.
[d] Mean of 9 samples taken between Sept. 21, 1906, and June 21, 1907, at Port Huron.
[e] Mean of 11 samples taken between Sept. 19, 1906, and Aug. 28, 1907, at Buffalo.
[f] Mean of 11 samples taken between Sept. 18, 1906, and Aug. 19, 1907, from the St. Lawrence River at Ogdensburg, 64 miles downstream from Lake Ontario.

There are distinct resemblances between the waters from the various lake basins, especially in that calcium is the chief metal and carbonates are the predominating salts. The water is alkaline in reaction. Observations made in the northeastern part of Lake Michigan during the summer months show that the pH of the water approximates 8.1.[6] In Lake Erie, during the summer of 1928, the surface water maintained a pH close to 8.2 while the extreme range in the open lake was from 7.8 to 8.4. The lower values were found in deeper water (Burkholder, 1929, pp. 70–71). The bottom water in Lake Superior, at depths of 700 to 900 feet in July, 1954, was only slightly alkaline, with a pH of 7.1 to 7.4 (Swain and Prokopovich, 1957, p. 536).

[6] Matteson, M. R., Dept. of Zoology, Univ. of Illinois, Urbana, personal communication.

The relatively high percentage of magnesium in the water of Lake Michigan probably is derived in great part from a large area of magnesian rocks in southeastern Wisconsin. An increase in sulfates is noted in the more southerly lake basins. This may be due to the leaching of sulfate-bearing shales.

Silica is present in all of the Great Lakes in minute quantities, the water being undersaturated with regard to the solubility of quartz and far undersaturated with regard to the solubility of amorphous silica. Table 4 shows that the highest relative concentrations occur in Lakes Superior and Huron, which receive drainage from extensive areas of crystalline Precambrian rocks.

Iron, reported as Fe_2O_3, is present in very minute quantities. It is interesting to note that its relative concentration in Lake Superior is two to three times that in the other lakes. The relative decrease of both silica and iron in the lower lakes apparently is a result mainly of the addition of other materials to the total dissolved-solids content in these lakes.

The total dissolved solids of the lake waters, ranging from 60 to 134 parts per million, is low compared to the total for sea water, which is 35,000 parts per million. Rain water usually contains from 30 to 40 parts per million dissolved solids (Welch, 1935, p. 141); thus the water of Lake Superior is nearly as dilute as rain water. The increase in concentration from Lake Superior to the St. Lawrence River is noteworthy. Several factors influencing the concentration are mentioned in the following paragraphs.

The percentage of lake-surface area to area of land draining into a basin has an important bearing on the quantity of dissolved solids in the water. The drainage areas of the various lakes are shown in Fig. 1. In the Lake Superior basin the ratio of lake surface to entire area (lake plus land) is 0.4 to 1.0. In other words, assuming that precipitation is evenly distributed over the basin, 40 per cent of the precipitation falls directly into the lake and thus has no opportunity to dissolve materials from the watershed. In the Michigan and Huron basins the lake surfaces make up 32 and 31 per cent, respectively, of the entire areas. In the Erie basin, the lake surface is 28 per cent, and in the Ontario basin the lake surface is 22 per cent of the entire area. There thus seems to be some degree of correlation between the decreasing percentage of lake-surface area receiving direct precipitation and the increasing concentration of dissolved solids. The slightly lower concentration in Lake Huron than in Lake Michigan (Table 4) indicates dilution of the Michigan and Huron waters by inflow from Lake Superior.

Evaporation from free water surfaces causes concentration of dissolved solids. The losses of water, shown in Table 3, are nearly the same in all of the basins. There is, however, a cumulative effect as the waters flow through the Great Lakes system.

The addition of municipal and industrial wastes to the lake waters is appreciable in the heavily settled areas of southern Lake Michigan and especially of Lake Erie. Data on dissolved solids in Lake Erie in 1951 and 1952, at the lake-water intakes of several Ohio cities and towns, show ranges of concentration from 148 to 367 parts per million, and an average concentration for all intakes of 182 parts per million (Youngquist and others, 1953).

Oxygen is present in the surface waters of the lakes, as shown by a number of analyses, and it is assumed that it is always present there. The principal sources of oxygen are (1) directly from the atmosphere through the lake surface and (2) from the photosynthesis of chlorophyll-bearing plants. Absorption of oxygen from the air is accomplished by direct diffusion at the surface and through the mixing in of air bubbles during wave action. It is assumed that absorption from the air is the principal source in the Great Lakes, because rooted vegetation is practically absent from the open lake basins and free-floating chlorophyll-bearing plants are not abundant. An additional source of some importance in the Great Lakes may be the inflowing streams, which have acquired oxygen from inland waters where green vegetation is more abundant.

Whether oxygen is present generally in the deeper waters of all of the lakes is not known from direct evidence.[7] In Lake Erie during the summer stationary period in 1928, the degree of oxygen saturation was found never to go below about 50 per cent. At the bottom of the lake in a depth of 197 feet the water was 59 per cent saturated in August. At no time was there any depletion of

---

[7] See n. 8.

dissolved oxygen in such degree as might menace the life of crustacea or fish (Burkholder, 1929, p. 68). The thermal stratification of the lake indicates that there is no transfer of water from the surface to the isolated deep water mass; yet the deep water mass does not become stagnant. It was suggested that the oxygen in the deep water is produced during the summer stationary period by green plants, which receive adequate light to carry on photosynthesis in the relatively shallow basin of Lake Erie (Parmenter, 1929, pp. 41-43).

The presence of certain species of fish in the deep water of Lake Michigan, where they have been caught at various times of the year, has been cited as an indication that oxygen is present at all times, and that Lake Michigan therefore must have a deep circulation at all times (Shelford, 1937, p. 60). The studies made by Church and reviewed earlier in this chapter have shown that this lake has a deep circulation only during the fall and spring periods of overturn, when the oxygen of the deep water can be replenished. The studies of Lake Erie have shown that a usable oxygen supply is maintained in deep water in spite of the absence of overturn, probably because of photosynthetic activity of plants. The deep basins of the other four lakes, however, presumably cannot receive sufficient light to support the growth of plants. In the oceans it has been found that no effective plant production can take place below a depth of about 260 feet (Sverdrup, Johnson, and Fleming, 1942, p. 774). If there is sufficient oxygen in the deep water of Lake Michigan to support fish throughout the long summer stable period, the oxygen brought down during the spring overturn must not be entirely exhausted. If this is true, then the number of oxygen-using organisms present and the quantity of organic matter settling into the deeper water must be insufficient to use up the oxygen.[8]

All the Great Lakes may be classified as oligotrophic or meso-

---

[8] As this book was going to press, the author was informed by D. C. Chandler (Great Lakes Res. Inst., Univ. of Michigan, Ann Arbor, personal communication, Aug. 11, 1958) that unpublished data obtained by the U.S. Fish and Wildlife Service show that oxygen is present in the deep water of the Great Lakes, in amounts approaching saturation, during the summer stable period.

trophic, or poor in nutrients. It is likely that organic matter produced in the lakes plus that brought in by streams constitutes a quantity which is too small to consume all of the available oxygen in the water.

Core samples of the deep-water sediments of four of the Great Lakes provide some indications of the probable nature of water circulation, not only at the present time but in early stages of the lake histories as well. These are discussed in the following chapter.

chapter three

SEDIMENTS OF THE GREAT LAKES

A wide variety of sedimentary materials occurs in the Great Lakes. Because of the size and varied topography of the lakes and the incompleteness of our knowledge, bottom-sediment maps of the entire lake basins cannot be drawn with any reasonable degree of completeness. Several maps showing sediment distribution in various portions of the lakes have been published. Following is a partial list of these:

Lake Michigan, entire (Emery, 1951, fig. 4).
Lake Michigan, southern (Hough, 1935, figs. 1 and 4).
Lake Michigan, Illinois shore (Division of Waterways, State of Illinois, 1952).
Lake Erie, eastern (Pegrum, 1929, fig. 4).
Lake Erie, western (Pincus, Roseboom, and Humphris, 1951, figs. 4 and 5).
Lake Erie, western (Pincus, Metter, Humphris, Kleinhampl, and Bowman, 1953, figs. 2-36, 3-34).
Lake Ontario, various parts (Kindle, 1925, figs. 2, 3, 4).

The U.S. Lake Survey published notations of bottom character on its navigation charts of the Great Lakes prior to 1935. When a detailed survey (Hough, 1935, pp. 72-75, fig. 4) showed that the materials varied considerably within a short distance, the Survey deleted all references to bottom materials on its charts, because the scattered notations might be more confusing than helpful to navigators.

66 GEOLOGY OF THE GREAT LAKES

No sediment-distribution maps are included in the present book. The distribution of the lake sediments is described in general terms, however, in the following paragraphs. Because all of the lakes have undergone extreme changes in the depth of their waters, only the present deep-water zones will contain a complete record of sedimentation. The descriptions of the shallow-water sediments are, therefore, brief.

GRAVEL

Exposed beaches of the lakes in the vicinity of rock cliffs and bluffs cut in glacial till are frequently composed of gravel (Fig. 24). Detailed studies of the size and shape of beach gravels have been made by Grogan (1945), and Krumbein and Griffith (1938). Gravel occurs on the lake bottom in depths as great as 250 feet (refer to the discussion of depth of wave action, Chap. 2).

Fig. 24. A gravel-and-boulder beach on the eastern shore of Lake Superior. The beach materials were derived from erosion of glacial till. Abandoned beaches at higher levels appear in the middle distance.

SAND

The most common material of the beaches is sand (Fig. 25), and sand is found on the lake bottoms in some areas to depths of more than 300 feet (Hough, 1952, pp. 164–65). The results of two detailed studies show that the beach sands of southern Lake Michigan and Cedar Point in Lake Erie commonly have a median grain diameter of from 0.25 to 0.35 mm., but individual samples have median diameters ranging from 0.139 to 1.520 mm. (Hough, 1935, p. 64, and Pettijohn and Ridge, 1932, p. 83). These sands generally are somewhat finer than those on the outer beach of Cape Cod, on the northern Atlantic shore of the United States; Schalk (1938, pp. 42–43) found the median diameters there to range generally from 0.30 to 1.90 mm. The Great Lakes beach sands compare closely, however, with the beach sand on the southern Atlantic shore between Charleston, South Carolina, and Miami, Florida, where Martens (1935, p. 1568) found a range of median diameters of individual samples from 0.108 to 0.690 mm., and an average median of 0.253 mm.

On the bottom, in Lake Michigan, there is generally first a de-

Fig. 25. A wide sand beach on the eastern shore of Lake Michigan, bordered by stabilized sand dunes.

crease in the size of the sand grains with increasing distance from shore and depth of water, then an increase to sizes considerably coarser than those of the beach sands, occurring in depths of 60 to 100 feet (Hough, 1935, figs. 2 and 3). The reason for the presence of a zone of coarser sand offshore in relatively deep water is unknown. A similar occurrence of coarse sand on the outer part of the Continental Shelf in the oceans has been observed (Stetson, 1938, pp. 34-36, and Shepard, 1948, pp. 153-54).

The sands of the Great Lakes beaches and bottoms, insofar as they have been studied, are relatively immature sediments. The size range of the grains generally is narrow, or the "sorting" is good. According to a classification by Folk (1951, p. 128), this indicates textural "maturity," and it is what would be expected as an adjustment to the strength of wave and current action in large bodies of water. The degree of rounding of the grains, however, generally is very low, and this excludes them from Folk's "supermature" textural class. The significance of the angularity of the grains is that they have not been in the present environment long enough to become rounded. Various experimental studies have shown that rounding of sand grains requires a very long time (Pettijohn, 1949, pp. 400-415). Mineralogically the sands are quite "immature"; while they are composed predominantly of quartz, they also contain significant amounts of feldspar and various other relatively unstable minerals such as augite, hornblende, hypersthene, and olivine. These statements on the texture and composition of the sands are based mainly on publications by Hough (1935), Pettijohn (1931), Pettijohn and Lundahl (1943), Pettijohn and Ridge (1932 and 1933), Pincus, Roseboom, and Humphris (1951), Pincus, Metter, Humphris, Kleinhampl, and Bowman (1953), and on unpublished work by the author.

The relative immaturity of the sands described in the foregoing paragraph reflects their main derivation from the mantle of glacial drift which occurs throughout the region; the glacial deposits consist mainly of unweathered debris derived from the various rocks of the region and transported by glacial ice. Both streams and shore processes erode the material from the drift and deliver it to the lakes. The resulting deposits are quite young or "immature." A certain amount of sorting as to size, shape, and specific gravity has taken place but the materials generally are but slightly affected by abrasion and weathering. The finer-grained constituents have been winnowed out and transported to deep water. One indication of the importance of this process is the occurrence of a microscopic Devonian plant fossil in great numbers in the Lake Michigan bottom clays of the southern basin. The fossil, a waxy spore case, occurs in the Devonian black shales and is also a common constituent of the glacial till around the southern end of Lake Michigan. Because the Devonian shales are nowhere known to be exposed at the surface in the area, it is apparent that the glacial till is the immediate source of the fossil for the present-day lake sediments. This fossil, once named *Sporangites* and now called *Tasmanites,* is thus observed in three cycles of deposition; in the original sediment, the Devonian shale, then in glacial till, and finally in present-day lake clays (Hough, 1934, pp. 646–48).

Sand dunes occur in many places along the shores of the Great Lakes (Fig. 25). The most spectacular dunes form a nearly continuous belt along the east coast of Lake Michigan from the Illinois-Indiana state line northward to Grand Traverse Bay. The prevailing westerly winds have removed sand from the beach and blown it inland where it generally accumulates in ridges parallel to the shore. Easterly winds are more commonly accompanied by rain, and the wet sand can not be returned to the lake by the wind. The western shore of Lake Michigan has very few dunes, probably because the prevailing westerlies return to shore any sand removed by winds from other directions. The eastern side of Lake Huron south of the entrance to Georgian Bay has dunes along the shore at intervals, but they are interrupted by extensive sections of wave-cut cliffs.

SILT

A relatively minor type in the Great Lakes, silt is occasionally found between the zones of sand and the deep-water clay deposits. It ranges from sandy silt to clayey silt, and generally occurs on gently sloping bottoms as in the area on the eastern side of Lake Michigan, between three and ten miles offshore from Grand Haven.

CLAY

Sediments of clay size are common in all of the deep basins of the Great Lakes, as was shown by notations on the U.S. Lake Survey charts published before 1935. Smith, in 1871 (pp. 373–74), stated that "in all the deeper parts of the lake [Superior], the bottom . . . is covered with a uniform deposit of clay, or clayey mud, usually very soft and bluish or drab in color." Kindle, who was the first geologist to study the deep-water sediments, reported the following information for Lake Ontario:

The sediments which are accumulating on the deep water portion of the lake bottom, remote from the shore line where depths ranging from 200 to 700 feet prevail, are distinguished from the inshore deposits by the fineness of their materials. This deep water area is surrounded by a belt of coarse sediment. In place of the sand gravel and muddy sand which comprises much of the sediments of the inshore zone only the finest mud of jell-like consistence is generally met with in the greater depths. . . . At station 270 [depth 615 feet] a sample 3 feet 6 inches in length was taken. This sample is of similar texture throughout. It is composed of soft fine textured ooze with the consistency of thin mush. The colour is buff grey and black in alternating bands. It is stratified throughout in alternating bands of black and buffish grey mud which apparently differ only in colour. When dry the colour is a uniform ash grey. The banding disappears after a half hour's exposure to the air [Kindle, 1925, pp. 47–48].

In Lake Erie, Pegrum (1929, p. 22) found "mud and clay" of a dull gray color in the deep eastern basin and extending nearly to the shore in the protected waters of Long Point Bay. All of the western part of Lake Erie was floored by irregular areas of sand and gravel with a few patches of mud.

In Lake Huron no extensive study has been made of the deep-water sediments, but the author arranged for a bottom-sampling expedition which was carried out in September, 1947, by cooperation between the Woods Hole Oceanographic Institution and the U.S. Lake Survey. Eleven core samples, ranging in length from three feet to six feet, three inches, were obtained. Ten of the cores were from depths between 234 and 270 feet in the southwestern basin of the lake, and one was from a depth of 528 feet on the western flank of the deep main basin of the lake. They contained mainly clay-sized material and showed colors ranging from a "rusty" dark brown at the top to light brown or gray at deeper levels. Two of the longer cores contained pink-colored clay at their bottom ends (Weinberg, 1948). None of the cores was from the bottom of the deepest part of the lake.[1]

The deep-water areas of Lake Michigan have been studied in greatest detail. Nearly 100 core samples have been taken from this lake (Hough, 1935, pp. 77–78, and 1955, p. 957), and the information obtained from them is summarized in the following paragraphs. One core sample 35 feet long, taken at the deepest point in Lake Michigan (923 feet), contains the entire column of lake sediment from the glacial-till bottom to the present surface of the deposit (see log of core M-1 in Fig. 26). At the bottom of this core there is a compact, pebbly, sandy clay which undoubtedly is glacial till. Because of its red color, this till is believed to be of Valders age and to correlate with the red till found onshore in northeastern Wisconsin. Above the compact till is a 16-inch-thick layer of red clay containing some sand and a few pebbles but showing no lamination. Above this is one foot of laminated red clay and gritty red clay. This is followed by two feet, two inches of finely laminated red clay and silt which appear to be glacial lake varves (annual layers). A few isolated pebbles or granules occur within the "varved" clay. These fine laminations become indistinct in the upper part of the zone, and the zone thus grades into unlaminated, fine-grained red clay above. The unlaminated, fine-grained red clay, which begins 4.5 feet above the till, extends upward for approximately 19 feet and then grades into a gray clay. The gray clay extends to the top of the core, through a

---

[1] Since the above paragraph was written the author has taken several core samples in the northwestern part of Lake Huron, in association with the Great Lakes Research Institute. These will be described in a separate publication after detailed studies have been completed. Preliminary examination of the cores indicates that there will be no conflict with the statements made here.

72 GEOLOGY OF THE GREAT LAKES

LOG OF CORE SAMPLE M-1 FROM DEEPEST POINT IN LAKE MICHIGAN

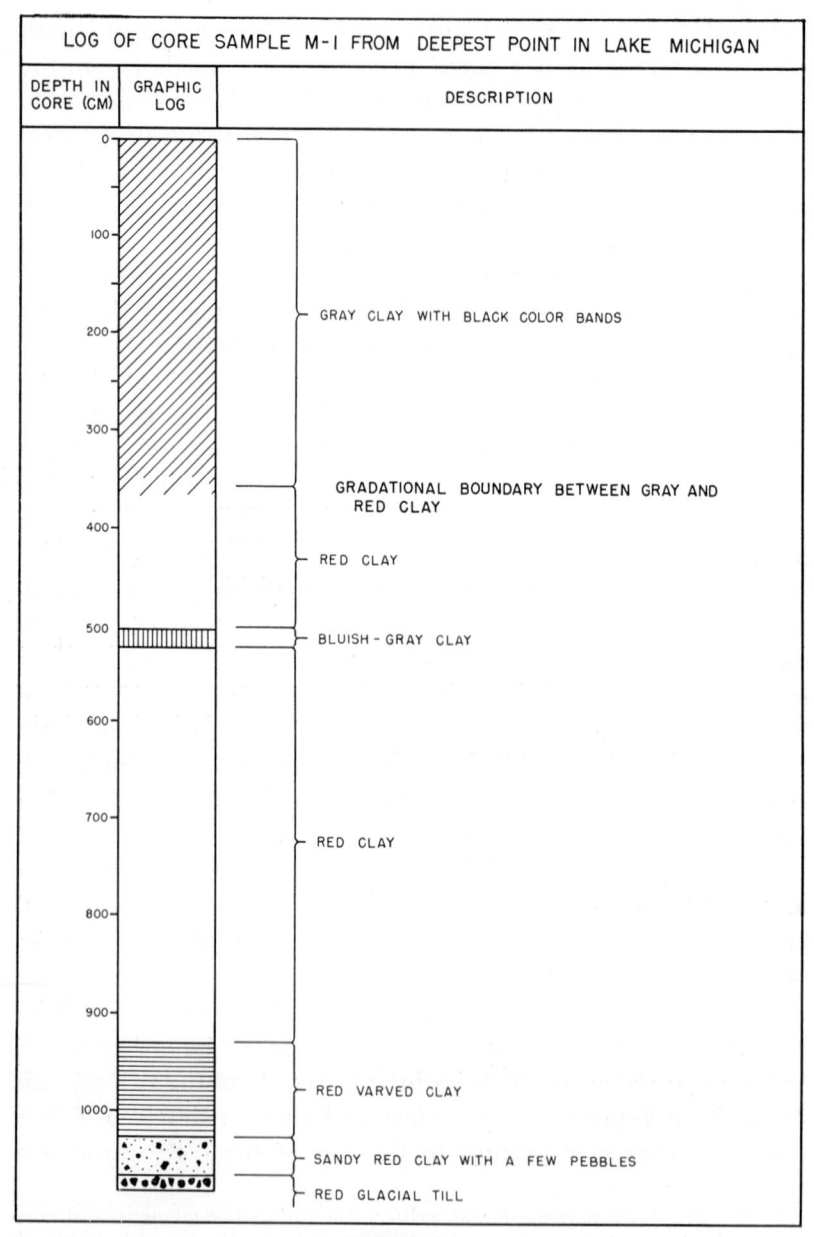

DEPTH IN CORE (CM)	GRAPHIC LOG	DESCRIPTION
0–360		GRAY CLAY WITH BLACK COLOR BANDS
360–430		GRADATIONAL BOUNDARY BETWEEN GRAY AND RED CLAY
430–500		RED CLAY
500–530		BLUISH-GRAY CLAY
530–940		RED CLAY
940–1030		RED VARVED CLAY
1030–1050		SANDY RED CLAY WITH A FEW PEBBLES
1050+		RED GLACIAL TILL

Fig. 26. Log of core sample from the deepest point (923 feet) in Lake Michigan.

distance of approximately 10 feet. The red clay and the gray clay, which together have a thickness of 29 feet, are extremely fine-grained material which has obviously been deposited in deep water in the lake. The median diameter of this material ranges from 0.0003 to 0.0042 mm. and averages approximately 0.0016 mm. throughout its thickness. The clay averages 96 per cent finer than sand size.

The gray clay, comprising approximately the upper ten feet of the deposit, is essentially homogeneous in texture but it contains color bands in which jet-black zones alternate with gray zones. The jet-black zones are generally two to eight mm. in thickness and are separated by from one to four cm. of gray clay. The black color disappears within an hour after exposure to the air. It will be recalled that Kindle (1925, p. 48) observed a similar phenomenon in the Lake Ontario clays. Differential thermal analysis of the black material from Lake Michigan core M-1, while it was still fresh, showed no appreciable organic matter to be present, but an exothermic reaction at about 450° C. indicated the presence of iron sulfide. Thermal analysis of the black zones after they had lost their color on exposure to the air indicated the absence of sulfide, and X-ray analysis showed the presence of the iron oxide, hematite. Evidently the jet-black color is due to a state of reduction of the materials present. The underlying red clays contain hematite.

All of the fine-grained clays, including the color-banded black and gray clays and the underlying red clays, are similar in their content of quartz, clay minerals (illite is predominant, kaolinite is present), calcite, dolomite, and siderite.[2]

The colors of the sediments in core M-1 have been recorded in terms of the notations of the Goldman-Merwin color chart prepared under the auspices of the National Research Council and based on Ridgeway (1912). The colors were noted immediately after the core was sectioned while the material was still moist and

---

[2] The data on differential thermal analysis and X-ray studies were obtained by R. E. Grim, Dept. of Geology, Univ. of Illinois, Urbana, personal communication. Detailed mechanical analyses of the sediments were made by Snodgrass (1952).

unaltered by exposure to the air. The colors of the materials were recorded as follows:

"Gray" clay: 5/23/6, 5/23/o, 5/23/i
(with jet-black zones)
"Red" clay: 3+/1/a, 3/1/a, 3/1/o, 3/i
"Bluish-gray" clay: 4/44/o
"Red" varved clay: 3+/1/a

Because the sandy red clay with pebbles and the red till, at the bottom end of the core, had dried out before the core was opened, their colors were not recorded.

The characteristics of the materials in this deep-water core sample are interpreted as a possible record of changing limnological conditions, as follows: after the front of a glacier (presumably the Valders ice) retreated north of this position, there was deposition of sediment from the melt waters of the wasting ice, and some coarser material was rafted in blocks of glacial ice and dropped to the bottom. A little later, but while glacial ice was still present in the northern end of the Michigan basin, seasonal variations in the rate of melting resulted in the deposition of the laminated red clay and silt, or "varves." The unlaminated red clay suggests that following this, and possibly beginning with the disappearance of glacial ice from the Michigan basin, there was a long period during which no appreciable amount of organic matter was deposited with the sediment. The dearth of organic matter may have been due to a limited supply from the land and a limited growth of aquatic organisms; it may also reflect greater destruction of the existing organic matter by oxidation in the deep water and on the surface of the bottom, before burial. The latter process may reflect a generally less stable condition of the water column which permitted more vertical circulation during the year and no oxygen depletion of the deeper waters. The gray clay with black zones, in the upper part of the core, records reducing conditions in the bottom sediment. This seems to indicate burial of an amount of organic matter sufficient to cause reduction of the iron. The increase in amount of organic matter buried in the bottom may have been due to an increase in the supply from the land and in the growth of aquatic organisms; it may also reflect an appreciable decrease in the oxygen saturation of the deeper waters during the summer seasons. The development of the marked stability of the water column during the summer seasons, which exists at the present time, may have coincided with the beginning of deposition of the gray clay.

No other core samples obtained from Lake Michigan contain the complete sequence shown by core M-1, but all of them contain parts of the sequence and they indicate the widespread distribution of the red clay to gray clay sequence and the occurrence of the color banding in the gray clay. One detail which has not previously been discussed is a bluish-gray zone within the red clay, occurring well below the bottom of the gray clay (Fig. 26). This is present in the red clays of many of the core samples and therefore records some general condition. The significance of the bluish-gray zone is not known, but it may represent a temporary reducing condition whose cause cannot as yet be assigned to any known events of the lake history.

All of the core samples taken from depths greater than 350 feet show at least the upper parts of the sequence illustrated in Fig. 26. Cores taken from depths of less than 350 feet contain a record of a temporary lowering of the lake surface to 350 feet below the present level. This is indicated by the truncation of some of the normal deep-water layers and the deposition of a thin layer of sand and shallow-water shells, followed by resumption of deposition of the normal deep-water clays. This will be described in detail in Chapter 13, in the discussion of the Lake Chippewa low-water stage.

chapter four

PREGLACIAL HISTORY OF THE REGION

HISTORY OF THE BEDROCK FORMATIONS

Introduction

The distribution of the major types of bedrock in the Great Lakes region has been described in Chapter 1, where it was shown that the locations of the lakes and many details of their topography are closely related to the type of rock and to the structural framework of the region. It has been pointed out that the Great Lakes are known to have existed for only about 20,000 years, and this period dates from the time when a continental ice sheet, which had covered all of the lake basins, first shrank back from the southern divides and allowed water to accumulate between the divides and the glacial fronts. The basins themselves have been excavated in certain zones, generally coinciding with the outcrop belts of the less resistant rocks. The detailed history of the formation of the bedrock is of very little importance in connection with the history of the *lakes* of the Great Lakes system. As a part of the history of the Great Lakes *region,* however, a brief summary of the history of the bedrock formations will be of interest.

The Geologic Time Scale

The standard geologic time scale as used in the Great Lakes region is shown in Table 5. For a detailed discussion of the geologic history represented by this scale, the reader is referred to textbooks on historical geology.[1] The approximate-age figures were determined from the radioactivity of the rocks (Moore, 1949, pp. 31–33). An age of approximately 2700 million years has been obtained for a rock occurring in Manitoba, northwest of the Great Lakes (Ahrens, 1955, p. 159).

The Precambrian Era

Most of the details of Precambrian history can never be deciphered because their record has been largely obliterated by metamorphism of the rocks and by removal of great volumes of rock by erosion. The Precambrian rocks have been classified as shown in Fig. 27.

Keewatin. The oldest rocks (Keewatin of Table 5 and Fig. 27) occurring in the Great Lakes region are composed of great thicknesses of lava flows, volcanic ash beds, and sedimentary deposits. These were buried deeply in the earth, where they were recrystallized and complexly folded during a mountain-building period, and they now are found as isolated patches of a few square miles up to a few thousand square miles, distributed between Lake Superior and Hudson Bay. The Keewatin rocks are surrounded by the Laurentian granites which are obviously younger because they cut through and across the layered rocks. The granites were formed at considerable depth in the earth. Following this stage there was a time of general uplift of the region and profound erosion which removed large volumes of rock and produced smoothly beveled surfaces on the Laurentian granites and on the Keewatin stratified rocks.

Timiskaming. The Timiskaming rocks were largely sandstones, shales, and limestones, which were deposited over the beveled surfaces of the Laurentian granites. Sedimentation apparently continued without much interruption during many millions of years; then another period of widespread mountain building occurred. The sedimentary rocks were metamorphosed, and folded and intruded by the Algoman granites. Another long period of

---

[1] Moore (1949) gives an excellent summary of the Precambrian events in the Canadian shield area (pp. 71–81), and a number of paleogeographic maps showing the distribution of Paleozoic seas in North America (figs. 64, 82, 99, 113, 129, 144, 161).

Table 5. *The geologic time scale.*[a]

Era	Period		Approximate number of million years ago
Cenozoic	Quaternary	Recent	
		Pleistocene	0–1
	Tertiary	Pliocene	1–12
		Miocene	12–28
		Oligocene	28–40
		Eocene	40–60
		Paleocene	
Mesozoic	Cretaceous		60–130
	Jurassic		130–155
	Triassic		155–185
Paleozoic	Permian		185–210
	Pennsylvanian		210–235
	Mississippian		235–265
	Devonian		265–320
	Silurian		320–360
	Ordovician		360–440
	Cambrian		440–520
Precambrian	Killarney Granite		
	Keweenawan		
	Huronian		
	Algoman Granite		520–2700 [b]
	Timiskaming		
	Laurentian Granite		
	Keewatin		

[a] Based on the time scale recognized by the U.S. Geol. Survey as quoted by Ladd (1957, p. vii), with the addition of subdivisions of the Precambrian as recognized in Canada.

[b] The date, 2700 million years, is from Ahrens (1955).

erosion then ensued, during which the Timiskaming sediments and the Algoman granites were beveled. The Timiskaming sediments are not recognized south of Lake Superior.

Huronian. During the Huronian division of time, great thicknesses of sandstones, shales, and carbonate rocks were deposited, and there was also some volcanic activity which is represented by

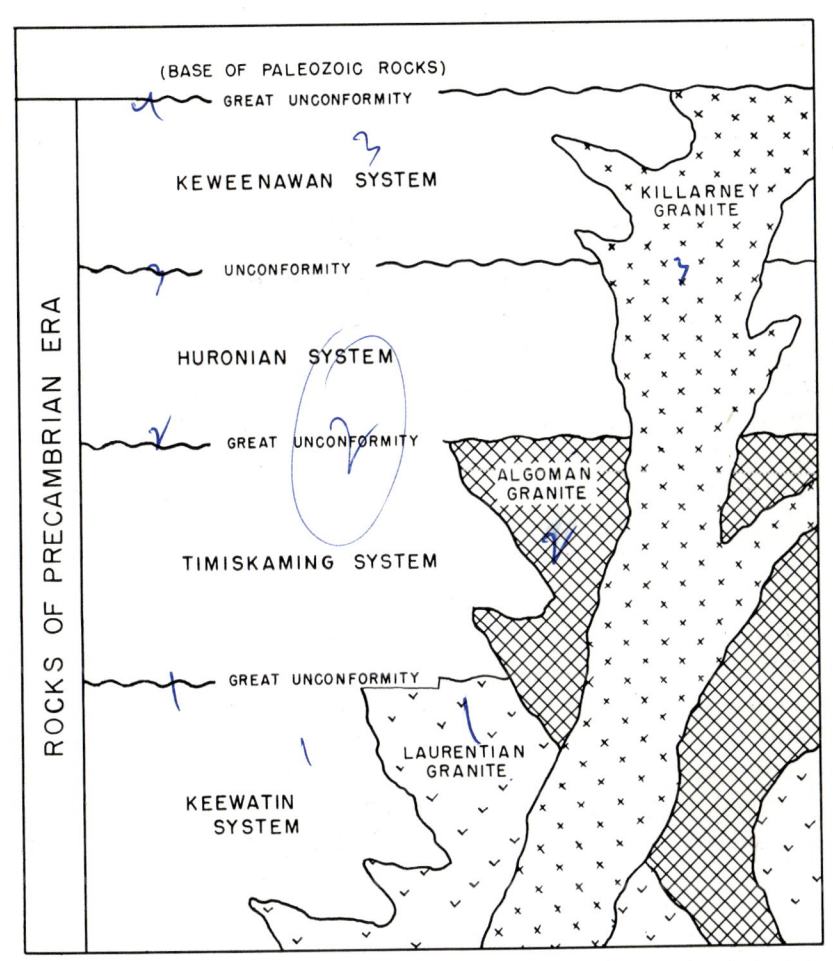

Fig. 27. Diagrammatic summary of the classification of Precambrian rocks of the Lake Superior region. (Modified from Moore, 1949, fig. 49.)

lava flows. Of particular interest are the iron formations of the Huronian, occurring in Minnesota, northern Michigan, and northern Wisconsin. These contain the iron ores which have been mined and transported by lake boats to steel plants in the Chicago and Pittsburgh districts. In Ontario, north of Lake Huron, there are thick deposits in the Huronian which consist of boulder conglomerates bearing what appear to be well-defined glacial scratches and showing the heterogeneous mixture of materials characteristic of glacial till. These presumably record an ancient ice age of considerable extent and they have been correlated with similar deposits in China and other parts of the world. The Huronian sediments and associated volcanic rocks were uplifted and eroded to some extent, but there was no great mountain-building activity before the next period of deposition began.

Keweenawan. The later part of the Precambrian, the Keweenawan, began with the deposition of thin conglomerates, sandstones, and arkoses in the vicinity of the south side of Lake Superior, and a thicker series of reddish and white detrital rocks with considerable dolomitic and calcareous material in the vicinity of Black and Nipigon bays on the north side. This was followed by a great outpouring of lavas. It is probable that the structural form of the Lake Superior basin was developed at this time, as the lavas apparently came from extensive fissures near the present center of Lake Superior. A detailed study of structural features found in the Middle Keweenawan rocks on the south shore led Hotchkiss (1923, pp. 669–78) to build up the theory that the main center of outflow lay near the present center of the lake and was related to a deep-seated batholithic intrusion. As thousands of cubic miles of lava were expelled through extensive fissures, the roof of the chamber of molten material gradually collapsed, forming, eventually, the structural basin now occupied by the lake. Observations made by Tanton (1920, p. 3e, and 1931, pp. 64, 86–87) on the north shore of Lake Superior indicate that the lava flows in that region were derived from the south, thus corroborating the theory of Hotchkiss. It is not supposed that the Lake Superior basin has been a topographic basin throughout all the time which has passed since its formation. The central part of the synclinal structure was occupied by lava as it formed, and Upper Keweenawan sediments were deposited over the lava. Further downwarping of the basin apparently occurred, to give the sediments a synclinal structure, and the topographic basin was excavated at some later time by removal of the soft sediments. The processes and time of excavation will be discussed in a later paragraph.

The Keweenawan lavas, which cover thousands of square miles in the land areas around Lake Superior, attain an apparent thickness of 60,000 feet in northern Wisconsin and Michigan (Van

Hise and Leith, 1911, pp. 418–19). The lavas constitute the central range of Keweenaw Peninsula, where the beds are tilted steeply to the northwest and are thrust upward over the edges of Upper Keweenawan sediments to the southeast (Fig. 9). The principal copper deposits of the Lake Superior region occur in the Middle Keweenawan rocks of the peninsula.

Lying above the lavas are the Upper Keweenawan conglomerates, sandstones, arkoses, and shales. These lie essentially conformably upon the older lavas where structural displacement has not obscured the relationship.

The Killarney Granite (Table 5) is found intruding various Keweenawan rocks in some places, and Butler and Burbank (1929, p. 47) suggest that it also cuts parts of the Upper Keweenawan sediments.

The end of Precambrian time is marked by granitic intrusion and by strong structural deformation of the Keweenawan rocks in certain areas, especially in the country south of Lake Superior and extending a considerable distance northeastward from Lake Huron. Outside of the belt of strong folding and metamorphism, the late Keweenawan rocks are only gently tilted or they remain nearly horizontal, as in the southeastern portion of the Keweenaw Bay area (Fig. 9). In such areas it is the presence of an erosional unconformity between the uppermost Keweenawan and the lowest Upper Cambrian sandstone which marks the end of Precambrian time.

As a generalization, it may be stated that the break between the Precambrian rocks and the Paleozoic rocks is the most pronounced stratigraphic boundary in the Great Lakes region. The Precambrian rocks generally have been more or less highly metamorphosed and are dense and crystalline. The various rock formations have been truncated by erosion, and the Precambrian terrane apparently was reduced to a surface of low relief, characterized as a peneplain. This surface of the Canadian shield descends beneath the Paleozoic rocks and is buried under the greater part of the Great Lakes region, outside of the Lake Superior basin.

The Paleozoic Era

The first record of Paleozoic deposition is found outside of the Great Lakes region. The Appalachian and Cordilleran geosynclines are downwarped tracts which were flooded by the Lower and Middle Cambrian seas (Table 5), while the central parts of the continent remained above water. The Upper Cambrian seas then spread across the interior of the United States and covered southern Ontario, but probably did not extend very far into Canada in the Great Lakes region. The Upper Cambrian rocks which crop out in the region generally are relatively thin formations of clean-washed sandstones which indicate deposition in shallow water and an absence of mountainous terrane in their watersheds.

Throughout the Paleozoic era the pattern of land and sea, or of shallow-water and deep-water areas, was set by the distribution of stable or "positive" areas and of "negative" areas (Fig. 28). The geosynclines on the eastern and western sides of the continent generally maintained their strong downsinking trends and accumulated great thicknesses of sediment, measured in tens of thousands of feet. Other areas, such as the Michigan basin (centered in the Lower Peninsula of Michigan) and the Eastern Interior basin (centered in Illinois), also showed a distinct tendency to subside. Over 10,000 feet of Paleozoic sediment accumulated in each of these. In contrast to the "negative" areas, other parts of the region remained nearly stationary or showed a tendency to rise at intervals. These stable or positive areas include the Canadian shield and its extension into Wisconsin, and the Cincinnati arch which is located in northern Kentucky, southwestern Ohio, and southeastern Indiana. The Cincinnati arch is linked to the Wisconsin positive area by the Kankakee arch extending across northern Indiana and northeastern Illinois. The Findlay arch extends from the Cincinnati arch northeastward through western Ohio and into southern Ontario.

The pattern of marine invasion of the Cambrian was repeated in the Ordovician and Silurian (Table 5), with somewhat greater flooding of the continent. Apparently no earth movements of importance occurred in the central part of the continent, for limestones and shales were deposited almost continuously. The structural framework which was outlined in the foregoing paragraph was not strongly expressed in the Ordovician, but in Silurian time the basin-and-arch pattern was quite marked.

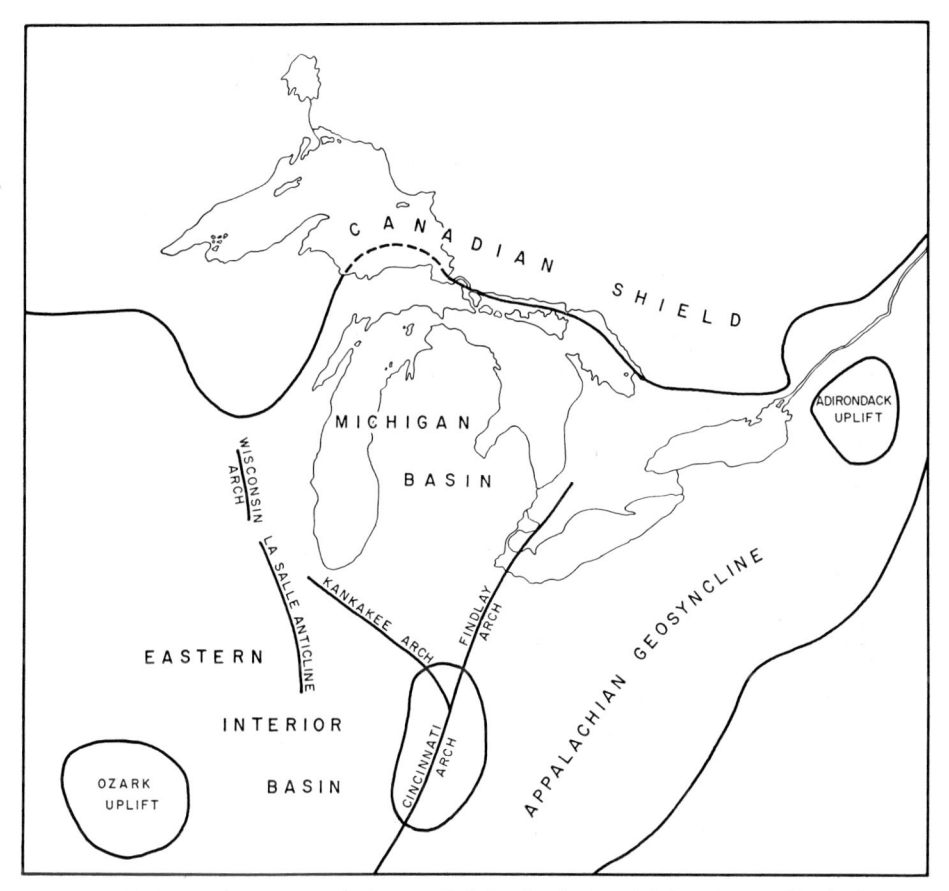

Fig. 28. Regional structures which controlled the distribution of Paleozoic seas. The basin areas accumulated thousands of feet of sediments (refer to Fig. 10 for a cross section of the Michigan basin).

In Middle Silurian time a number of coral reefs [2] existed in parts of Indiana, Illinois, and Wisconsin. These formed a zone paralleling the edges of the deeper parts of the Central Interior and Michigan basins, which were connected at the time as part of the extensive epicontinental sea. The reefs later were buried under younger formations, but because of their greater resistance they since have been exhumed by erosion of the overlying and

---

[2] Coral reefs, both ancient and modern, actually are composed mainly of many other types of calcareous organic remains besides corals.

surrounding rocks, and some of them form topographic features at the present time. Some of the shoals along the western side of Lake Michigan, from Chicago north to Racine, are Silurian coral reefs.

Later in the Silurian period the seas were more restricted and an area comprising the Michigan basin and the northern Appalachian geosyncline apparently was somewhat isolated from the open sea. This is inferred from the extensive deposits of salt and gypsum of the Salina Series. Toward the end of the Silurian period the seas generally were more extensive, and normal marine sediments, mainly dolomites, were deposited.

From the Devonian through the Permian (Table 5), the Paleozoic era can be generalized as a time of alternate flooding and draining, with a distinct trend toward less extensive flooding each time the seas advanced. During a time of emergence of the Great Lakes region within the Devonian there was considerable erosion, and leaching of parts of the Salina salt beds to produce the Mackinac breccia which has been described in Chapter 1. Following this, black shales were deposited in widespread Upper Devonian seas.

Through Mississippian time shales, siltstones, and sandstones were the most common sediments deposited in the Great Lakes region, though limestone continued to be the predominant rock type formed farther south in the Eastern Interior basin. By Pennsylvanian time the Appalachian geosyncline was filled, and almost the entire northeastern United States became an area characterized by cyclical sedimentation in which there were many alternations between marine conditions and emergence of the land. During a part of each cycle there generally was a time during which extensive swamps existed, and the peat accumulations in these swamps have since been altered to coal. The Pennsylvanian-age coals, together with the Huronian iron ores, form the basis for the industrial development of the Great Lakes region.

During the Permian, the last period of the Paleozoic era, there was some deposition of continental sediments, generally red in color, in the Appalachian area. The entire eastern part of the continent apparently had risen above the sea. In the Great Lakes region there is no trace of Permian deposits, and it may be inferred that the long period of emergence and erosion, which has continued to the present time, actually began within the late Paleozoic.

The Paleozoic was brought to a close by the building of the Appalachian Mountains. In the eastern part of the Appalachian geosyncline the Paleozoic rocks were uplifted, folded, and overthrust in an intense structural deformation. There is little evidence of this in the essentially undisturbed interior district which includes the Great Lakes, except for some gentle warping of the strata.

The foregoing discussion has dealt only with the major aspects of the Paleozoic history of the Great Lakes region. It would be impossible to give a fair picture of the physical history without a large amount of detailed description, and this is beyond the scope of the present book. A more detailed review of general Paleozoic history may be found in several textbooks of historical geology, and the details of stratigraphy are contained in the compilation by Schuchert (1943) and in the literature with which geologists are familiar.

Post-Paleozoic Time

Late in the Paleozoic era the sea withdrew from the Great Lakes region, apparently never to return (except for a brief postglacial flooding of the St. Lawrence valley, in which brackish water reached Lake Ontario and up the Ottawa River valley). The nearest approach of the sea in the Mesozoic is recorded by marine Cretaceous rocks of the Gulf of Mexico embayment which extend only as far north as the southern tip of Illinois. Two small areas of Cretaceous sediments which are regarded as continental deposits (von Engeln and Caster, 1952, fig. 247) have been found in Ontario along the Missinaibi and Moose rivers, which are tributary to Hudson Bay.

During the Mesozoic and Cenozoic eras, which together represent approximately the last 200 million years, the Great Lakes region apparently remained above sea level. With the exception of some continental sedimentation in the Tertiary, represented by scattered terrace-gravel deposits, and the glacial deposition of the Pleistocene epoch, the predominant process throughout the region since the Paleozoic era has been erosion.

PREGLACIAL EROSION AND DRAINAGE SYSTEMS

Erosion Surfaces

After the withdrawal of the Paleozoic seas, streams became established and began their normal work of dissecting the land. The oldest recognizable erosion surfaces on the Paleozoic rocks are, however, of late Tertiary age (Table 5) so that the long time lapse between the end of the Pennsylvanian and the late Tertiary is not recorded either by sediments or erosion surfaces.

A generally accepted principle pertaining to the sculpturing of an extensive land area by streams is that erosion is cyclical. The cycle begins with an uplifted, uneroded land surface and proceeds through the stages of youth, maturity, and old age. The lowest level to which an area can be reduced, the base level, is determined by sea level. In the final stage of old age a surface of low relief, or peneplain, extends over a wide area. Uplift of the land, with consequent rejuvenation of the streams, will initiate a new cycle. The cycle may be interrupted at any stage by further uplift, and if the final cycle does not reach completion, two or more erosion surfaces produced by different cycles may be preserved.

Remnants of uplifted peneplains or extensive erosion surfaces were recognized in southwestern Wisconsin at an early date. Horberg (1950, pp. 87–96) has extended the study of these surfaces into Illinois, and he was able to demonstrate the existence of some of them beneath the cover of Pleistocene glacial deposits. In very brief summary, it may be stated that the earliest recognizable erosion surface is the Dodgeville peneplain, which is represented by accordant summit levels of all but the highest hills in the driftless area of southwestern Wisconsin and northwestern Illinois. This surface probably was completed during the later half of the Tertiary and extended an unknown distance southward. The surface developed during the next cycle of erosion is more widely preserved and is represented by the Lancaster

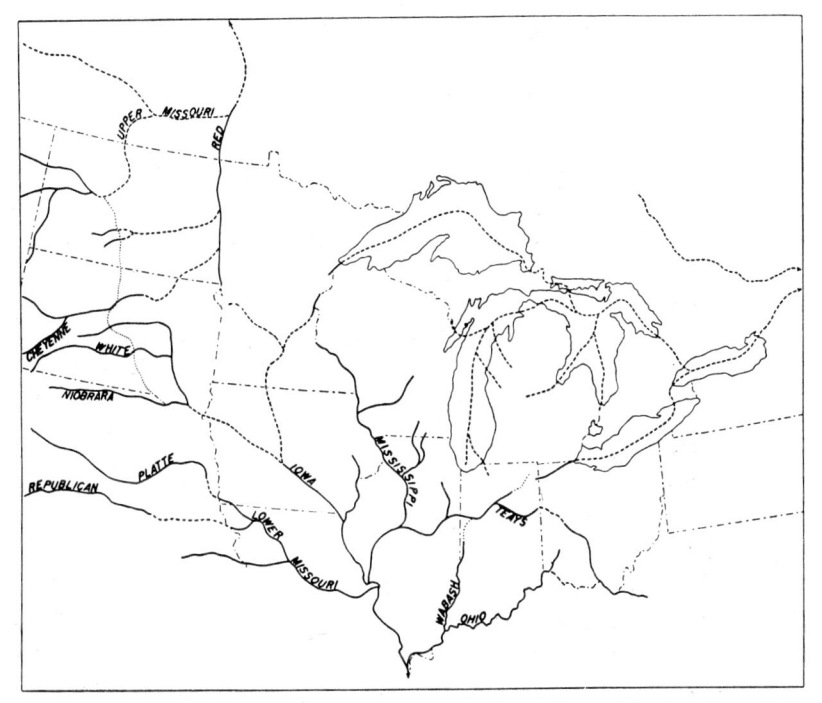

Fig. 29. Preglacial drainage systems in the central United States. The locations of the preglacial rivers outside of the Great Lakes region are based on well records and maps of the bedrock topography, and are fairly well established. The drainage pattern in the Great Lakes region closely follows the early views of Spencer (1891a), and is highly conjectural. (Reproduced from fig. 2 of Horberg and Anderson, 1956, Univ. of Chicago Press, by permission.)

peneplain in Wisconsin and northern Illinois, the Calhoun peneplain in western Illinois, and the Ozark peneplain in southern Illinois. Somewhat more obscure lower erosion surfaces have been suggested. These various surfaces can not be traced into the drainage basin of the Great Lakes region with any degree of confidence, because modification of the topography by glacial scour and deposition has obscured them if they existed there in preglacial time.

Drainage Systems

The reconstruction of preglacial drainage systems has been attempted by various writers over a period of more than 50 years. The most recent summary and synthesis of these studies, by Horberg and Anderson (1956, pp. 103–7), gives references to the extensive literature on the subject. Figure 29 shows the inferred preglacial drainage as published by Horberg and Anderson (1956, fig. 2). In this figure it is seen that the discharge of the upper Missouri River system is believed to have gone to Hudson Bay, and that other parts of the Missouri system were connected to the Mississippi system by routes which were different from the present ones. The upper Mississippi River, in preglacial time, did not extend so far north, and it included a route from the northwestern corner of Illinois into the central part of that state. The preglacial master stream in the region now drained by the Ohio River system was the Teays River, whose headwaters drained the Appalachian plateau and whose lower course ran through north-central Indiana and east-central Illinois. The changes in these systems from their preglacial to their present locations were caused by Pleistocene glaciation. The glacial ice actually covered much of the region involved and thus directly diverted the river flow to new channels. Upon its retreat it left deposits in many of the old valleys so that they could not be reoccupied by the major streams.

The interpretation of the preglacial drainage system in the Great Lakes region which is shown in Fig. 29 (after Horberg and Anderson, 1956) closely follows the early views of Spencer (1891a). The present author considers this interpretation to be highly conjectural. In some of its details, the drainage system is plotted through major ridges on the lake bottoms which almost certainly are composed of Paleozoic bedrock. Adjustments could be made to avoid these ridges without changing the major pattern, but there still would remain considerable doubt as to the validity of some major parts of the pattern. Because there almost certainly has been a large amount of glacial scour of the lake basins (see Chap. 5), and because thick deposits of glacial drift (amounting to as much as 700 feet) cover the bedrock in some places, the possible connections of major parts of the ancient stream system are obscure. No attempt is made in the present book to revise the interpretation, but the subject is worthy of further research.

The most important point regarding the preglacial drainage system of the Great Lakes is that the master valleys obviously were located in the outcrop belts of the rock formations which were least resistant to erosion, and that the divides generally were formed by the more resistant rocks. The various parts of the system which were located in broad belts of the less resistant rocks probably were broad lowland areas. The gross features of the preglacial landscape must have exerted a strong influence on the direction of ice movement when the continental ice sheets of the Pleistocene epoch invaded the region.

chapter five

GLACIAL HISTORY OF THE REGION

RÉSUMÉ OF PLEISTOCENE GLACIATION

Introduction

The occurrence of multiple stages of glaciation has been demonstrated mainly by work during the past 50 years in the Great Lakes and Mississippi valley regions,[1] but it is well shown in Europe also (Flint, 1947). Recent work on Antarctic and Pacific ocean-bottom sediments has revealed multiple layers of sediment types which apparently correlate with the glacial stages of the northern hemisphere (Hough, 1950 and 1953b); this is reviewed in Chapter 6.

There are four distinct glacial ages within the Pleistocene epoch, as shown by the record in North America. Table 6 lists these and the intervening interglacial ages. Figure 30 shows the positions of the margins of the various glacial stages in the central and northeastern United States. Each glacial age is represented by an extensive deposit of till or boulder clay. Each till deposit is distinguished by the presence of a weathered zone at its top and by the occurrence of nonglacial deposits of various kinds lying over it. These are not present everywhere but they have been

---

[1] Much of the early work done in these regions is reported in detail in the U.S. Geol. Survey monogs. by Leverett (1899, 1902, 1932), and by Leverett and Taylor (1915). The classification of the Pleistocene is discussed by Kay (1931) and Leighton (1933).

Table 6. *The Pleistocene time scale for North America.*[a]

Glacial ages	Interglacial ages
Wisconsin	
	Sangamon
Illinoian	
	Yarmouth
Kansan	
	Aftonian
Nebraskan	

[a] Only the major subdivisions are given in this table. Further subdivisions of the Illinoian are mentioned in the text and those of the Wisconsin are listed in Table 7.

found in a sufficient number of places to establish the age relationships.

The length of the Pleistocene epoch appears to be not less than one million, certainly not more than five million, and probably not more than two million years (Flint, 1947, pp. 405–6). The later parts of the epoch have been dated more accurately by modern radioactivity methods. A detailed discussion of the dating methods and a summary of the results obtained is presented in Chapter 6. Meanwhile, to provide points of reference for a résumé of Pleistocene events, the most generally accepted estimates of age are given in the following tabulation:

Glacial Stage	*Approximate Age in Years*
Wisconsin	10,000–50,000
Illinoian	300,000
Kansan	700,000
Nebraskan	1,000,000

The duration of each glacial stage is considered to have been relatively short, of the magnitude of 50,000 years, while the interglacial stages were relatively long.

Nebraskan Age

The outer margin of the earliest known Pleistocene glacial stage, the Nebraskan, has been mapped in eastern Nebraska, northeastern Kansas, and northern Missouri. It undoubtedly extended throughout the Great Lakes region to the northeast of its

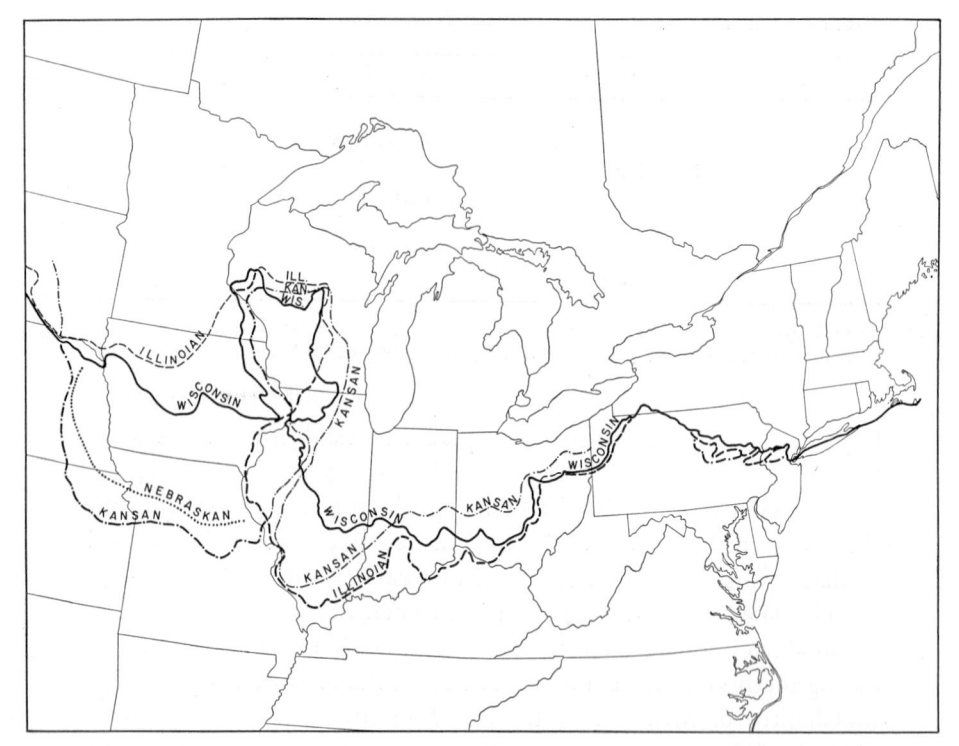

Fig. 30. Maximum extent of the four principal glacial stages in the central and northeastern United States. (Compiled from Flint and others, 1945, and various other sources.)

known exposures, but later advances of the ice have obliterated the record in that region. The upper part of the Nebraskan till is deeply weathered, with a zone of gumbotil [2] eight to nine feet thick, which represents a very long period of weathering. In places beds of peat and water-laid sediments containing remains of animals that indicate a cool, temperate climate are associated with the gumbotil. This deeply weathered zone is buried beneath the next younger till, the Kansan, which is unweathered at its base. The long period of weathering which formed the Nebraskan gumbotil thus occurred before the next glacial stage began, and it represents the Aftonian interglacial stage.

---

[2] Gumbotil is a gray, leached, deoxidized clay which is very sticky when wet and extremely firm when dry. The clay contains stones, but most of them are of siliceous types highly resistant to decomposition.

Kansan Age

The Kansan glacial ice occupied about the same area as the Nebraskan, in the region where material of that age is known. To the northeast, in parts of Pennsylvania and New Jersey, the Jersey drift is believed to be of Kansan age. The Kansan till, like the Nebraskan, has in places a well-developed gumbotil horizon and this attains a thickness of as much as 12 feet. This, along with deposits of peat, loess,[3] and gravel, represents the Yarmouth interglacial stage, which is believed to be the longest and warmest of the interglacial stages (Flint, 1947, pp. 281-82).

Illinoian Age

A third ice sheet, the Illinoian, extended over a large part of Illinois, Indiana, and Ohio, where it deposited an extensive sheet of till. Correlative deposits have been mapped in eastern Iowa and in Pennsylvania and New Jersey. Three fairly distinct substages of the Illinoian are represented by the Payson, Jacksonville, and Buffalo Hart moraines in Illinois.

During the Sangamon interglacial stage that followed, the Illinoian till was weathered to form four to six feet of gumbotil in many places, and peat, loess, and volcanic ash have been found associated with or lying over the gumbotil. The loess was leached to a depth of three to five feet before glacial ice again advanced over the region.

Wisconsin Age

The Wisconsin drift sheet is the youngest and most completely preserved of the Pleistocene glacial deposits. It exhibits only slight weathering and an almost unmodified glacial topography, and because of its excellent state of preservation, many more details of its history can be deciphered than are possible to discover in the older drift sheets.

The complexity of detail still is in process of being worked out by a number of investigators, so that any summary of the classification of the Wisconsin deposits which is given at the present time may be subject to revision in the light of future discoveries. The

---

[3] "Loess," as used here, means a wind-transported silt.

presently accepted classification of the Wisconsin has been established by Leighton (1933, p. 168), expanded by Leighton and Willman (1950, pp. 602–3), and further revised by Leighton (1957, pp. 108–9). This is shown in the first column of Table 7. A revised scale, adapted for use in the present book, is shown in the second column of Table 7. The classification of Wisconsin

Table 7. *Wisconsin-age time scales for the Great Lakes region.*

Time scale of M. M. Leighton [a]	Time scale used in the present book [b]
glacial substages	*glacial substages*
	7. Cochrane [c]
6. Valders	6. Valders
5. Mankato	5. Port Huron (Mankato)
4. Cary	4. Cary
3. Tazewell	3. Bloomington (late Tazewell)
2. Iowan	2. Shelbyville-Iowan (early Tazewell)
1. Farmdale	1. Farmdale

[a] The middle four divisions were proposed by Leighton (1933, p. 168), and the lowest, the Farmdale, by Leighton and Willman (1950, pp. 602–3). Farmdale till was described by Shaffer (1956). The Valders drift was named by Thwaites (1943, p. 121), and the Valders was recognized as a substage by Leighton (1957, p. 109).
[b] Refer to text for justification of this revision.
[c] The Cochrane was proposed by Karlstrom (1956, p. 326).

events is discussed in the following paragraphs. The Wisconsin moraines have a very marked pattern which indicates that the ice sheets were strongly lobate and were influenced by the Great Lakes basins in their movements.

Substages of the Wisconsin

Farmdale. The first substage of the Wisconsin is one which has been discovered recently. Leighton and Willman (1950, pp. 602–3) inferred the existence of the Farmdale glacial substage from the occurrence of the Farmdale loess. This loess, from its characteristics and relationship to the valley of the ancestral Mississippi River, was judged as requiring a glacial valley train source. This, in turn, implies an extension of an ice sheet during the Farmdale substage into the drainage basin of the ancestral Mississippi River. Farmdale till has since been discovered and mapped in northwestern Illinois (Shaffer, 1954b, pp. 693-94, and 1956, pp. 1-25).

Iowan. The second substage is the Iowan. Before the Farmdale drift was recognized, the Iowan was considered the oldest Wisconsin substage. It has not yet been unequivocally identified outside of Iowa and Minnesota (Flint, 1947, p. 246). In Illinois, before discovery of the Farmdale, the Shelbyville drift of the Tazewell substage was considered the oldest Wisconsin substage. This drift has not been recognized west of the Mississippi River, except for a short extension reported by Shaffer (1954a, p. 448). It has been considered younger than the Iowan because a thin sheet of loess which presumably is younger than the Iowan covers much of the Tazewell drift in Illinois. According to Flint (1947), "the fact that the Iowan drift is most extensive in Iowa whereas the Tazewell drift is most widespread in Illinois and Indiana suggests that the effective center of radial outflow in this sector of the ice sheet shifted eastward between the Iowan sub-age and the Tazewell sub-age" (p. 250). An alternative interpretation is being considered: that the early Tazewell Shelbyville drift is correlative with the Iowan and the two were deposited at approximately the same time (Shaffer, 1954a, p. 455).[4] If this is true, the early Tazewell (Shelbyville) is Iowan and the later Tazewell (Bloomington) may represent a later substage.

Tazewell. Assuming the Farmdale to be first and the Iowan to be second (as shown in Table 7), the third substage which is generally recognized is the Tazewell (Flint, 1947, p. 249; Leighton and Willman, 1950, fig. 2; Ruhe, 1952, fig. 1). Several conspicuous end moraines are present in the Tazewell drift, the two most prominent being the Shelbyville at the margin of the drift and the younger Bloomington moraine. Each of these moraines records a readvance of the ice sheet after a retreat of unknown amount. The Bloomington and several younger moraines which are con-

---

[4] Also Ekblaw, G. E., geologist, Illinois Geol. Survey, personal communication, 1956.

centric with it form a pattern which is distinctly discordant with the pattern of the Shelbyville drift. The discordance suggests a considerable interval of time and possibly a large amount of shrinkage of the ice between the Shelbyville and Bloomington advances. This is in accord with an observation of weathered Shelbyville overlain by Bloomington, reported by G. E. Ekblaw.[5] If further study indicates that the Shelbyville is Iowan, the Bloomington may be retained as the representative of the Tazewell or third Wisconsin substage.

Cary. The fourth substage according to the present classification is the Cary, represented by a number of well-developed morainic systems. The pre-Cary stages which have been described all involved advances of the ice through and beyond the Great Lakes basins, so that they have no relationship to known lake stages. Early in the Cary, however, the intervals between the glacial advances involved retreats of the ice front into the lake basins and the history of the Great Lakes thus began at that time. Various events of lake history may be used in deciphering and classifying the details of the Cary and later substages, and a discussion of many of those details will be given in the chapters dealing with lake history.

In broad outline, the Cary of the Lake Michigan lobe began with the formation of a group of minor moraines, the Minooka, Rockdale, and Manhattan, which occur in northeastern Illinois. The first major morainic system is the Valparaiso, which marks the southern rim of the Lake Michigan basin and the continental divide in this area. Younger Cary moraines which are more or less concentric with the Valparaiso are the Tinley and the Lake Border system. The next younger, the Port Huron system, occurs farther north in the Lake Michigan basin. It has been considered a part of the Cary but more recently it has been separated as a later glacial substage. This will be discussed in connection with the Mankato substage. The Cary is represented in northeastern Indiana by a group of moraines which include the Mississinawa to Fort Wayne series, and in northwestern Ohio and southeastern Michigan by these and the Defiance moraine. All of these are concentric about the western end of the Erie basin. Detailed corre-

---

[5] Personal communication, 1956.

lation has not been made between these and the moraines of the Michigan basin. Recent work by White (1951) has delineated the Cary and older drifts in eastern Ohio but these have not as yet been correlated in detail with the moraines to the west.

The oldest morainic system which can be traced continuously from the Michigan to the Huron basin is the Port Huron. This extends from the vicinity of Ludington on the eastern shore of Lake Michigan northeastward, around the north end of the interlobate area, down the western side of the Huron basin passing around the head of the Saginaw Bay embayment, and to the vicinity of Port Huron, Michigan (its type locality), at the south end of Lake Huron (Martin, 1955). From there it swings around the south end of the Huron basin and passes around the north end of the southern Ontario upland, where it extends southward west of Lake Ontario and reaches the Erie basin north of Long Point. It may be represented by a ridge on the bottom of Lake Erie (Fig. 6), and an apparently correlative moraine has been mapped south of the eastern end of Lake Erie (Flint and others, 1945).

Using the Port Huron morainic system as a reference datum between the Michigan, Huron, and Erie basins, it should be possible to work backward through the sequence of moraines and select some probable correlatives in those basins. In the Michigan basin the next older system is the Lake Border, and north of Chicago this system is composed of four moraines (Bretz, 1955, p. 28); north of the west end of Lake Erie the next older moraines include the Adair, Emmett, Mount Clemens, and Detroit in Michigan (Leverett and Taylor, 1915, pl. 32) and the Seaforth, Leamington, Charing Cross, and Blenheim in Ontario (Chapman and Putnam, 1951, fig. 11, pp. 40–41); south of Lake Erie to the east of Cleveland the next older moraines are designated the Erie Lake Border system and this is composed of four moraines where the entire system occurs on land in northeastern Ohio (White, 1953, pl. 26). It appears quite possible that all of these various moraines are a correlative system and they will be so considered, on a tentative basis, in the present book.

At the south end of Lake Michigan the moraines of the Lake Border system, the Tinley moraine, and those of the Valparaiso system are all rather closely bunched. The Tinley is geographically separate from the Lake Border in the Chicago region, however, and relations between the moraines and valley trains indicate an appreciable time interval between the formation of the Tinley and the earliest Lake Border moraine (Bretz, 1955, p. 108). In the Erie basin the Defiance moraine appears to be correlative with the Tinley of the Michigan basin because the Defiance is separated from the Erie Lake Border system by a considerable distance.

The Tinley moraine is built against the Valparaiso system, at the south end of Lake Michigan, but it is recognizable as a distinct feature because it is built across the heads of the Sag and Des Plaines valleys which were cut through the Valparaiso system in post-Valparaiso time. A post-Valparaiso, pre-Tinley "Lake Chicago" is inferred from this, the discharge from which was sufficiently large and long-continued to cut the transmorainic channels (Bretz, 1955, p. 107). The Erie basin Defiance moraine, which has been selected as a correlative of the Tinley, is separated from the next older moraine, the Fort Wayne, by a considerable distance at the extreme western end of the basin.

The early Cary Valparaiso system is composed of five moraines north of the Chicago region (Bretz, 1955, p. 107). At the west end of the Erie basin in eastern Indiana and in the Miami and Scioto lobes in western Ohio there are five principal moraines in the early Cary drift (Flint and others, 1945). Leverett (1902, p. 51) named the first two of these, the Union and Mississinawa, as probable correlatives of the Valparaiso system. In the present book all five of the early Cary moraines in the western Erie basin are considered as equivalent to the Valparaiso system. The youngest of the group, the Fort Wayne, is thus considered equivalent to the youngest Valparaiso moraine. This stage of history is shown in Fig. 53.[6]

The position of the outer border of the Cary drift east of the Erie basin has not been determined with certainty. Flint (1953, pl. 1) suggests a probable position which lies well outside of the Great Lakes region so that we are not concerned with its accurate delineation in the present study. Likewise, the Cary border in

---

[6] Figures 53–75, lake stage maps, follow Chapter 16.

the Lake Superior district (Flint and others, 1945, and Wright, 1955, p. 407) lies far enough beyond the lake margin for its exact position to be of no immediate concern in the present study.

Mankato. The fifth subdivision of the standard time scale (Table 7) takes its name from Mankato, Minnesota (Leighton, 1933, p. 168), in the Des Moines lobe drift which lies to the west of the Great Lakes. The late Wisconsin drift which was included in this substage had been described in detail by Leverett (1932, pp. 8, 56–105), who stated that the correlative morainic system in the Great Lakes region appeared to be the Port Huron system. The Port Huron system is described in detail by Leverett and Taylor (1915, pp. 293–315). They stated that "It is of the substage order of magnitude and is, in fact, one of the best developed and most clearly defined moraines in the Great Lakes region. . . . The main moraine of the Port Huron system marks a pronounced readvance of the ice front and records one of the simplest of these changes, uncomplicated by any later or earlier ones" (p. 293). The foregoing information seems to indicate that the Mankato–Port Huron drift constitutes a distinct substage which is later than the Cary.

The simple interpretation given in the foregoing paragraph has been thrown into a state of confusion by the interpretations placed on an extensive body of red glacial drift which occurs throughout the upper Great Lakes region from the head of the Superior basin through the northern Michigan basin to the northwestern Huron basin. This red drift in northeastern Wisconsin was named the Valders by Thwaites (1943, p. 121) from a type locality at Valders, Manitowoc County, Wisconsin. At the type locality the red Valders drift lies in part on gray till which is correlated as Cary, and in part on an adjacent bedrock surface which bears two sets of striae crossing at nearly a right angle. Thwaites stated that no definite time relation of the Valders to the Mankato of Minnesota and Iowa was implied, and pointed out that such a vast expanse of unsettled, little-known territory separated the Mankato and the Valders drifts that their equivalence is far from assured. The red color of the till generally is believed to have resulted from the deposition of red lake clays (probably derived from red Precambrian iron-bearing rocks) in the lake basins during an interglacial time and the incorporation of those clays into the till during a subsequent advance of the ice. Various writers have stated that the red (Valders) drift of northeastern Wisconsin correlates with the Port Huron morainic system of Michigan. Leverett (Leverett and Taylor, 1915, pp. 309–10) made this correlation, not on the basis of composition or topographic form of the drift but on its relationship to lake beaches. Only the third or Toleston beach of Lake Chicago is present on the drift on both sides of Lake Michigan, north of a certain parallel in the areas of occurrence of the Port Huron and the Valders, whereas all three of the Lake Chicago beaches are present on both sides of the lake farther south. Leverett stated, "There could scarcely be better grounds for correlation than are presented in the relation of these moraines to the Lake Chicago beaches" (Leverett and Taylor, 1915, p. 310). There are better grounds for correlation, as detailed study of the drift has shown. These were summarized by Bretz (1951b, pp. 418–24), who enumerated the following significant points: (1) The red Valders drift in Wisconsin has no end moraines and scarcely any topographic expression of its own but rather is a thin veneer over the glacial topography of the underlying gray drift. (2) The Port Huron, in Michigan, is composed of gray till which has a strong morainic expression. (3) There is a thin red till in western Michigan which laps onto the margin of the Port Huron morainic belt. It appears obvious that the red till of Michigan and the red till of Wisconsin are correlative, and that both are younger than the gray Port Huron till. Further, the present author points out the fact that all lake beaches above the present beach of Lake Michigan have been destroyed by shore erosion for a distance of tens of miles on both sides of the lake in the areas of occurrence of the Port Huron and Valders tills, and that therefore the absence of the higher Lake Chicago beaches cannot be cited as a basis for correlation of the two tills (Chap. 9).

Alden, however, had accepted Leverett's argument, and in 1918 (pp. 135–36) stated, "the red lacustrine silts were being deposited during the readvance of the ice to the main moraine of the Port Huron morainic system in Michigan and during the deposition of the red till in Wisconsin," and showed a correlation of the Port Huron moraines with the red drift of Wisconsin on his plate 23. Various students of the Pleistocene came to accept this correlation without question. Both the Port Huron and the Valders drifts became identified as Mankato in age, and they were so shown on the "Glacial Map of North America" (Flint and others, 1945). When Bretz (1951b, pp. 423-24) demonstrated that the Valders was younger than the Port Huron, he referred the Port Huron to a late Cary age and retained the Valders as the representative of the Mankato substage. Melhorn (1954) mapped the red drift of the Lower Peninsula of Michigan in detail, assigned it to the Valders, and confirmed its status as a drift which is distinct from the Port Huron and probably of substage rank. Zumberge (1956) has expressed serious doubt of the Valders-Mankato correlation, and has recommended that the term "Mankato" be suspended from use in connection with features of the Michigan and Huron basins. The author has followed this recommendation in the preparation of the present book.

Recent dating by the radiocarbon method (refer to Chap. 6) has indicated that the Mankato drift in Minnesota is significantly older than the Valders of northeastern Wisconsin; it therefore now appears reasonable to separate the Valders drift as the representative of a separate Valders substage. This will be discussed in further detail in a subsequent paragraph. Meanwhile, the Port Huron drift may be returned to the Mankato substage classification. This is done simply because there is a good basis for its separation from the Cary as a separate substage (as stated by Leverett and Taylor, quoted in the foregoing discussion), and because the Mankato of Minnesota and Iowa is younger than Cary and older than Valders. Leighton recently has accepted this reclassification and has proposed (1957, p. 109) that both the Mankato and the Valders be recognized as successive substages following the Cary, as is shown in Table 7. The present author believes, however, that the name "Port Huron" should be used to designate the Port Huron-age drift in the Great Lakes region, because the name "Mankato" has been used so extensively in the past to designate Valders-age drift. In the Michigan lobe the sequence of Cary–Port Huron–Valders is well established and it is recommended that these terms be used in any correlations that are

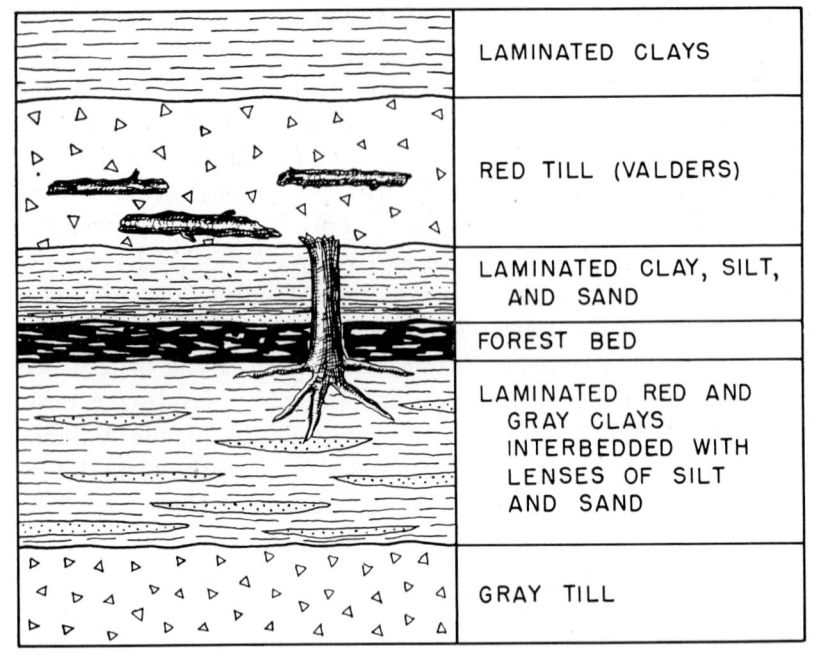

Fig. 31. The Two Creeks forest bed and associated deposits near Manitowoc, Wisconsin. Trees in growth position were broken off by Valders ice and buried in the red till.

made between the drift of this area and other drifts to the east, as was done by Flint (1956, pl. 1 and p. 275).

Two Creeks interval. One of the major points of reference in the Wisconsin stage in the Great Lakes region is the Two Creeks forest bed near Manitowoc, Wisconsin (Fig. 31). The bed marks an interval between gray till, presumably of Port Huron age, and the overlying red till of Valders age. Wood from the forest bed has been dated by the radiocarbon method (see Chaps. 6 and 16) and has been found to have an age of 11,400 years before the present. Spruce, pine, and birch trees have been identified in the bed, and mosses and several kinds of mollusks are present also (Wilson, 1932 and 1936). This assemblage suggests a climate like the present one in Ontario north of the Great Lakes. The Two Creeks forest bed is represented by correlative material in an area of 500 square miles in the Fox River valley in Wisconsin.

The sequence of deposits illustrated in Fig. 31 record the following events: (1) Deposition of gray till by glacial ice. (2) Retreat of the ice from this locality allowing the waters of Lake

Chicago to flood the area, and deposition of laminated clays and lenses of silt and sand in the lake. The lake was 60 or 40 feet deep at first, because either the Glenwood or the Calumet stage (more probably the former) existed at this time and discharged southward through the Chicago outlet. (3) Drainage of the lake, caused by further retreat of the ice front to the north at least far enough to open a lower outlet; this outlet could have been through the lowland extending eastward from Little Traverse Bay to Lake Huron, or it may have been the Straits of Mackinac. (4) Growth of the forest. The forest bed extends down to an elevation at least as low as the present surface of Lake Michigan. This indicates that the lake must have drained through an outlet considerably lower than the one at Chicago, and the only available dischargeway must have been at the northeastern edge of the Michigan lake basin. (5) Readvance of the ice over the low outlet to the northeast, causing flooding of the forest and deposition of sand, silt, and clay over the forest floor. (6) Advance of the ice over the forest, breaking off the trees and laying them over in a southwesterly direction and depositing red till. (7) Retreat of the ice from the locality, allowing lake water to flood it, and deposition of laminated clays. (8) Further retreat of the ice front to uncover a lower outlet to the northeast and allow drainage of the area.

The most significant aspect of this sequence of events is not the probable length of time required for establishment and growth of a forest while the locality was above water (this period could have been measured in only hundreds of years), but the extent of retreat of the ice which was required to open a new, lower outlet to the northeast.

The Two Creeks interval separates the Valders from the Port Huron. If the Port Huron is Mankato in age, the Two Creeks interval is not pre-Mankato, but is post-Mankato and pre-Valders. Several writers have correlated the Two Creeks low-water stage with the Bowmanville low-water stage, which is indicated by deposits found in the Chicago region. The question of the validity of this correlation is discussed in Chapter 9. Meanwhile, the name "Two Creeks" will be used to designate the Port Huron–Valders interval.

Valders. The sixth substage of the Wisconsin stage, as shown in the time scale of Table 7, is the Valders (named by Thwaites [1943, p. 121]), which is represented by the red Valders drift in the northern part of the Lake Michigan basin. A summary of how the Valders came to be classified as Mankato and is now identified as younger than the Mankato has been given in the preceding pages. The Port Huron system is retained in the Mankato-age classification, as it was shown on the "Glacial Map of North America" (Flint and others, 1945).

The present author believes that the magnitude of the break between the Port Huron and the Valders is great enough to warrant a separation into two substages. This is shown in one way by the Two Creeks forest bed, which represents a retreat of the ice front sufficient to drain the lake waters of the Michigan basin through an available outlet channel at the northeastern corner of the lake. This in turn suggests a general retreat of the ice margin in the eastern lake basins sufficient to permit drainage to the east and to lower the lake surfaces in the Huron, Erie, and Ontario basins. Otherwise the water surface in the Michigan basin could not have been drawn down to permit growth of the Two Creeks forest at a level below the sill of the Chicago outlet channel. The discharge in Two Creeks time may have gone down the Champlain–Hudson River valley, or down the St. Lawrence valley to the sea, if that part of the continent were free of ice.

Additional evidence which strongly suggests a retreat of large magnitude is the color and composition of the Valders till and its associated lake deposits. Some of the older tills in the region are slightly pink (for example, the Bloomington) or contain occasional reddish patches, but generally they are not red or even pink. Older lake clays almost without exception are described as grayish or bluish in color. The Valders drift is characteristically a strong red in color and this seems to require some special explanation. Murray (1953, pp. 148–54) has made a detailed petrographic and textural examination of the Valders material and has found that the red color is due to fragments of hematite in the clay size range and to a hematite stain on the larger grains. The Valders till has a much finer texture than the Cary. Murray's conclusions as to the source of the Valders till embrace the suggestions of Alden (1918) and Thwaites (1943) that the Valders ice ploughed up red lake clays deposited in the pre-Valders interval. Because the Valders ice followed a course from a center to the west of that from which the Port Huron and Cary ice came, and because the mineral composition of the coarser fractions of the Valders is identical with that of the older till, it appears that the Valders ice accomplished little scour of the bedrock but derived the coarser parts of its load from the glacial drift over which it moved. Further, because the finer fractions of the Valders appear identical with the red lake clays of the region, it is concluded that the ice mixed the lake clays with the older till to produce the red Valders till. Murray believes that the pre-Valders ice retreated far enough to permit the formation of a large lake in the Superior basin, which he named "Lake Keweenaw," and that this lake drained to Lake Chicago in the Michigan basin. He proposes that the fine sediments produced by glacial erosion in the Superior basin, which is underlain by red sedimentary rocks, were in part carried through to Lake Chicago to give a red color to the clays deposited there. If these conclusions are correct, there was a marked retreat of the ice between the Port Huron and Valders advances, indicating a distinctly warm interval.

Another feature of the Valders drift which distinguishes it from the deposits of the Port Huron and most of the older Wisconsin drifts is its general lack of well-developed morainic topography and relatively slight thickness. These aspects of the drift, together with Murray's indications that little scour of the bedrock was accomplished by the Valders ice, suggest that the ice advanced as a thin and relatively fluid mass. This is further suggested by the shape of the ice margin, deduced from the extent of the red drift. The Valders lobe of the Michigan basin extended much farther south than did the Port Huron, as shown by Bretz (1951b, fig. 1) and confirmed by the present author in the discovery of red till on the lake bottom at two points southwest of Grand Haven, well out in the southern basin of the lake. In the area east of Grand Traverse Bay in Michigan, the red drift extends into several embayments of the inner margin of the Port Huron morainic system but it does not occur very high on the slopes. All of these features indicate that the Valders ice was a thin and mobile mass.

In view of the information reviewed in the foregoing paragraphs, the author proposes that the Valders drift represents a distinct glacial substage. The duration of the Two Creeks interval, which is post-Mankato and pre-Valders, was relatively short. There may have been a period of only about 2000 years between the maximum of the latest Port Huron advance and the Valders maximum. The extreme mobility of the Valders ice, as inferred in the foregoing paragraph, would account for the rapidity of its advance. Further discussion of the late Wisconsin time scale is presented in Chapter 6. Whether the Valders drift represents a separate, formally recognized substage or whether it is considered a late phase of the Mankato substage is a matter of small importance, so long as the Valders is recognized as a distinct part of the glacial succession which is younger than the Port Huron and the type Mankato drift.

The Valders glaciation, as recorded in the Lake Michigan basin, was not a simple single advance and retreat. After the retreat from its maximum there were at least two halts in the recession, or minor advances, recorded by end moraines. These occur in the Upper Peninsula of Michigan in the part which separates Lake Michigan from Lake Superior. They were mapped and described by Leverett (1929a, pl. 1, pp. 49-53), and have been found by the present author to extend northeastward into Ontario. These will be discussed in further detail in Chapter 9, in connection with the Algonquin stage of the Great Lakes.

The outer margin of the Valders red drift has been traced eastward from the Michigan basin to the Huron basin, where it extends along the inner border of the Port Huron morainic system to within 12 miles of Thunder Bay, then runs northward and offshore into Lake Huron at a point about two miles northwest of Rogers City (Melhorn, 1954). It is possible that non-red Valders-age drift exists in Michigan southeast of the margin of the red drift and was deposited by ice of a different glacial lobe. It appears certain, however, that the margin of Valders-age ice did not extend to the southern end of Lake Huron.

In the southern Ontario upland east of the south end of Lake Huron there is glacial drift which is undoubtedly of Valders age. Older drift which is confidently correlated with the Port Huron occurs in a strongly developed morainic system of two or more prominent ridges similar to that of the Port Huron in Michigan, and which is traceable on land to a junction with the type Port Huron. Chapman and Putnam (1951, p. 30) show the margin of the Port Huron stage as extending from this morainic system southward west of Lake Ontario and across the eastern end of Lake Erie. They show (p. 31) a later stage in which the Georgian Bay glacial lobe stood against the northern edge of the southern Ontario upland and the Ontario basin lobe stood on the Niagara escarpment between Lakes Erie and Ontario. This later "stage" logically is the Valders substage of the present author's revised classification.

In the Niagara River area the Whirlpool–St. David buried gorge, which extends from the "whirlpool" of the present Niagara gorge to the vicinity of St. David, is filled with glacial drift. The crest of a faint moraine which coincides with the position of the later-stage ice margin of Chapman and Putnam (Valders of the present author) crosses the northern part of the filled gorge and forms the highest part of the filling (Kindle and Taylor, 1913, p. 17). This filled gorge was regarded as pre-Pleistocene by early writers (Pohlman, 1884, p. 202, and Grabau, 1901, p. 51), but was dated as interglacial and pre-Wisconsin by Kindle and Taylor (1913, p. 17). The present author regards the buried gorge as post–Port Huron, pre-Valders in age; it logically records erosion by discharge from the Erie basin during the Two Creeks interval, when the glacial lake waters were drained down from the early Lake Warren stage (see Chap. 8) and a low-water stage ensued. The length of the buried gorge is approximately two miles, and if we allow 2000 years for the cutting of the gorge, the rate of recession of the falls which poured into it becomes approximately five feet per year; this is in accord with the present rate of recession of Niagara Falls. Flint (1953, pp. 905–6) presented an interpretation which is similar to this one, except that he then regarded the Valders as Mankato. If one inserts Valders in place of Mankato in his discussion of the St. David gorge, his interpretation would coincide with the present one. Elsewhere in the paper cited, "Mankato" is the equivalent of Port Huron.

The position of the Valders margin eastward from Lake Ontario has not been mapped in detail as yet.

Correlation of the Valders drift of the Michigan basin with

deposits west of Lake Superior has not been accomplished by direct tracing. When Thwaites proposed the name "Valders" in 1943 he cautiously refused to make such a correlation. Later, however, Thwaites (1946, pp. 86–87) proposed that the Valders was an early stage of the Mankato, because gray till of the Des Moines lobe, presumably Mankato in age, overlies red till of the Superior lobe, which he presumed to be Valders. Wright (1955, pp. 405–11) has shown that red and gray till interfinger and that there are two red tills in the area west of Lake Superior. The upper red till is very reasonably correlated with the Valders on the basis of its high clay content and the fact that it is a thin sheet which mantles previously formed topography. The lower red till, which is sandy, is regarded as late Cary; the present author suggests Mankato–Port Huron as the proper identification of this material, keeping in mind his designation of Valders as a post-Mankato, post–Port Huron substage. A gray till of the St. Louis sublobe of the Des Moines lobe, which is in part overlain by the younger, clayey Valders red till and in part intertongued with it, presumably is of Valders age. The older sandy red till thus may be regarded as Mankato- (and Port Huron-) age drift of the Superior lobe, which should be correlated with gray Mankato till which occurs farther west and south in the Des Moines lobe.

Cochrane. The Cochrane glacial age is represented by three separate readvances of the ice margin in the vicinity of Cochrane, Ontario, 200 miles north of Georgian Bay. These are recorded by an outwash apron near Nellie Lake, 23 miles south of Cochrane; by remnants of an end moraine at Cochrane and in the Suskequa and Frederick House rivers; and by the Turgeon Bend moraine, which lies about 50 miles north of Cochrane.

The age of the Cochrane is still highly controversial. Because the Cochrane events occurred far to the north of the Great Lakes and had little effect on those lakes, the controversy is not reviewed here. The interested reader is referred to Antevs (1928 and 1953), Bryan and Ray (1940), Flint (1953), and Karlstrom (1956). The last-named writer (Karlstrom, 1956, p. 327) presents a convincing argument that the Cochrane glaciation is a post-Valders event which occurred before the "altithermal," or "climatic optimum," warm period of 4000 to 6000 years ago. He also suggests that the Cochrane should be ranked as a substage of the Wisconsin glacial stage. It is so considered in the tentative revision of the Wisconsin time scale by the present author, in Table 7.

Altithermal. A postglacial period of warm and relatively dry climate occurred between 6000 and 4000 years ago, according to various kinds of evidence which are summarized by Flint (1947, p. 487); since then the climate has become cooler and more moist down to the present time. This "altithermal," or "climatic optimum," period has been correlated with certain low-water stages in the Great Lakes, to be described in a subsequent chapter. It is not suggested, however, that the low-water stages resulted directly from excessive evaporation or even from diminished precipitation; it is believed that in all of the stages of the Great Lakes the basins were filled with water to the level of some available outlet. An indirect effect of a warm climate with subnormal precipitation would be the rapid wasting of an ice sheet, and this would result in the uncovering of new outlets which would serve as dischargeways to drain the lakes to lower levels.

CENTERS OF GLACIATION

According to the most widely held view, the glacial lobes of the Great Lakes region originated from two major centers, the Labradorean center east of Hudson Bay and the Keewatin center west of Hudson Bay, and from a secondary center, the Patrician, southwest of Hudson Bay. A review of the development of this concept is given in the following paragraphs.

Dawson (1886, p. 58R) apparently was the first writer to recognize that ice had moved radially outward from the region of the Canadian shield rather than from the Polar region. He later (1890, p. 162) proposed the name "Laurentide glacier" for a "sea of ice" occupying the region between the Great Plains and the Atlantic Ocean. T. C. Chamberlin (1895, p. 725) adopted Dawson's name "Laurentide glacier" and applied it to an inferred continental glacier covering the land east of the Rocky Mountains. Tyrell is the one who is primarily responsible for establishing the concept of separate major centers of glaciation north of the Great Lakes. This concept was developed in a series of publications on the Keewatin and Labradorean centers (Tyrell, 1894, p. 397; 1895, p. 439; 1896, p. 812; 1898, p. 150), and later the Patrician secondary center was elaborated on (Tyrell, 1913, p. 197; 1914, p. 523; 1935, p. 1). The evidence on which these centers of glaciation were based was largely radiating groups of striae made by late Wisconsin-age ice, though till lithology also was cited.

Following the usage proposed by Tyrell, the terms "Labradorean," "Keewatin," and "Patrician" have been quoted extensively in professional journals and textbooks and they have been applied to the supposed chief centers of glaciation throughout the Wisconsin age and even quoted in connection with pre-Wisconsin glacial events. Some writers, while recognizing the flow of ice from the general areas represented by these named centers, have preferred not to emphasize the importance of the centers (Upham, 1914, p. 515; Johnston, 1933, p. 16; Leverett, 1929b, p. 749; 1932, p. 3; 1935, p. 7).

Flint (1943, p. 325) has reviewed the evidence for the centers of glaciation, and has concluded that the Keewatin and Labradorean centers of radial flow were not certainly in existence before the latest fraction of the Wisconsin age, were not fixed in position even during that time, and were accompanied by other centers identified from evidence of a similar kind. He states (p. 332) that the concept of a single Laurentide ice sheet is most adequate when used for a continuous ice sheet occupying, at its maximum, most of Canada east of the Rocky Mountains and having a geographic center in the general region of Hudson Bay. In the same publication (p. 356) Flint develops the concept that if the Laurentide ice sheet were nourished chiefly by air masses moving northward, eastward, and northeastward from the Atlantic, the Gulf of Mexico, and the Pacific, the most active part of the ice should have been a broad marginal zone where relatively large accumulation induced relatively rapid flow. At the maximum the marginal zone of most active flow would have extended from Great Bear Lake southeastward through the Great Plains of Canada and across the region of the Great Lakes. Accumulation in this zone would have thickened the ice and induced flow toward areas of thinner ice, that is, toward the ice-sheet margin and also toward the central region of the ice sheet. Further, it may be supposed that accumulation was not uniform throughout the length of the marginal belt but that some sectors were provided with greater increments than others. The result would have been broad, very low domes on the upper surface of the ice sheet from beneath which the basal ice would have flowed slowly away in all directions. The present author believes that this concept is worthy of serious consideration, particularly in connection with the interpretation of the late Wisconsin events of Great Lakes history.

Holmes (1952, p. 1009), in a study of the drift in west-central New York, concluded that the Valley Heads (Cary) glacier evidently formed and spread radially from the deepest part of the Ontario basin, and that its deposits contain no record of glacial transportation from the Precambrian rock area north of Lake Ontario during Valley Heads time.

The flow of glacial ice out of the Lake Erie basin northward (Dreimanis and Reavely, 1953, p. 238) may better be explained by Flint's mechanism of flow from a thickened marginal belt of a Laurentide ice sheet than by flow of ice in an extremely elongate lobe from a distant center to the northeast.

Horberg and Anderson (1956), in a study of the relationship between bedrock topography and the Pleistocene lobes in the central United States, have concluded that "on the evidence of the bedrock topography alone, the case against major centers of glaciation is probably strengthened rather than weakened. The conformity between bedrock lowlands and glacial lobes shows that the bedrock topography was the dominant controlling factor in their development and that over-all relations can be explained by topographically controlled marginal drainage from a single ice sheet centered over Hudson Bay and from minor local centers" (p. 114). They also state, however, that factors of drift composition and distribution still require some consideration of the concept of major centers, perhaps in a modified form.

GLACIAL SCOUR OF THE LAKE BASINS

The origin of the Great Lakes basins has been a subject of interest to geologists for more than a hundred years. C. F. Volney, in 1803, stated that Lake Ontario lies along a great subterranean earthquake belt and that because the basin is deep and is almost completely [?] encircled by an escarpment, the lake basin must be a volcanic crater (Volney, 1803, pp. 129–31). Volney's earthquake belt has not been substantiated, and the rocks of the basin rim have been found to consist of Paleozoic sedimentary rocks (Fig. 13).

In the middle of the nineteenth century Agassiz's glacial theory was rapidly winning support, and when William E. Logan, first director of the Canadian Geological Survey, noted glacially scoured bedrock along the shores of the Great Lakes, he concluded that "These great lake-basins are depressions, not of geological structure, but of denudation; and the grooves on the surfaces of the rocks, which descend under their waters, appear to point to glacial action as one of the great causes which have produced these depressions" (Logan and others, 1863, p. 889).

In 1874, Newberry claimed the existence of an ancient river flowing from Lake Superior through the lake basins and down the Mohawk valley into the trough of the Hudson River, and thence to the ocean by New York. He believed that the valley of this stream, locally expanded into boat-shaped basins by glacial action, formed the basins of the Great Lakes (Newberry, 1874a, pp. 136–38). In another publication in the same year, Newberry (1874b, pp. 72–74) again stated his concept that the Great Lakes "are excavated basins, wrought out of once continuous sheets of sedimentary strata by a mechanical agent," and that ice or water or both were involved in the excavation. He emphasized the thought that no other agent than glacial ice is capable of producing broad, deep, boat-shaped basins.

Crustal warping was suggested as a cause of the Great Lakes basins by Spencer (1891a, pp. 86–97), who stated that the valleys of the Great Lakes are the result of stream erosion carried to a mature stage, followed by the development of barriers formed in late glacial times. In small part these barriers were due to drift fillings of portions of the original valleys, but more especially they were due to warping of the earth's crust, raising the downstream ends of the lake basins sufficiently to pond the waters. Deepening of the lake basins by glacial scour was not included in Spencer's hypothesis; in fact, in this regard he stated, "At the present stage in the investigation this subject can be quickly dismissed. The question whether glaciers can erode great lake-basins is hardly pertinent, for nowhere about the lakes is the glaciation parallel to the shores or vertical escarpments which are associated with the lakes" (Spencer, 1891a, pp. 93–94). A paper by Fairchild (1905, pp. 13–74) entitled "Ice Erosion Theory a Fallacy" gave support to Spencer's views by generally discounting the erosive powers of continental glacial ice.

Glacial scour was emphasized again by Shepard (1937, pp. 76–88) in a critical review of the characteristics of the Great Lakes, and he concluded that glacial erosion was the principal cause of the basins.

The close relationship between topography of the lake basins and bedrock geology which was illustrated in Chapter 1 leaves little room for doubt that the Great Lakes were excavated by erosional processes and that they are not primarily the result of diastrophic action. The relative importance of preglacial stream erosion and of glacial scour cannot be evaluated closely, but it appears that both were important. The fact that upwarp of the land to the northeast has occurred, raising the outlets of several of the lakes, does not disprove the probable occurrence of a great amount of glacial scour in shaping of the basins.

A detailed analysis of this question of the importance of glacial erosion has been made by Horberg and Anderson (1956, pp. 102–3), who have concluded, "The preglacial topography was extensively modified by glacial scour in the Canadian Shield and Superior Upland and by profound deepening of preglacial lowlands along the axes of the Great Lakes. Even in regions of scour, however, it is probable that the major topographic elements are inherited from the preglacial land surface."

INTERGLACIAL LAKES

The probable existence of lakes in the Great Lakes basins in interglacial times may only be inferred. The oldest lakes which have left any direct traces, in the forms of shore terraces, beaches, or other erosional or depositional features, are of late Wisconsin age. Because the early Wisconsin ice expanded through and stood with its outer margin beyond the lake basins, it overrode and apparently destroyed all direct evidence of any lakes which may have existed previously. The strong similarity of pattern of the early Wisconsin glacial lobes to those of the later Wisconsin glaciers suggests that the lake basins existed in essentially their present forms in early Wisconsin time. If this were true, we might expect that lakes would have existed in the basins at the various times when the ice front stood between the southern rims of the basins and any possible outlets to the north. How far back into the Pleistocene epoch this supposition may be projected is unknown.

A possible indirect indication of the existence of lakes in interglacial (or interstadial) times is a high percentage of clay- and silt-sized material in the till deposits, if such material can be presumed to be of lacustrine origin. It may be assumed, however, that the preglacial landscape was mantled to a large extent with residual soil, which would contain clays resulting from the weathering of the bedrock. The first glacier to advance over a maturely weathered land surface would be expected to pick up a considerable load of clay-sized materials, including clay minerals, and this material would not have been derived from lake deposits. Successive advances of ice over the same terrane, provided that the intervals between advances were of short duration, might be expected to acquire progressively smaller amounts of clay derived from weathering, and the tills deposited by those ice sheets then might be expected to contain smaller percentages of clay-sized material.

In a study of Wisconsin tills of northeastern Ohio, Shepps (1953, p. 46) found that the successively younger tills (Tazewell, early Cary, late Cary) contained progressively larger percentages of clay-sized material, and he quoted similar findings by other investigators of Wisconsin tills. In discussing these findings, Shepps pointed out that due to a protective coating of the bedrock by earlier tills, progressively less local bedrock material was made available for incorporation into the tills of the later ice advances. Hence, sources of coarse material for the later tills would be limited largely to the till which mantles the surface, to interstadial deposits such as dunes and local stream deposits, and to the scattered bedrock exposures which escaped the mantling by earlier tills or from which this mantle was removed by later erosion. In contrast, an increasing source of fine materials might be introduced during the interstadial period in the form of lake clays and silts and loess.

From the factors reviewed in the foregoing discussion it is apparent that the interpretation of the size composition of tills must be undertaken with great care. The inference of early Pleistocene lakes in the Great Lakes basins from any considerations of mechanical composition of the tills lying to the south probably would be very questionable.

chapter six

DATING THE EVENTS OF LAKE HISTORY

RELATIVE AGES

The sequence in which many geological events occurred can be determined directly from the order of superposition of deposits and of other features such as soil horizons and erosion surfaces. The Pleistocene glacial history of the Great Lakes region has been pieced together largely on this basis. Lake stages, where they are represented by beaches, sometimes can be dated as to relative age by observing the termination of the beaches against glacial moraines whose relative ages are known.

Abandoned beaches lying at various elevations in a given area generally are considered to have been formed in sequence from the highest to the lowest, so long as there is no evidence of reworking of the material. An example of a reworked beach is that of the Lake Wayne stage in the Erie and Huron basins, which "shows clear evidence of having been submerged and greatly modified" (Leverett and Taylor, 1915, p. 389). Because of this, the Wayne stage is dated as older than the late Warren stage, which is represented by a strongly developed and sharply defined beach lying 20 feet higher up the slope. In a few places older beach materials are actually overlain by younger shore deposits. An example of this is a group of impounded Early Algonquin beach spits, which were built into Sucker Creek Bay in Grey County, Ontario (Georgian Bay area), during a low-level stage and which were covered by a strong Algonquin bar built across the bay mouth at a later and higher stage of the lake (Stanley, 1938a, p. 483). A further indication of relative age is found where warping of the land surface has tilted one group of beaches, and a later beach has been formed which cuts across the earlier group, as occurred where the "lower Algonquin" beaches are transected by the Nipissing beach in Georgian Bay (Stanley, 1936, p. 1948, and 1937, p. 1679). Another relationship, which has given evidence for the occurrence of a low-water stage followed by a higher stage, is the occurrence of essentially undisturbed Nipissing beach deposits extending across truncated spurs and across intervening gulleys or small valleys which extend from a generally eroded terrane down to the Nipissing level and apparently extended to still lower levels before water of the Nipissing stage rose to drown the valleys (see p. 250 and Hough, 1953a, pp. 84–85). This low-water stage is correlated with Lake Chippewa, which was inferred from the stratigraphy and fossils of core samples taken in deep water in Lake Michigan (described in Chap. 13 and in Hough, 1955, p. 957).

ESTIMATES AND MEASUREMENTS OF AGE

Rate of Soil Formation on Glacial Deposits

Many estimates of the ages of geological events in the Pleistocene epoch have been based on considerations of the rate at which some process has been supposed to have operated. Among these are the following: (1) rate of change of animal and plant populations, due to climatic changes; (2) rate of migration of animals and plants; (3) rate of accumulation of humus and other organic deposits; (4) rate of deposition of stratified and unstratified glacial drift; (5) rate of formation of beach features such as terraces and cliffs; (6) rate of erosion of pre-existing topography by ice; (7) rate of surface erosion of glacial deposits; (8) rate of cutting of gorges and recession of falls; (9) rate of decomposition of pebbles and boulders in glacial drift; (10) rate of oxidation of drift; (11) rate of leaching of calcium carbonate from drift. These were tabulated by Kay (1931, pp. 453–54), and he recognized in connection with each that the rates are unknown or imperfectly known. He then stated, "The estimates which have been made by investigators by using the different criteria to which reference has just been made will be introduced here only insofar as they contribute to the judgments now to be developed by the writer regarding Pleistocene duration." Kay then reviewed data on depth of leaching of the various Pleistocene deposits of Iowa, and set up a relative age scale based on unity for the time since the retreat of the latest Wisconsin ice sheet from Iowa.

The figure for the duration of post-Wisconsin time was derived by Kay (1931) as follows: "It would seem to be a safe judgement in the light of recent investigations to assign 25,000 years to the time since the retreat of the Late Wisconsin ice-sheet from Iowa. When this figure was submitted recently to Antevs and to Leverett for their judgement as to the advisability of my adopting it in this paper both of them expressed the view that, based on present evidence, 25,000 years is, perhaps, the time estimate that can best be defended" (p. 454). It would be of interest to know the nature of the evidence on which the 25,000-year estimate was based, and to hear a defense of it. It is probable that this figure was derived in part from estimates of the age of the Niagara gorge, to be discussed in subsequent pages.

By using 25,000 years for the value of unity in his ratios for the relative ages of the Pleistocene interglacial ages, and by making certain assumptions regarding the rates of advance and retreat of the ice sheets, Kay derived a time scale for the Pleistocene which is shown in Table 8. This time scale has been quoted extensively for the ages of Pleistocene events in North America (e.g., Thornbury, 1940, p. 471; Piggot and Urry, 1941, p. 1197; Flint, 1947, p. 399; Moore, 1949, p. 463).

Kay's ratios for relative ages of materials were based mainly on the observed depths of leaching of carbonates. Dreimanis (1957, p. 403) has pointed out that the thickness of a carbonate-leached zone is a very unreliable index of the amount of leaching which has occurred, unless the total carbonate content of the original unleached material is known. For example, if ten feet of gravel, containing 10 per cent carbonates, is leached, nine feet of noncalcareous gravel will remain; while if ten feet of gravel, containing 90 per cent carbonates, is leached, only one foot of noncalcareous gravel will remain. The great difference in the result-

Table 8. *Minimum duration of the Pleistocene period in Iowa.*[a]

Interglacial age	Years	Glacial age
	0	
(Post–late Wisconsin)		
	25,000	
		Late Wisconsin
	28,000	
Peorian		
	55,000	
		Iowan
	58,000	
Sangamon		
	178,000	
		Illinoian
	187,000	
Yarmouth		
	487,000	
		Kansan
	495,000	
Aftonian		
	695,000	
		Nebraskan
	705,000	

[a] After Kay (1931, pp. 460–61).

ing measurable thickness (nine feet and one foot) does not mean that the nine-foot weathering profile is older than the one one-foot deep. The opposite may be closer to a correct answer. In view of this, Kay's age estimates are considered unreliable.

Age of Niagara Gorge

The first serious attempt to place a date in terms of years on events of lake history was the calculation of the age of the Niagara River gorge. This feature almost certainly has been cut since the last ice sheet receded from the locality, exposing the Niagaran escarpment and allowing the discharge from the Lake Erie basin to pour over it. The age of the gorge may be calculated by taking the length of the gorge and dividing it by the average amount of recession of the falls per year.

Upham, in 1896, stated:

dividing the length of the Niagara gorge (about six and a half miles) by the recent rate of average annual recession of the falls (nearly five feet), we have approximately 7,000 years, as announced by Gilbert at the meeting of the American Association in Buffalo in 1886, as the probable time required for the erosion of the gorge.

This measure, which (not to be too exact in figures depending on somewhat varying conditions of the Niagara history) we may place in round numbers as between 5,000 and 10,000 years, is of great interest to geologists because it is at the same time the duration of the period since the end of the Ice Age, or, speaking more definitely, since the retreat of the continental glacier from the northern United States and southern Canada. It may be so accepted with confidence, for it agrees with the estimates and computations independently made for the same period by Professor N. H. Winchell, from the recession of the Falls of St. Anthony; by Professor G. Frederick Wright from the filling of depressions among kames and eskers, and from erosion by streams tributary to Lake Erie; and by Professor B. K. Emerson, from post glacial deposition in the valley of the Connecticut River. In Europe, likewise, numerous estimates of the lapse of time since the Glacial period, as collated by Hansen, are found to be comprised between the limits of 5,000 and 12,000 years, thus being well harmonious with the measure given us by Niagara Falls [pp. 176–77].

The validity of the age calculated solely from a recently determined average annual rate of recession is affected, however, by several factors. Probably the most important of these is that the entire discharge of the upper Great Lakes did not pass through the Niagara gorge during its entire history. Spencer (1894, p. 472), taking this into account, estimated the life of the Niagara gorge and river as 32,000 years. Taylor (1895a) stated that the calculated age of the gorge (7000 years) should be multiplied by tens in arriving at an age estimate, and that "As a measure of the duration of post-glacial time, therefore, I do not see how the gorge can have any value worth mentioning" (p. 70). Kindle and Taylor, however, in 1913 (p. 24) state that the Niagara gorge cutting may have taken between 20,000 and 35,000 years. Sometime during the next 18 years the age of the Niagara gorge came to be accepted as 25,000 years. A search of the literature of this period has failed to reveal any substantial basis for this age. Leverett, in 1930, stated, "From a study of Niagara Falls by Spencer, Taylor and others [see Kindle and Taylor, 1913] it appears that the Port Huron morainic system, which marks the limits of the late Wisconsin drift, was formed some 25,000 to 30,000 years ago" (p. 193). Then Kay (1931), as quoted on a foregoing page, selected 25,000 years as the date of the retreat of the latest Wisconsin ice sheet from Iowa and, in personal correspondence with Leverett and Antevs, obtained their approval of the date but gave no evidence for it. The beginning of the cutting of the Niagara gorge was associated with the retreat of the "latest Wisconsin ice sheet," and within a few years the familiar figure, Kay's 25,000 years, came to be accepted as a convenient one for the age of the Niagara gorge. Meanwhile, however, Taylor (1929, p. 261), on the basis of new evidence, had revised his estimate of the age of the gorge to 18,000 to 20,000 years.

More recent studies of the late Wisconsin drift have shown that the Valders glacial advance occurred after the "late Wisconsin" of the writers quoted in the foregoing paragraph. The Valders has been dated at about 11,000 years before the present, by the radiocarbon method (see a subsequent part of this chapter). It is thought by some, including the present author, that the Valders ice margin lay upon the Niagaran escarpment in the vicinity of the Niagara gorge and that the present gorge, therefore, is post-Valders in age. An age of something less than 11,000 years thus is indicated. This interpretation will be discussed further in sections on the history of the lake stages.

The age of the entire postglacial Niagara gorge cannot be calculated from rates of erosion with any confidence, because of uncertainty regarding variations in discharge from the upper lakes and because of the complicating effect of the presence of an earlier, drift-filled gorge which is, in part, occupied by the present gorge. The age of the present "Upper Great Gorge," upstream from the reopened buried-gorge section, may be evaluated somewhat more closely, however. From an evaluation of lake history presented in a subsequent chapter, it appears certain that the volume of discharge from the Erie basin was augmented by the discharge from all of the upper lakes beginning with the Nipissing stage, and that this large volume of discharge has continued uninterruptedly to the present. The cutting of the wide "Upper Great Gorge" is correlated with this entire post-Nipissing period. Kindle and Taylor (1913, p. 24) calculated the age of the upper gorge as 3000 to 3500 years, based on the current rate of recession of the Horseshoe Falls. A later revision of the rate of recession (Johnston, 1928, p. 29) raised the figure to 4000 years or more. Radiocarbon dates for the Nipissing stage, as reviewed by Hough (1953c, p. 138), indicate the age of the Nipissing to be approximately 4000 years.

Varve Correlation

A method of time measurement which has a potentially high degree of accuracy is the counting of the seasonally banded layers, or varves, of clay or fine silt deposited in glacial lakes. It is generally believed that rapid melting of glacial ice in the summer season, and the resulting high volume of water discharge with a high quantity and coarseness of sediment load, give the relatively coarser layers of glacial lake sediments, and that the finer-grained layers result from settling out of material from suspension in the lake water during the winter, when inflow has been reduced or stopped and when freezing over of the lake surface prevents agitation which would keep the fine material in suspension. The winter band is sharply separated from the summer band above, but grades into the summer band below. Chemically the winter bands contain more ferric oxide, alumina, and potash, and somewhat less lime. The winter bands generally are darker in color.

One pair of layers, a coarser summer band and the overlying finer winter band, is generally considered to represent the deposit of one year. Assuming this to be true in all cases, a count of the pairs of bands in any one deposit gives a measure of the years represented in making the deposit. Fluctuations in the quantity and coarseness of material supplied to a lake from one year to the next, and in the relative lengths of the summer and winter seasons, cause a variation in the thicknesses of the annual layers and in the relative thicknesses of the summer and winter layers in a deposit. Because such variations are related to climate, the individual layers in deposits of separate glacial lake basins in the same region may be correlated by comparison of their varve-thickness measurements. Many such correlations have been made in the Scandinavian countries and in the northeastern United States and eastern Canada.

The most noteworthy varve study, in relation to Great Lakes history, is that of Antevs (1953), in tracing and dating the retreat of the glacial ice margin from New Jersey and southern New England to the region between Cochrane, Ontario, and James Bay. By making correlations between the contemporaneous portions of the deposits of a succession of glacial lakes which followed the ice margin northward, Antevs established a late glacial chronology for eastern North America. This is based mainly on varved glacial clays and partly on time estimates of gaps in the varve data. As stated in 1953 it comprises 15,700 years, of which 12,055 are represented by counted varves and 3645 by estimates. Antevs joined his North American varve chronology with a Finno-Swedish varve chronology by assuming a transatlantic correlation between a late ice advance in the Cochrane area and a late ice advance in southern Finland. Because the Finno-Swedish chronology is tied to the present day with only a small degree of uncertainty, Antevs believed that he had tied his North American varve chronology to the present day. His transatlantic correlation is based on considerably more detail than is reviewed here, and his argument for the correlation is logical, but the correlation is not established beyond reasonable doubt.

On the basis of his varve chronology, Antevs (1953, p. 212) derived a date of approximately 19,000 years for the Valders maximum, which was correlated with the St. Johnsbury moraine in Vermont. In 1955 (pp. 497–99) Antevs correlated the Valders maximum with a moraine at the northern end of Lake Willoughby in Vermont and derived its age as approximately 18,500 years. In earlier publications (Antevs, 1928 and 1931) considerably greater ages apparently were derived for more or less correlative events (though the terms "Mankato" and "Valders" were not in use as yet). As Karlstrom (1956) has stated, "This progressive downward revision in the length of the varve chronology is not the product of new varve data so much as of re-estimates of the durations represented by the gaps in the varve sequences, and of changes in correlations between moraines on the east coast and type sections in the Midwest and in northern Europe. The delineation of the precise ice boundaries for Wisconsin events from the Midwest to the east coast, as well as the transatlantic correlations, remains a subject of some conjecture and of no small controversy. Because of these uncertainties, Antev's figures may be accepted as shrewd first approximations but not necessarily as precise figures" (p. 310). The ages of late glacial and postglacial events as determined by the radiocarbon method (discussed in a subsequent part of this chapter) are appreciably less than those estimated from the varve chronology.

Uranium-Ionium-Radium Method of Age Determination

A method of dating marine sediments by the per cent of equilibrium between uranium, ionium, and radium has been developed by Urry (1942). Though the method is not applicable to fresh-water or terrestrial sediments, it has been used to date the material in various sea-bottom core samples which include a record of Pleistocene glacial events.

The longest detailed record available is contained in a southeastern Pacific Ocean core sample (Hough, 1953b). This core contains several layers of red clay separated by globigerina ooze. Because calcium carbonate is more soluble in cold water, the red clay layers (with low carbonate content) are interpreted as records of cold-water deposition and the globigerina ooze layers (with high carbonate content) are interpreted as records of warmer-water deposition. These indications of water temperature at time of deposition may, further, be indications of colder and warmer climatic conditions. The author's interpretation of this Pacific Ocean core is given in Fig. 32, where its presumed climatic record is compared with the North American Pleistocene chronology.

Other ocean-bottom core samples which have been dated by the uranium-ionium-radium method include three from the North Atlantic Ocean collected by C. S. Piggot (Bramlette and Bradley, 1940, and Piggot and Urry, 1941) and three from the Ross Sea, Antarctica, collected by J. L. Hough (Urry, 1949, and Hough, 1950). A comparison of the last 70,000 years' record of cores selected from each of these groups, and the last 70,000 years' record of the southeastern Pacific core is made in Fig. 33.

The North Atlantic cores contain glacial marine (ice-rafted) sediment in the layers shown in black. Cold-water foraminiferal faunas occur in these layers and in some other zones marked "C" in Fig. 33. A warm-water faunal zone centered 6000 years ago presumably is a representative of the "altithermal" or "climatic optimum" which is generally recognized by students of postglacial events (Flint, 1947, p. 487). This warmer period is represented in the southeastern Pacific core by a layer of globigerina ooze centered at 6000 years ago. The generally cooler climate which has prevailed in the northern hemisphere since the altithermal is recorded in the southern hemisphere by glacial marine sediment in the Ross Sea core from the present back to nearly 6000 years ago. Dated zones in this and the other Ross Sea cores indicate that the northern limit of pack ice in the sea moved northward more than 100 miles between 6000 and 4000 years ago (Hough, 1950, p. 259). Of particular interest in the North Atlantic core is the zone between 10,800 and 11,800 years ago; this is composed of mud comparatively low in calcium carbonate and containing a cold-water fauna, and within this zone at 11,000 years there is a coarser-grained zone which was characterized as "anomalous," and was attributed to submarine slumping, by Piggot and Urry (1941, p. 1198). This description was written before the Valders drift was named by Thwaites (1943) and several years before the Valders was dated as 11,000 years old (Flint and Deevey, 1951). In the light of present knowledge this cold-water zone with the coarser-grained material at 11,000 years may be confidently identified as a representative of the Valders substage of the Wisconsin. This is indicated as "Substage VI" in Fig. 33 and is correlated with the lower part of a cool-water zone in the southeastern Pacific core and with an occasional ice-rafted pebble occurring in the Ross Sea core at the 11,000-year-old level. The remaining cold-water zone correlations of the cores illustrated in Fig. 33 are fairly obvious. It has been suggested that the six numbered substages of those cores correspond with six Wisconsin glacial substages of the Great Lakes region (Fig. 32 and Hough, 1953b, p. 258).

The accuracy of the percentage of equilibrium of the uranium-

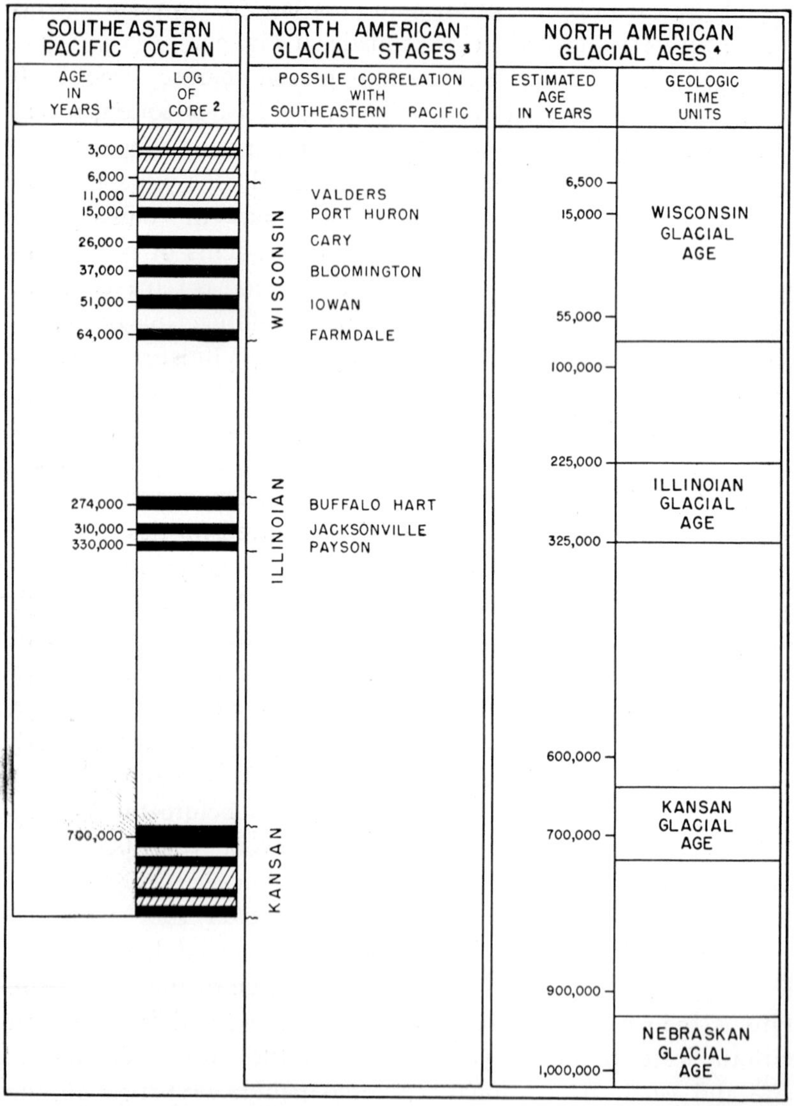

126 GEOLOGY OF THE GREAT LAKES

SOUTHEASTERN PACIFIC OCEAN		NORTH AMERICAN GLACIAL STAGES [3]	NORTH AMERICAN GLACIAL AGES [4]	
AGE IN YEARS [1]	LOG OF CORE [2]	POSSIBLE CORRELATION WITH SOUTHEASTERN PACIFIC	ESTIMATED AGE IN YEARS	GEOLOGIC TIME UNITS
3,000				
6,000			6,500	
11,000		VALDERS		
15,000		PORT HURON	15,000	WISCONSIN GLACIAL AGE
26,000		CARY		
37,000		BLOOMINGTON		
51,000		IOWAN	55,000	
64,000		FARMDALE		
			100,000	
			225,000	
274,000		BUFFALO HART		ILLINOIAN GLACIAL AGE
310,000		JACKSONVILLE		
330,000		PAYSON	325,000	
			600,000	
700,000			700,000	KANSAN GLACIAL AGE
			900,000	
			1,000,000	NEBRASKAN GLACIAL AGE

Fig. 32. A dated climatic record in a southeastern Pacific Ocean core sample, compared with the North American Pleistocene chronology. Legend: [1] Ages by the uranium-ionium-radium method, by W. D. Urry (from Hough, 1953b, fig. 2). [2] Solid black = red clay (cold-climate deposits); diagonal lines = calcareous red clay (cool-climate deposits); white = globigerina ooze (warm-climate deposits). [3] Stages from Tables 6 and 7; subdivisions of Illinoian from Leighton and Willman (1950, p. 602). [4] Time scale from Flint (1947, p. 532).

ionium-radium method has not been discussed critically in the literature. Theoretically, the method is valid to an age of about 300,000 years, but the percentage of error of determinations within this age are not known to the author. Inaccuracies in the dates assigned to events in Figs. 32 and 33 are present because several of those dates have been derived by interpolation between the dates for the sampled points in the core samples. Because of these considerations, the time scale for events dated by the uranium-ionium-radium method should not be considered as absolute. It is of interest, however, to note the correspondence in date (11,000 years) between the Valders of the Great Lakes region, dated by the radiocarbon method, and the events identified as Valders in age in Fig. 33, dated by the uranium-ionium-radium method; also the general correspondence between climatic events in the widely separated oceanic regions represented in Fig. 33. At the very least, it appears reasonable to assume that the four major glacial marine zones of the North Atlantic contained in cores P-124 (3) and P-126 (5) of Fig. 33 are representatives of the Wisconsin age (as stated by Piggot and Urry, 1941, p. 1196) rather than representatives of the four major Pleistocene glaciations, which was one possibility suggested by Bramlette and Bradley (1940, p. 10) before the dates were obtained. The detailed correlations indicated in Fig. 32 are suggestions only, and it is hoped that further investigations of oceanic sediments will be made, for comparison with the growing body of information on the dates of late glacial events in the Great Lakes region.

Radiocarbon Method of Age Determination

A method of dating materials containing carbon has been developed by W. F. Libby. This was first described in the literature by Anderson and others in 1947, and the first tests of the method against archeological samples of known age were reported by Arnold and Libby (1949, p. 678). A brief history of the development and early applications of the method, and an appraisal of it, are given by Johnson and others (1951). The method has been developed further, and it has been applied to a large number of archeological and geological specimens. The results generally are considered valuable and reasonably accurate back to ages of 20,000

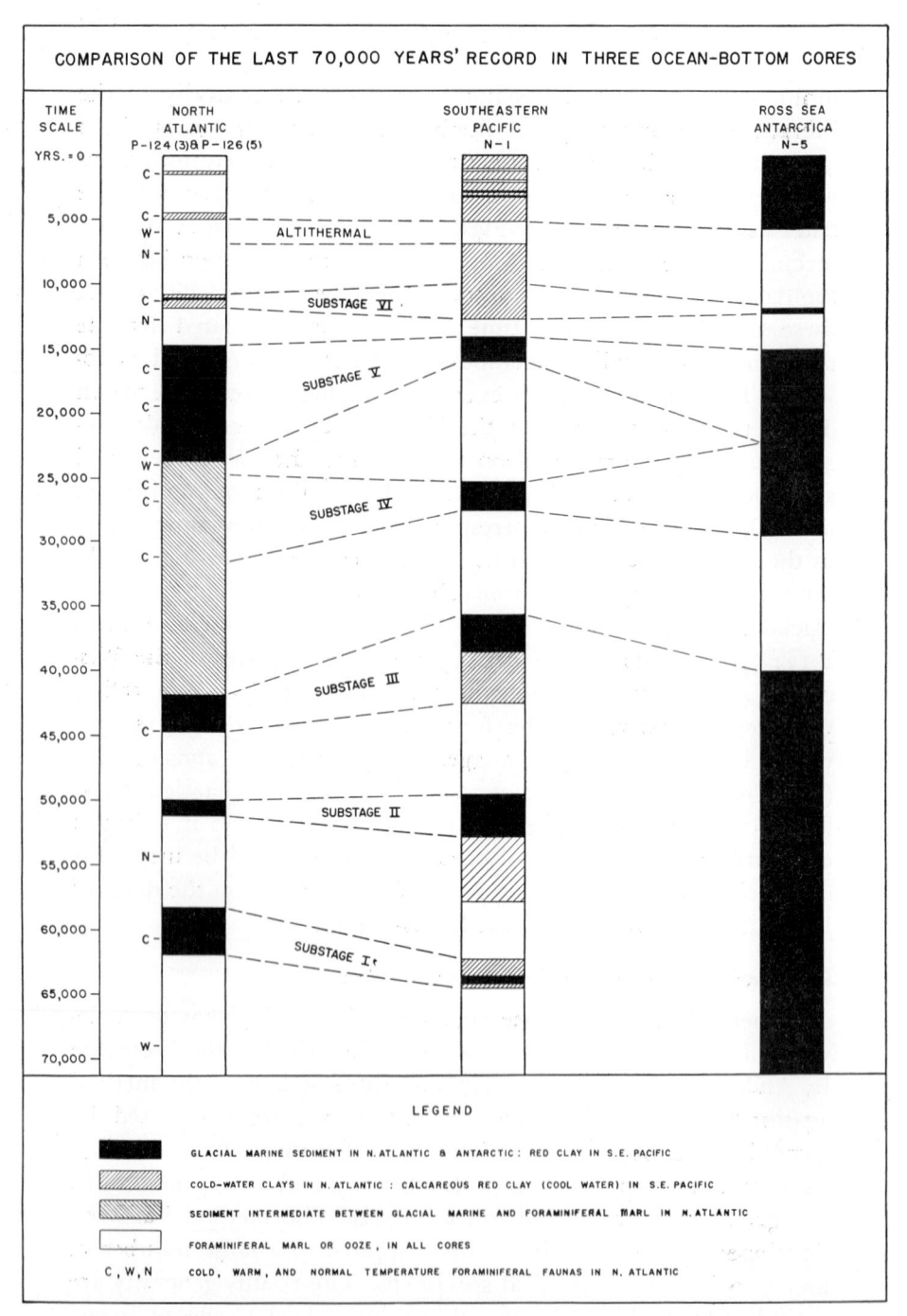

Fig. 33. Climatic record of the last 70,000 years in three ocean-bottom cores, dated by the uranium-ionium-radium method. The Valders glacial substage of the Great Lakes region, dated at 11,000 years by the radiocarbon method, may be correlated with Substage VI. (Reproduced from fig. 3 of Hough, 1953b, Univ. of Chicago Press, by permission.)

or 30,000 years, and the method shows promise of being extended to 50,000 years. This makes it particularly useful in dating the late glacial and postglacial events of the Great Lakes region.

A critical evaluation of the radiocarbon dating method is beyond the scope of the present book. It is noted that the results obtained thus far generally have appeared to be trustworthy within fairly narrow limits because the dates obtained generally have been consistent in their sequence with those expected from the known stratigraphic positions of the dated material, and, after the first shock of the relatively recent ages indicated for late glacial events, they have appeared to be fairly consistent with values predicted. The most outspoken critic of the radiocarbon method has been Antevs (1953, 1955, 1957), whose varve-based chronology gives substantially greater ages for most of the late glacial events in the Great Lakes region.

It must be admitted that any single radiocarbon age determination may be erroneous because of a number of factors, such as contamination of the sample by either older or younger material. It is possible that there is a bias in the method which gives progressively greater errors with increasing age of the materials. Relatively few radiocarbon dates of Great Lakes events are at present available, and some of them obviously are erroneous by comparison with the dates of correlative events. For these reasons a radiocarbon time scale is not applied generally in the discussions of lake history throughout Part 2 of this book. Radiocarbon dates are cited in Chapters 8 through 15, for the purpose of indicating relative ages rather than assigning absolute dates to the events discussed. An absolute time scale, based on the radiocarbon dates, is presented in Chapter 16 (Table 22) along with lists of the dates (Tables 17–21) used in preparing the time scale.

PART TWO

HISTORY OF LAKE STAGES

chapter seven

INTRODUCTION TO HISTORY OF THE LAKES

EVIDENCES OF FORMER LAKE STAGES

The evidence for the many lake stages which have occurred in various parts of the Great Lakes system has been reported in detail in other publications. The principal reference is the monograph by Leverett and Taylor (1915), which summarizes and interprets the information known at the time, including the results of several years of study by those authors. The purpose of the present book is to present a revision of the history of the lakes which takes cognizance of new information and which includes some reinterpretation of the old data. The detailed evidence for each of the lake stages is not restated here unless a revision of the published interpretations is involved.

The most common evidence for a former lake stage is a sand or gravel beach deposit, similar to the beaches of the present shores of the Great Lakes but lying inland from them at a higher altitude. Some of these abandoned beaches extend for many miles and they have been used as routes for highways and railways. Conspicuous beach ridges near Lakes Erie and Ontario were among the first to receive careful study. In Ohio, Colonel Charles Whittlesey, in 1838, reported a line of beaches skirting Lake Erie at a maximum distance of five miles and at an altitude of 90 to 120 feet above the lake (Whittlesey, 1838, p. 55). In Michigan a corresponding shoreline north of Lake Erie was traced for 60 miles by Bela Hubbard, and described by him in the *Annual Report* of the Geological Survey of Michigan (1840, p. 104). Sir Charles Lyell visited the eastern Great Lakes in 1841, and observed 11 distinct terraces near Toronto, the highest of which was 680 feet above sea level. His impressions were expressed as follows: "with the exception of the parallel roads or shelves in Glen Roy, and some neighboring glens of the Western Highlands of Scotland, I never saw so remarkable an example of banks, terraces, and accumulations of stratified gravel, sand and clay maintaining, over wide areas, so perfect a horizontality, as in this district north of Toronto" (Lyell, 1845, p. 106).

Abandoned beaches are bordered in some areas by old sand dunes, many of which are now stabilized by a cover of vegetation. Wave-cut terraces and cliffs are next in abundance among the evidences of former stands of the lakes. In some places more recent wave action or stream erosion has cut into the land, leaving only a fragmentary record of the older shorelines.

A large stream entering a lake may deposit a delta which is graded nearly to lake level, and such deltas now found above water record earlier lake levels. Probably the most extensive feature of this type is the Allendale delta on the Grand River in western Michigan, which covers nearly 100 square miles (Bretz, 1951b, p. 422). The top surface of this deposit records a lake level 60 feet above the present surface of Lake Michigan. The most spectacular delta of the entire region is on the Magpie River near Michipicoten, Ontario, east of Lake Superior. Its top surface stands about 400 feet above Lake Superior and its steeply sloping front stands as a bold break in the valley topography.

Outlet channels of some of the lake stages may still be observed, notably the Lake Chicago outlets west of Chicago and the Lake Maumee outlet at Fort Wayne, Indiana. The elevations of the channel bottoms record the minimum depths of the lakes draining through them, unless upwarp of the land has since raised the outlet area as is the case at North Bay, Ontario.

Other less common evidences of former lake levels are described in the detailed discussion of lake history.

Throughout the discussion of lake history, the various lake stages are assigned specific elevations. It is to be understood that the figures given are average values representing a range of elevations, generally through a vertical distance of about five feet. The elevation of the crest of a depositional beach ridge varies from place to place, depending on the height of storm waves which reach the shore and on the amount and size of debris available for beach construction. Constructional forms such as ridges and bars generally have crest elevations which are higher than wave-cut terraces formed elsewhere in the same lake. The elevations of the present Great Lakes have varied as much as five feet through a period of a few years (Figs. 20A and 20B), and it is reasonable to suppose that the earlier lake stages also had appreciable ranges in elevation.

WARPING OF THE SHORELINE FEATURES

Many of the shoreline features formed by older stages of the lakes are found to rise as they are traced northward around the lake basins. This rise indicates that the region has been tilted and warped slightly since the beaches were formed. The cause of the tilting is generally considered to be the result of isostatic rebound as the glacial ice was removed by melting. Presumably the land was depressed as the continental ice sheets advanced into the region, and when the ice disappeared the land tended to rise back approximately to its previous altitude.

Because there were several advances and retreats of the ice, it seems likely that the land was depressed and then was elevated several times. A study of the configuration of the beaches indicates that there were several distinct periods of tilting of the land, separated by periods of no movement. During the later part of the Wisconsin glacial stage, each advance of the ice margin generally was to a line which lay farther north than that marking the previous advance. Some of the older beaches (for example, the Maumee and Whittlesey of the Erie basin) are observed to be horizontal in their extreme southern areas and to rise to the north to their points of termination against younger glacial deposits. Somewhat younger beaches occurring farther north (for example, the Algonquin of the Huron basin) are observed to be horizontal in their southernmost areas and to be tilted farther north. This indicates that the tilting which had deformed the

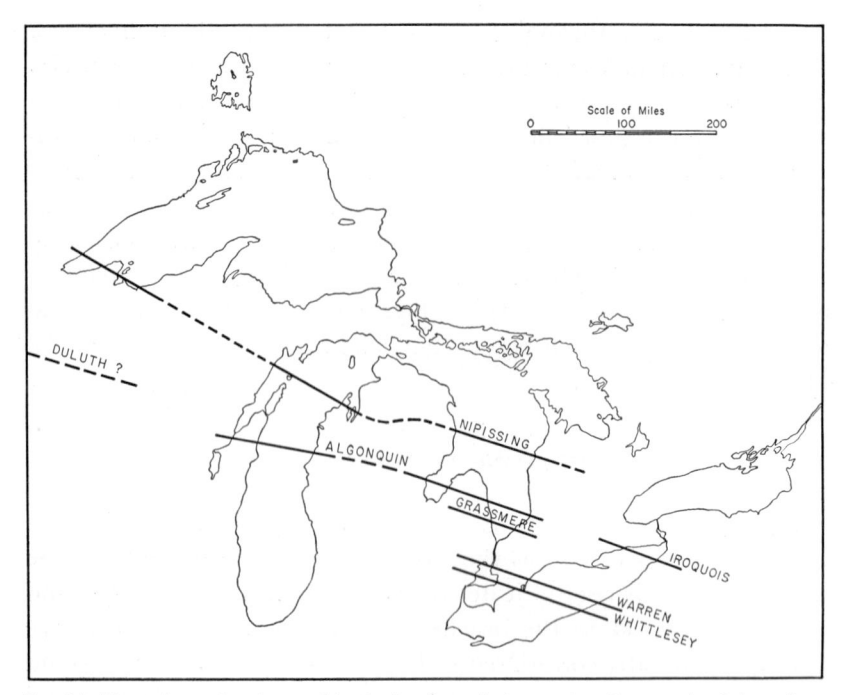

Fig. 34. Hinge lines of various uplifts in the Great Lakes region. From each of these lines the beaches of the lake stage named rise to the northward.

earlier beaches had ceased, and that a later episode of tilting occurred after the later beaches were formed; also that the later tilting was hinged on a line lying north of the hinge of the earlier tilting.

Figure 34 shows the "hinge lines," or lines of zero uplift, of several of the beaches in the Great Lakes region. From each of these hinge lines one of the beaches rises to the north. The details of the slope of the tilted planes of two of the individual beaches are shown by Fig. 42 (Chap. 12) and by Fig. 51 (Chap. 14). Warping of beaches is shown diagrammatically in Figs. 43 and 44 (Chap. 13).

A discussion of the various possible causes of crustal deformation in the Great Lakes region was given by Leverett and Taylor (1915, pp. 502–18), who concluded that isostatic rebound resulting from the melting of glacial ice was the principal cause. Gutenberg (1933) reached the same conclusion. Deformation of the region is still in progress, as shown by the change in elevations of lake gage markers reported by Gutenberg (1933) and by Moore (1948). Moore concluded that the deformation is not primarily a result of the disappearance of the ice, but is a tectonic movement of the region which dates from earlier geologic eras. Gutenberg, however, has assured the author [1] that a calculation of the possible isostatic rebound due to removal of the ice load is quite adequate to explain the observed deformation.

INTERPRETATION OF THE EVIDENCE

The evidence for each one of the lake stages, within a single lake basin, is sufficiently clear in most cases to be beyond reasonable doubt. The order in which the stages occurred, within a single basin, generally is well established, but there are some portions of the history in which the exact sequence of events is not certainly known. The correlation of events in two or more lake basins is considerably less well proven, and some of the details of the history, such as the particular outlet channel used by a given stage, are matters of inference which are not supported by evidence in all cases.

The history of the entire region has been synthesized by reasoning from the well-known events to the most probable sequences of happenings and from there to relationships and correlations, some of which are merely conjectures. When more information has become available, it has too often been interpreted within the framework of an existing hypothesis, without reviewing all the evidence which may be relevant.

In the present synthesis of the history of the Great Lakes, an attempt is made to present a review which proceeds from the more probably correct interpretations to inferences that are less well founded, and to indicate the degree of doubt or certainty existing in the conclusions which have been drawn.

Because the history of the lakes is long and very detailed, all of the alternative interpretations which are possible cannot be presented without making the story unduly long and complicated.

---

[1] Personal communication, 1948.

It should be understood that the synthesis given here is subject to revision, especially in the light of any additional facts which are discovered.

It is not feasible to review all of the various interpretations of lake history which have been published, or to acknowledge the source of all information and ideas which have been contributed. Leverett and Taylor, in their monograph, assembled much of the data which were available at the time and made their own interpretations of it. In general, the present book is based on the publication of Leverett and Taylor (1915) and on the material which has been published since 1915.

OUTLINE OF DISCUSSION OF THE LAKE HISTORY

From the earliest known lakes to the time of Lake Algonquin, each lake basin generally had a distinctive series of lake stages which were more or less independent of events in the other basins. For this reason the earlier part of the history is most conveniently described by subdividing the material into chapters dealing with the separate basins. This is presented in Chapters 7 through 10. The entire history of the Ontario basin is given in Chapter 11. From the time of Lake Algonquin to the present several of the lake basins had the same lake stage in common; it is, therefore, more convenient to discuss each of these stages in order, subdividing the material into chapters according to stage rather than by lake basin. This later part of the history is presented in Chapters 12 through 15.

The final chapter, 16, includes a summary of the major events of the history of the Great Lakes and presents a tentative time scale based on radiocarbon dates.

chapter eight

EARLY LAKE STAGES IN THE ERIE AND HURON BASINS

INTRODUCTION

The history of lake stages in the Erie and Huron basins, from the earliest known lakes to the Algonquin stage, is presented first because the record of those events is probably more fully known and less controversial than the record of the earlier stages in the other basins. Further, the Erie and Huron basins are centrally located in the region and the outflow from them to the other basins during the early stages generally can be established without reference to specific events in the other basins.

The Great Lakes region is shown in Fig. 53 [1] at a time just before the first known lakes were formed. This time was in a waning stage of the Valparaiso phase of the Cary glacial substage in the Michigan basin and a waning stage of the Fort Wayne phase of the Cary glacial substage in the Erie basin.

Several small lakes were formed at intervals along the margin of the glacier wherever the ice front withdrew to the north side of the southern boundary of the Great Lakes watershed. These earliest small lakes discharged southward independently through low points in the divide; later, upon further retreat of the ice margin, they were joined with the first stage of glacial Lake Maumee. Descriptions of these early independent lakes have been given by Leverett (1902, pp. 610–11).

The major lake stages of the Erie basin are listed in Table 9,

---

[1] Figures 53–75, lake stage maps, follow Chapter 16.

140 GEOLOGY OF THE GREAT LAKES

Table 9. *Lake stages of the Erie basin.*

Name	Altitude above sea level (feet)	Outlet
Highest Maumee	800	Ft. Wayne–Wabash River
Middle Maumee	790	Ft. Wayne–Wabash River
Lowest Maumee	760	Huron basin
Whittlesey	738	Huron basin
Highest Arkona	710	Huron basin
Middle Arkona	700	Huron basin
Lowest Arkona	695	Huron basin
Highest Warren	690	Huron basin
Middle Warren	682	Huron basin
Lowest Warren	675	Huron basin
Wayne	655	Syracuse–Hudson River
Grassmere	640	Huron basin?
Lundy	620	Huron basin?
Early Algonquin	605	Huron basin?
Erie	573	Niagara River
Early Erie	<573	Niagara River
Two Creeks–interval low stage	?	Niagara River
Cary–Port Huron–interval low stage	?	Unknown—eastern

in the order of their altitudes above sea level. This is not the order of their occurrence in all cases. Figure 35 shows the lake stages of the Erie basin in the order of their occurrence, according to the present author.

The major lake stages of the Huron basin, many of which were continuous with those of the Erie basin, are listed in Table 10, in the order of their altitudes above sea level. Again, this is not the order of their occurrence in all cases. Figure 36 shows the stages of the Huron basin in the order of their occurrence according to the present author.

HIGHEST LAKE MAUMEE

The first major lake stage in the Erie basin was formed by recession of the margin of the glacial ice eastward from the south-

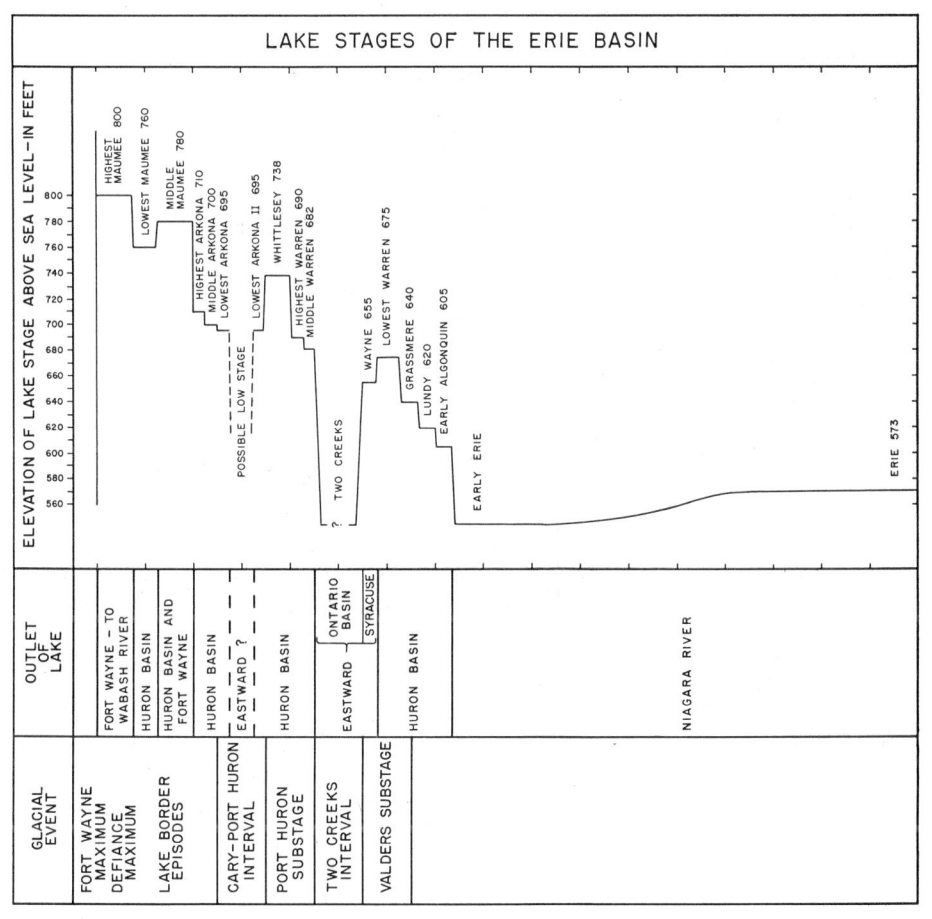

EARLY LAKE STAGES IN THE ERIE AND HURON BASINS 141

Fig. 35. Lake stages of the Erie basin. Lake levels are plotted against an unspecified time scale in the upper part of the figure, and lake outlets and correlative glacial events are shown in the lower part.

western end of the basin, leaving a low area between the ice and the Fort Wayne moraine, which is of Cary age. Water ponded in that area discharged over a low part of the moraine and soon established an outlet channel, which is located at Fort Wayne, Indiana. This stage, the Highest Maumee (see Fig. 54), stood at an elevation close to 800 feet above sea level and is represented by well-developed beaches which have been described by Leverett

Table 10. *Lake stages of the Huron basin.*

Name	Altitude above sea level (feet)	Outlet
Lowest Maumee	760	Erie basin
Whittlesey	738	Grand River
Early Saginaw (in Saginaw Bay)	735	Grand River
Highest Arkona	710	Grand River
Middle Arkona	700	Grand River
Lowest Arkona	695	Grand River
Highest Warren	690	Grand River
Middle Warren	682	Grand River
Lowest Warren	675	Grand River
Wayne	655	Unknown—eastern
Grassmere	640	Northwest to Michigan basin?
Lundy	620	Northwest to Michigan basin?
Early Algonquin	605	Northwest to Michigan basin and (later) to Erie basin
Algonquin	605	Northwest to Michigan basin and Port Huron
Nipissing	605	First, three-outlet stage: North Bay, Port Huron, and Straits of Mackinac
		(Possible) second, two-outlet stage: Port Huron and Straits of Mackinac
Algoma	595	Port Huron [a]
Huron	580	Port Huron
Kirkfield	555	Kirkfield
Post-Algonquin "upper group"		Kirkfield?
Wyebridge	540	Unknown—eastern
Penetang	510	Unknown—eastern
Cedar Point	493	Unknown—eastern
Payette	465	Unknown—eastern
Sheguiandah	?	Unknown—eastern
Korah	?	Unknown—eastern
Stanley	230–	North Bay

[a] High-water phases also discharged through the Straits of Mackinac.

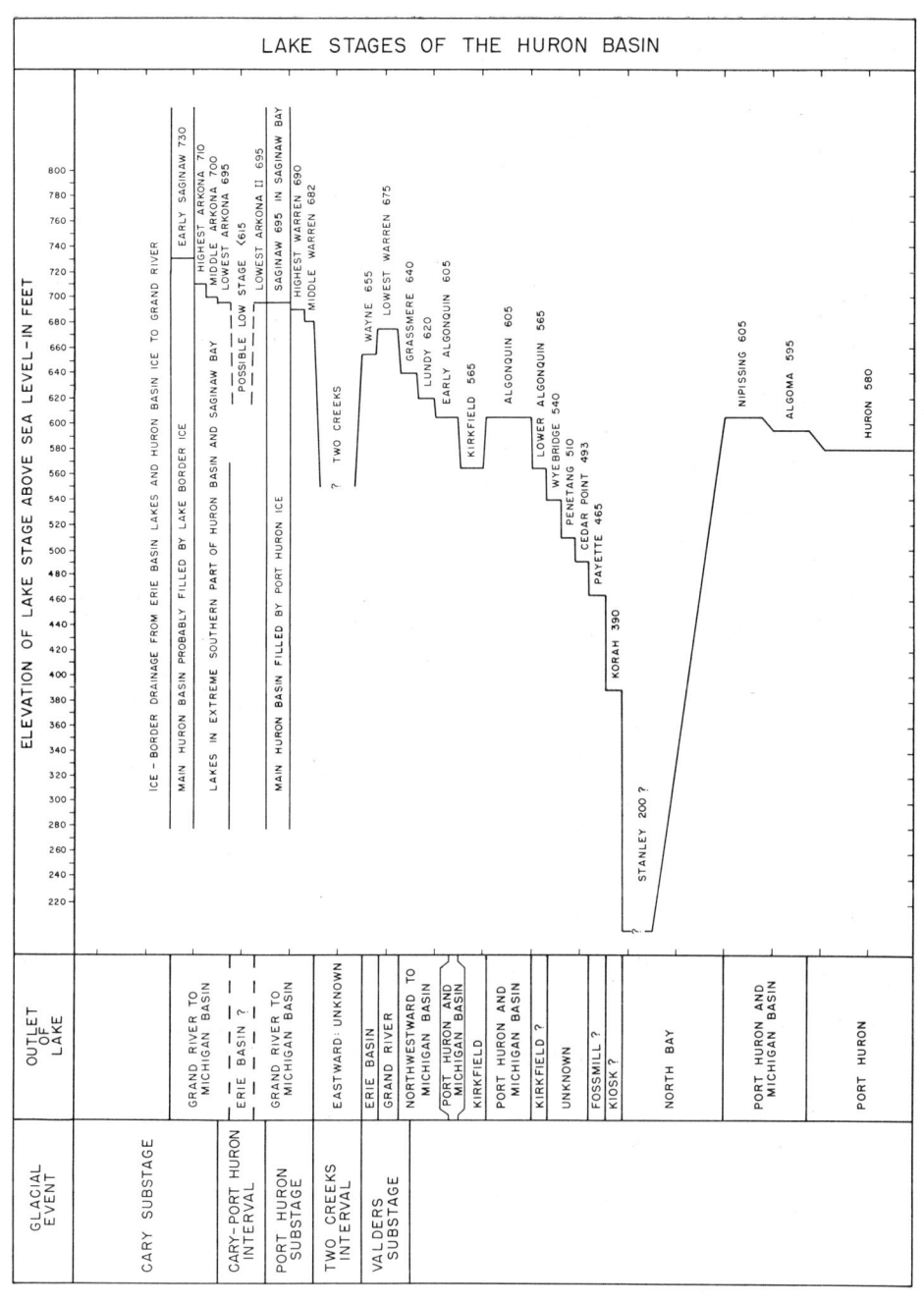

Fig. 36. Lake stages of the Huron basin. Lake levels are plotted against an unspecified time scale in the upper part of the figure, and lake outlets and correlative glacial events are shown in the lower part.

(1902, pp. 714–40). The first level of the ponded water may have been higher than 800 feet above tide for a brief period while the outlet channel was being cut and stabilized.

The Highest Maumee stage existed during the ice retreat from the Fort Wayne moraine, when the lake was expanding (Fig. 54), and during the subsequent ice advance which built the Defiance moraine, when the lake was reduced in area (Fig. 55). The extent of the retreat during the Fort Wayne–Defiance interval apparently was considerable. Leverett and Taylor (1915, p. 322) suggested that the ice front, in retreating from the Fort Wayne moraine, did not stop at the position of the Defiance moraine but receded to a line at least 25 or 30 miles east of that position before readvancing to build the Defiance moraine. Shepps (1953, p. 41) has shown that the Defiance-age till in eastern Ohio has a much higher content of silt and clay than the earlier till, and this is interpreted as an indication that the ice retreated from a large part of the Erie basin during the Fort Wayne–Defiance interval, and that fine-grained lake sediments were deposited there and then incorporated in the Defiance till during the next advance of the ice. The Highest Maumee level evidently was maintained during a part of the next ice retreat (from the Defiance moraine), because a Highest Maumee–level beach occurs on the inside of the Defiance moraine, in Michigan.

LOWEST LAKE MAUMEE

The next stage of the lake, the Lowest Maumee, is recorded by a beach at an average elevation of 760 feet A.T. This has been found at many places along the southern margin of the Erie basin in Ohio, along the northwestern margin of the Erie basin in Michigan, and extending into the Huron basin in Michigan (Leverett and Taylor, 1915, p. 334). This Lowest Maumee stage resulted from the retreat of the ice front northward down the "thumb" of Michigan, between the main Huron basin and the Saginaw Bay area, which allowed the lake waters to drain westward along the ice margin to the Grand River valley and thence to the Lake Michigan basin. This stage is shown in Fig. 56.

The Lowest Maumee beaches have a washed or submerged

character, which indicates that they were flooded by the next, or Middle Maumee, stage (Leverett and Taylor, 1915, p. 322).

MIDDLE LAKE MAUMEE

During the next advance of the ice, which built the Erie Lake Border moraines, the ice front moved up the slope of the "thumb" of Michigan, overriding the Lowest Lake Maumee outlet, and built the Flint moraine in that area. This raised the level of the lake to about 780 feet A.T. (see Fig. 57). According to the detailed description by Leverett and Taylor (1915, p. 322), this stage of Lake Maumee was barely too low to overflow at Fort Wayne; it discharged westward through the Imlay channel, alongside the Flint moraine, to the Grand River valley. The possibility of concurrent discharge through the Fort Wayne outlet is mentioned by Leverett and Taylor (1915) in their table on page 469. The author regards this as more than just a possibility, because the present elevation of the bed at the Fort Wayne outlet is 757 feet A.T., as recorded by Leverett (1902, p. 712), quite low enough to carry an appreciable part of the discharge from a lake standing at 780 feet A.T. It is suggested here that the Fort Wayne outlet was cut to this low level during the third or Middle Maumee stage, and that during this process most if not all of the discharge was transferred from the Imlay to the Fort Wayne outlet.

LAKE SAGINAW

When the ice front began to retreat from the Owosso moraine in the Saginaw Bay area, and presumably while the latest stage of Lake Maumee still existed to the south, water was ponded between the ice front and the divide around the southwestern edge of the Saginaw Bay embayment. This body of water was named "Early Lake Saginaw" by Leverett and Taylor (1915, p. 323), but its precise time of occurrence, as it is stated here, was redefined by Bretz (1951a, p. 245). Lake Saginaw stood at an elevation of 735 feet A.T. and discharged westward down the Grand River valley (Fig. 58).

Later, downcutting of the Grand River outlet channel lowered the lake in Saginaw Bay to an elevation of 710 feet A.T., and this stage has been referred to as a later Lake Saginaw. The next stage in the main basins of Huron and Erie, the Highest Arkona, was brought to this same level and later Lake Saginaw and Highest Arkona were joined to form a single lake. If the junction occurred prior to the lowering from the Early Lake Saginaw level, there is no need for formal recognition of this stage of Lake Saginaw because it was a part of the Highest Arkona stage.

A discussion of detailed correlations between periods of ice advance and retreat and the channels leading into the Grand River valley has been given by Bretz (1951a), and this discussion includes a consideration of the time of downcutting of the head of the outlet channel in relation to the junction of Lake Saginaw and Lake Arkona. A still later stage of Lake Saginaw occurred at 695 feet, when a portion of Lake Arkona was cut off by the advancing Port Huron ice sheet.

LAKE ARKONA

During retreat of the ice, later in the Erie Lake Border stage of glaciation, the ice barrier on the "thumb" of Michigan retreated sufficiently far northward to allow the waters of the southern Huron and the Erie basins to drain freely into Saginaw Bay. This may have occurred after the water level in that area had been lowered to an elevation of 710 feet A.T., or the lowering to 710 feet may have occurred at the time of the junction or a little later.

Highest Lake Arkona (Fig. 58) was dammed east of the Erie basin by the ice front, where it lay on high ground in New York south of the Lake Ontario basin. The Ontario basin was still nearly filled with glacial ice.

Lake Arkona stood for a time at each of three levels, 710, 700, and 695 feet A.T., while continuing to discharge down the Grand River. The static stage at 710 feet is correlated with the minor ice advance which built the Chesaning moraine in the Saginaw Bay area, and the lowering to the next level through downcutting of the outlet is correlated with ice retreat which delivered an increased amount of glacial melt water. The static stage at 700 feet is correlated with another minor advance (represented by the moraine north of Chesaning), and the lowering to the next level,

695 feet, is correlated with ice retreat which again delivered an increased amount of melt water and caused further downcutting of the outlet channel. The detailed correlations of lake stages in Saginaw Bay with times of ice advance are those of Bretz (1951a), and they are based on the premise that the discharge from a lake basin is less during times of growth of an ice sheet and is greater during the wasting of an ice sheet.

LOW-WATER STAGE DURING THE CARY–PORT HURON INTERVAL

Because the present author correlates the Port Huron glacial advances with the Mankato substage of the Des Moines glacial lobe, and because Leverett and Taylor (1915, p. 293) recognized the Port Huron as "of substage order of magnitude" and stated that it marks a pronounced readvance of the ice front, it appears likely that the ice front had retreated sufficiently far during the Cary–Port Huron (or Cary-Mankato) interval to uncover outlets to the east which were low enough to drain the lake waters down to a low level (Fig. 59). If this occurred, there was a Cary–Port Huron–interval low stage which has not yet been recognized in the Great Lakes region. This is noted on the lake stage diagrams for the Erie and Huron basins (Figs. 35 and 36) as a "possible low stage." When this possibility was pointed out to G. W. White, he recalled evidence for such a stage in the Erie basin, observed in a gravel pit located south of the Cleveland, Ohio, city limits, three and a quarter miles north of the center of Independence Village and one mile north of the Willow traffic interchange at Willow. White's unpublished field notes, taken during the course of an investigation which was not directly related to the matters of lake history, record a section extending well below the Lowest Arkona level (695 feet) and which includes deposits at elevations of 658 to 677 feet described as crossbedded and having the appearance of beach sand. These apparently record shallow-water or shore deposition, and they are overlain by finer-grained, presumably deeper-water deposits. Dr. White has pointed out the need for further study of this section with the low-water-stage hypothesis in mind, but has kindly allowed the present author to quote his observations.

An "episode of free eastward drainage and no lakes" in the

Ontario basin was recognized by Fairchild (1909) as occurring between the Vanuxem I and Vanuxem II stages (see Chap. 11). This is regarded here as a correlative of the Cary–Port Huron–interval low stage.

SECOND LAKE ARKONA

The Port Huron glacial advance, following the low-water stage of the Cary–Port Huron interval, closed off the eastern outlet in the Ontario basin area. It seems probable that this occurred before the advancing ice front reached the "thumb" of Michigan in the Huron basin, and that the waters of the Erie and Huron basin were raised as a confluent lake to spill down the Grand River. Because the Grand River outlet had been abandoned by the previous Lowest Arkona stage at 695 feet, it may be assumed that the rising water in early Port Huron time would come to the same level and constitute a second Lowest Arkona stage.

A lagoon deposit at an altitude of 690 feet at Cleveland, Ohio, is overlain by 10 to 12 feet of sand and silt. This apparently was deposited just before the rise of lake water to the Whittlesey stage, and it thus represents the second Lowest Arkona stage. A sample of broken twigs, roots, and leaves collected from the lagoon deposit by G. W. White has been dated by the radiocarbon method as 13,600 ± 500 years old (Suess, 1954, p. 469). This is the oldest radiocarbon date of materials directly related to a lake stage.

LAKE WHITTLESEY AND THE PORT HURON GLACIAL SUBSTAGE

Following the low stage of the Cary–Port Huron interval and the second Lowest Arkona stage, the advancing Port Huron ice rode up the "thumb" of Michigan to Ubly, and separated the waters of Saginaw Bay from those to the east and south, raising the level of the latter to the Whittlesey stage (Fig. 60). Lake Whittlesey stood at an elevation of 738 feet A.T. in the Erie basin while the ice occupied all of the main Huron basin and built the Port Huron moraine. Drainage from Lake Whittlesey passed along the western margin of the ice northward to the Ubly channel, and through it to Lake Saginaw, which had remained at an elevation of 695 feet A.T. and which continued to discharge westward down the Grand River valley.

The Port Huron moraine is the oldest one in the Great Lakes region which can be traced from one basin to another. It extends from the western shore of Lake Michigan, around the northern part of the Lower Peninsula of Michigan, around Saginaw Bay, and southward to Port Huron, its type locality at the south end of the Huron basin. From there it is traceable in Ontario northward around the southern Ontario highland and back southward to the north shore of Lake Erie near the base of Long Point. It may be projected through a distinct ridge on the bottom of Lake Erie lying west of the eastern deep basin (Fig. 6), and from there the moraine apparently emerges on the south shore of the lake and extends eastward into New York.

The Whittlesey beach is one of the strongest and best developed in the Great Lakes region. This may imply that it was formed by a lake which occupied one level for a long time, but the fact that the beach was built by a rise from a lower level must be considered. Various writers have stated that stronger beaches are formed by an encroaching lake, which sweeps shore debris up the slope and concentrates it in a zone at the highest level reached, and that weaker beaches are formed by a lowering lake, which leaves shore debris scattered over a broad zone as the water recedes.

A radiocarbon date of $12,800 \pm 250$ years has been obtained for a sample of wood imbedded in beach sediments of Lake Whittlesey, four and a half miles southeast of Bellevue, Ohio.

The Whittlesey stage was brought to a close by retreat of the ice from the outermost Port Huron moraine on the "thumb" of Michigan, and the lake surface in the Erie basin was lowered to the level of the water in Saginaw Bay.

HIGHEST AND MIDDLE WARREN STAGES

Highest Lake Warren stood at an elevation of 690 feet A.T., five feet lower than the Saginaw and Lowest Arkona level (Fig. 61). The drop from the Saginaw–Lowest Arkona level to the first Warren level was caused by downcutting of the outlet, and this probably occurred when Lake Whittlesey was lowered, delivering a great volume of water to the outlet channel. The total lowering from the Whittlesey to the Highest Warren stage was 48 feet.

A Middle Warren stage, represented by a weak, discontinuous beach, next occurred at an elevation of 682 feet A.T. The lowering to 682 feet is correlated with the retreat from the Bay City moraine, a second stand of the Port Huron ice, and the lake stage at 682 feet is correlated with a readvance to the Tawas (or latest Port Huron) moraine (Bretz, 1951a, p. 257). Middle Lake Warren continued to discharge down the Grand River. Retreat of the ice from the Tawas moraine, presumably accompanied by an increase in volume of melt water, may have caused further downcutting of the outlet to bring the Middle Warren stage to a close. The next lower beach is that of the Lowest Warren stage at 675 feet A.T., and the outlet may have been lowered and stabilized to hold the lake at this level, as a result of the discharge which occurred early in the retreat of the last Port Huron ice. The main Lowest Warren stage occurred much later, however, after a lower episode had occurred.

LOW-WATER STAGES OF THE TWO CREEKS INTERVAL

The Lake Wayne stage, at 655 feet A.T., occurred before the Lowest Warren–level beaches, at 675 feet A.T., were last occupied. This is indicated by the fact that the Lake Wayne beach "shows clear evidence of submergence and modification, and the Warren beach does not . . ." (Leverett and Taylor, 1915, p. 386). The Wayne stage is correlated with the Two Creeks interval, between the Port Huron and the Valders glacial substages.

A stage much lower than the Wayne occurred in the Huron and Erie basins during this interval, as may be deduced from evidence found in areas outside of those basins. This stage is shown in Fig. 62. The section at Two Creeks, Wisconsin, which has been described in a foregoing discussion of the Wisconsin-stage glaciation (Chap. 5), indicates that the water in the Michigan basin was drained down at least as low as the present lake surface, 580 feet A.T. Because the only outlet of the Michigan basin which is low enough to permit this is one into the Huron basin, it is necessary that the water surface in the Huron basin also was at least that low. This, in turn, requires a low outlet to the east for drainage of the Huron and Erie basins. The drift-filled St. David gorge, which extends from the "whirlpool" of the present Niagara gorge to the vicinity of St. David, Ontario, apparently was occupied by the discharge from the Erie basin during this low-water stage. Whether the water from the Huron basin drained to the Erie basin at this time, or discharged directly to the Ontario basin by a more northerly route, is not known. It might have discharged down the Trent valley to the Ontario basin, or down the Mattawa-Ottawa valleys to the St. Lawrence valley, depending upon the relative elevations of the divides and whether the Mattawa-Ottawa route was ice-free.

If the lower St. Lawrence River valley was blocked by ice at this time, and detailed studies in that area suggest that it was, the Two Creeks low-water stage must have discharged down the Lake Champlain–Hudson River lowland.

The beaches of the Lake Wayne stage, at 655 feet A.T., were formed during a static period which may have occurred either during the lowering of water surface to the extreme low position of the Two Creeks interval or during the subsequent rise. In the absence of any evidence bearing on this question, the author has arbitrarily assigned it to the period of rise, during the advance of the Valders ice (Fig. 63). The discharge of Lake Wayne is believed to have been eastward to the Mohawk valley through some of the channels south of Syracuse, New York (Leverett and Taylor, 1915, p. 386).

LOWEST LAKE WARREN AND THE VALDERS GLACIAL SUBSTAGE

When the advancing Valders ice reached its greatest extent in the eastern part of the Great Lakes region, it dammed the eastern outlets of the Huron and Erie basins, and raised their waters to the Lowest Lake Warren level, 675 feet A.T. (Fig. 64). This late Lake Warren stage spilled down the old Grand River outlet, which had been abandoned during the retreat of the last Port Huron ice, and thus discharged its water to Lake Chicago.

The Niagara Falls moraine probably marks the position of the Valders ice border between the Ontario and Erie basins; at least, it is reasonable to assume that the Valders ice advanced onto the Niagaran escarpment in this area because it stood well up on the slope south of Lake Ontario to form the eastern dam of late Lake Warren. The Valders margin to the west extended northward around the southern Ontario upland, possibly along the Gibraltar moraine as mapped by Chapman and Putnam (1951, p. 31). The border of the Valders ice probably stood well out in the lake in the southern part of the Huron basin, but crossed the Michigan shore near Rogers City according to Melhorn (1954).

The Lowest Warren stage persisted until a lower outlet was uncovered by retreat of the Valders ice.

THE GRASSMERE, LUNDY, AND EARLY ALGONQUIN STAGES

The Grassmere beach at an elevation of 640 feet and the Lundy beach at 620 feet A.T. record two rather brief static periods during the lowering of the Huron and Erie basin waters (Figs. 65A and 66A). Wood from Lake Grassmere deposits at Bellevue, Ohio, has been dated by the radiocarbon method and found to be 8513 ± 500 years old (Libby, 1951, p. 292).

Detailed studies in the Detroit area have shown that during Grassmere time the Huron and Erie basins were connected by a shallow strait over the Port Huron moraine, and that by Lundy time the connection through this strait must have had to be deepened slightly by erosion of the channel bottom before the lake surface could be brought to the same level on both sides of the barrier (Leverett and Taylor, 1915, p. 399). The direction in which the water flowed while deepening this channel is not specified at this point in the discussion, but it is deduced from certain factors which are discussed in a subsequent paragraph.

An Early Algonquin stage in the southern part of the Huron basin, at an elevation of approximately 605 feet A.T., was inferred by Leverett and Taylor (1915, p. 328). There are no shoreline features in the Huron basin which can be referred to this separate introductory stage of Lake Algonquin, but the stage was based primarily on the existence of several abandoned channels through the Port Huron moraine, the divide between the Huron and Erie basins, at levels just below 605 feet A.T. Having conceived of an Early Algonquin stage and believing that it must have discharged southward to the Erie basin, Leverett and Taylor correlated its discharge with the cutting of the wide lowest portion of the Niagara gorge. They then cited the features of the gorge as proof of the existence of the southward-discharging Early Algonquin stage.

The Grassmere, Lundy, and Early Algonquin stages were explained by Leverett and Taylor (1915, p. 399) as follows: retreat of the ice front, from its position holding up the Lowest Warren stage, occurred in a manner such that an outlet for Lake Warren was uncovered somewhere to the east of Lake Erie, and the waters of the Huron and Erie basins were thus drained down to the Grassmere level; further retreat then uncovered a new and lower outlet to the east, and the lake drained down to the Lundy level; as a consequence of this supposed eastward drainage from the Erie basin, the slight entrenchment of the channel through the Port Huron moraine which was necessary to bring the levels of both the Huron and Erie basins to the Lundy level was accomplished by water flowing southward from the Huron basin; following this stage the ice in the Ontario basin retreated far enough to permit the lake waters to spill over the Niagara escarpment, and the cutting of the Niagara gorge began; the water level in the Lake Erie basin was lowered well below the divide between the Erie and Huron basins, dropping rather abruptly from the Lundy to a low, early Lake Erie stage; the southward discharge from the Huron basin then cut a channel deeper into the Port Huron moraine but stopped its downcutting at a level which held the Huron water surface at the Early Algonquin level, about 605 feet A.T.

The history of the Grassmere to the Early Algonquin stage which was outlined in the foregoing paragraph may be disputed on several bases. First, the supposed eastern outlets for the Grassmere and Lundy stages have never been identified. The following quotations from Leverett and Taylor (1915), in which the italics are by the present author, will illustrate this:

the outlet of Lake Lundy has not yet been determined by continuous tracing of the beaches into close connection with it, but Fairchild has shown that during the time of the Lundy or Dana beach it was *in all probability* through the great Marcellus-Cedarvale channel which crosses the front of the hills southwest of Syracuse, New York. It is certain that the outlet was not by the Grand River channel to Lake Chicago, for the divide at the head of that channel has an altitude of 653 feet and was within the area of horizontality, in which the Grassmere beach has an altitude of 640 feet to 645 feet and the Lundy beach is about 20 feet lower. It *seems* clear also that the outlet was not northwestward to Lake Chicago along the hills northwest of Alpena, for when the lake fell to a level 20 feet lower than the Grassmere beach, the discharge from the Lake Huron basin was southward to the Lake Erie basin. The marks of that discharge are very plain. If the outlet of Lake Lundy had been northwestward to Lake Chicago at the time of the Grassmere beach the waters would hardly have flowed south when the ice front drew back far enough to let the lake level fall 20 feet lower [p. 400].

It should be pointed out that although the channels of the Lundy-stage discharge across the Port Huron moraine are "very plain," the direction of flow in these channels is not evident. They could have been cut by northward-flowing discharge rather than by southward-flowing discharge. Further, the Leverett and Taylor correlation of Lake Lundy with the Lake Dana stage of the Ontario basin is in error, because Fairchild (1909, p. 54) stated that the Dana beach and Marcellus-Cedarvale outlet were at a present elevation of 700 feet above sea level, and were this level restored to its unwarped elevation it would be at approximately 590 feet above sea level or well below the Lundy level.

The precise location of the ice barriers of Lake Lundy is not known. *It seems necessary to infer* the presence of an ice barrier on the middle slope of the highlands west of Alpena, for if there was an outlet northwestward for the Grassmere stage of Lake Lundy along that slope it would have continued to carry off the discharge during the time of the Lundy beach which is about 20 feet lower, *but this it did not do* [p. 406].

Again, it is pointed out that Leverett and Taylor had no proof for southward flow at Port Huron during the Lundy stage.

On the other hand, any ice barrier near Syracuse, N.Y., must also have rested at a certain definite place on the face of the northward-sloping escarpment. . . . in the Niagara region the indications are that the barrier of Lake Lundy stood somewhere between the Niagara escarpment and the present shore of Lake Ontario. Thus, while the precise positions of the barriers are not known in Michigan or in New York, they must have been within very narrow zones in each locality [p. 406].

The second questionable item in the Leverett and Taylor history is their interpretation of the manner in which the Early Algonquin stage came into existence. This is quoted, with italics by the present author:

The introductory stage of Lake Algonquin was short and was followed by one or two transitional steps or substages in which different basins and different outlets were involved. *These steps of transition are mainly matters of inference and are not based on observation. Nevertheless, their existence is not to be doubted.* . . . Its existence is based on evidence of the establishment and erosion of its outlet through the distributaries of the St. Clair River at St. Clair, and on the character of the Niagara River and gorge [p. 409].

In a discussion of the St. Clair River, Leverett and Taylor (1915) interpret the features of the river in terms of the preconceived history of the lake:

Early Lake Algonquin had a comparatively short life, for the discharge soon left the Port Huron route and went to Kirkfield, Ontario, where it *appears* to have remained for a relatively long time. During this time some deepening at the mouths and in the lower courses of the tributaries *may* have been accomplished, but it is *doubtful* whether any recognizable evidence of it now remains. At the close of early Algonquin, St. Clair and Detroit Rivers had *probably* only cut through the St. Clair and Detroit barriers and the softer parts of the Trenton barrier. . . . The general bottom levels of St. Clair and Detroit Rivers were at that time *probably* not much, if any, below the present water surfaces of these rivers. . . . The later or main period of southward discharge of Lake Algonquin *appears* to have endured for a relatively long time, during which St. Clair and Detroit Rivers *must have* greatly scoured their beds, probably enough to bring their level down . . . [pp. 498–99].

It is obvious that the "matters of inference" regarding the lake history are not based on evidence from the St. Clair River;

inferences regarding the St. Clair River are based on inferences from the lakes.

Let us now examine the character of the Niagara gorge, the other basis claimed in support of the lake history. Kindle and Taylor in the "Niagara Folio" (1913, p. 21) give an outline of a part of the Great Lakes history which is the same as that presented by Leverett and Taylor (1915). They then say, "The main facts of the above brief history of the lakes, such as the order of the lake stages, their changes of outlet, and the effects of those changes upon the volume of Niagara Falls, are firmly established. *The variations of volume of the Falls are fixed primarily by facts of the lake history, without reference to the characters displayed in the Niagara gorge.* The history of the lakes is therefore the key to the history of Niagara, for no matter what characters are found in the gorge, it is certain that the volume of the river and Falls has varied in accordance with the lake stages."

It is obvious that the "matters of inference" regarding the lake history are not "based on evidence . . . on the character of the Niagara River and gorge." Here, the matters of inference are stated to be "firmly established."

The previously quoted statement that "These steps of transition are mainly matters of inference and are not based on observation. Nevertheless, their existence is not to be doubted" seems very out-of-place in a scientific monograph. Because of a complete lack of factual evidence to support it, the Leverett and Taylor interpretations of the beginnings of Lake Algonquin may be discarded.

At this point it is necessary to refer to events in the history of Lake Chicago, in the Lake Michigan basin. Three early lake stages in that basin include the Glenwood, at 640 feet, the Calumet, at 620 feet, and the Toleston, at 605 feet A.T., all of which discharged through an outlet at Chicago. Leverett and Taylor believed that Early Lake Algonquin, at 605 feet in the Huron basin, had come into existence after Lake Chicago had been lowered to the 605-foot Toleston stage, and that each was formed independently of the other. Then, "when the ice sheet drew away from the high ground northwest of Alpena, Michigan, it allowed glacial Lake Chicago in the basin of Lake Michigan to unite with

Early Lake Algonquin. If this occurred, as it most probably did, before the opening of the Kirkfield outlet in Ontario, the two lakes were probably already at very nearly the same level at the time of union, and no considerable change of altitude occurred in either, the overflow probably being divided between the two original outlets at Port Huron and Chicago" (Leverett and Taylor, 1915, p. 409).

The present author was first led to formulate a new hypothesis for the origin of the Early Algonquin stage by the apparent need for an explanation of a delicate adjustment between the Port Huron and Chicago outlets. If the Leverett and Taylor history is correct, the Port Huron channel was first cut by southward-flowing discharge from the Huron basin to a lower-level lake in the Erie basin. Downcutting stopped in unconsolidated material, at a level which held the Huron waters at an elevation of 605 feet, for no known reason. It was only by chance that this coincided with the level of the Toleston stage in the Michigan basin. A more logical explanation is that the Huron basin was discharging to the Michigan basin while its waters were being lowered to the 605-foot level, and that the Erie basin was discharging to the Huron basin at the same time (Fig. 67A). When the water level in the Huron basin reached 605 feet (determined by the Chicago outlet which was on bedrock), the discharge from the Erie basin could cut a channel at Port Huron only deep enough to bring the Erie level down to 605 feet. Thus, the cessation of downcutting is accounted for. Throughout the various lake stages which occurred at 605 feet (the Toleston and the Early Algonquin, the main Algonquin and the Nipissing), the discharge at the Chicago outlet was through a channel with a bedrock bottom. This could not be cut lower. The Port Huron outlet, however, was cut entirely in glacial drift. This material was susceptible to further deepening by scour, but the channel was not cut down by its discharge during the early and the main Algonquin stages, nor while the Nipissing lake stood at 605 feet.

In order to explain this stability of the Port Huron outlet during several later stages it may be postulated that the channel had a slightly higher bed elevation and a considerably smaller capacity than those of the Chicago outlet. If this were true, during the various 605-foot stages the Chicago outlet would carry a considerably larger proportion of the total discharge, and the smaller part going through the Port Huron channel then would not necessarily accomplish any downcutting. The new hypothesis provides for a smaller-capacity, probably shallower, channel at Port Huron. At the time of its inception this channel carried only the northward discharge of the Erie basin while the Chicago outlet necessarily developed the capacity to carry the discharge of the Erie, Huron, and Michigan basins. Later, when the water of the Erie basin discharged eastward and the water surface there dropped below the level of the Port Huron outlet, the Michigan-Huron basins spilled through the Port Huron outlet as well as through the original Chicago outlet. The foregoing hypothesis explains the outlet-channel relationships on a cause-and-effect basis.

Leverett and Taylor (1915) stated dogmatically that the first flow through the barrier at Port Huron was to the south:

> There are one or two other remote alternatives as to the precise order of events at the beginning of Lake Algonquin. An outlet to Lake Chicago or to Lake Simcoe and the Trent Valley may possibly have occurred before the barrier at St. Clair was uncovered, so that when the waters fell to the level of the barrier's crest, there was no tendency to overflow toward the south. . . . the discharge of those waters would have had to pass northward across the barrier and would have cut channels in that direction. But the first flow over the moraine left a clear record in distributary channels which prove absolutely that the flow was southward [p. 407].

Leverett and Taylor were not so positive on their next page:

> Hence, although from present knowledge of the moraines it seems hard to believe that at so late a stage of the ice retreat the ice was still barring the passage to Lake Chicago west of Alpena . . . yet the character and relations of the distributaries at St. Clair seem to leave no other alternatives. It is therefore concluded, provisionally, that Lake Algonquin developed directly from the Lundy stage . . . and had an early stage during which its whole discharge passed southward to the Erie basin . . . [p. 408].

In the entire discussion of the Port Huron–St. Clair River outlet area by Leverett and Taylor (1915, pp. 470–501), the present author finds no proof that the first flow through the barrier between the Huron and Erie basins was southward. On the contrary, a feature is described which is best explained in terms of northward flow. A ridge of gravel 11 or 12 miles long extends from Roberts Landing north to St. Clair, on the west side of the present St. Clair channel. An abandoned channel borders the ridge on the west. The crest of the ridge is approximately level, and at its northern end it is closely associated with gravelly river bars. The present author regards this ridge as an erosional remnant of a more extensive channel bed which was graded to the Port Huron outlet and which was formed by northward-flowing discharge to the Early Algonquin lake of the Huron basin. Leverett and Taylor (1915) stated, "In its stronger southern part its crest happens to lie through much of its length almost exactly at the level of the first St. Clair beach [correlative of the Algonquin beach], faint markings of which . . . show the altitude at which the waves acted. This relation seems to be accidental . . ." (p. 477). Leverett and Taylor named the ridge the "St. Clair Esker," though they admitted "difficulties in finding a clear and satisfactory explanation" for its origin. There are several other remnants of a surface lying just above 600 feet in the St. Clair area. These are shown on the following U.S. Geological Survey topographic quadrangles: Marine City, Rattle Run, St. Clair, and Port Huron, Michigan, published in 1937–40. The surface here shown, extending 18 miles southward from the Port Huron barrier, may represent the Early Algonquin stage in the Erie basin.

Further support of the concept of discharge from the Erie and Huron basins to the Michigan basin in Early Algonquin time is found in a review of events preceding the Algonquin stage. Besides the coincidence in level between the Algonquin and Toleston stages, the Lundy stage of the Erie-Huron basins, at 620 feet, has a correlative in the Calumet stage of Lake Chicago, at 620 feet, and the Grassmere stage of the Erie-Huron basins, at 640 feet, has a correlative in the Glenwood stage of Lake Chicago, at 640 feet. It seems beyond the limits of probability that three lake stages occurred at the same levels in adjacent basins simply by chance.

It appears that the coincidence in elevation of three successive lake stages in the separate lake basins indicates some cause-and-effect relationship, such as a connection between the basins. It is proposed that the lakes in the Erie and Huron basins were connected with the Michigan basin by a channel or an open water route around the north end of the Lower Peninsula of Michigan, by Grassmere time, and that the successive elevations of the Chicago outlet sill, which determined the elevations of the three Lake Chicago stages, also determined the elevations of the Grassmere, Lundy, and Early Algonquin stages (Figs. 65A, 66A, 67A).

Leverett and Taylor, in their statements of the correlatives of the various early stages in the Huron and Michigan basins (1915, pp. 349, 376, 385, 398), mention only Lake Chicago, without naming specific stages of that lake. Bretz (1951b), however, has stated that Lake Chicago was lowered from the Glenwood stage to the Calumet stage during the Port Huron glacial substage, or during the existence of Lake Whittlesey in the Erie basin, and that Lake Chicago was lowered from the Calumet to the Toleston stage during the Valders glacial substage, or during the existence of Lowest Lake Warren in the Huron and Erie basins. His opinion is based mainly on the absence of the Glenwood beaches from the inner margins of the Valders drift on the shores of Lake Michigan, where he expected to find them if the Glenwood stage had persisted into the time of retreat of the Valders ice. This absence of the Glenwood beaches may be explained simply by the fact that recent shore erosion has removed all of the older beaches above the present level for several miles northward from the southern limits of the Valders drift, on both sides of the lake. This explanation of their absence does not, of course, prove that they once were there, but it does indicate that they might have been there. A further discussion of Bretz's explanations of the stages of Lake Chicago is given in a subsequent chapter on the Lake Michigan basin.

Other evidence of a positive nature is required to substantiate the proposal that the Huron basin discharged to the Michigan basin in Grassmere time. The only possible route for this discharge lies generally along the ice border on the northern slopes of the Port Huron morainic system. It will be recalled that

Leverett and Taylor considered this route as a "remote alternative" in the quotation from their page 407, and that on their page 408 they stated, ". . . it seems hard to believe that at so late a stage of the ice retreat the ice was still barring the passage to Lake Chicago west of Alpena. . . ." They rejected a northwestern outlet for the Huron basin in Early Algonquin time only because they believed that their interpretation of the St. Clair River features made it impossible. The distribution of the red Valders till in Wisconsin, and its distribution in Michigan east of Grand Traverse Bay, north of the Port Huron moraine, indicates that the Valders ice occupied the Michigan basin as an extremely elongated tongue which did not rise very high on the basin margins. Only a slight retreat from its maximum position in Michigan would be sufficient to uncover a channel low enough to drain the lakes down from the Warren level. Field evidences of such a channel have been observed by the author, and they are described in the subsequent chapter dealing with the features of the Lake Michigan basin.

FURTHER DEVELOPMENT OF THE EARLY ALGONQUIN STAGE AND EARLY LAKE ERIE

As proposed in the foregoing paragraphs, the Early Algonquin stage in the Huron and Erie basins came into existence because downcutting of the Chicago outlet channel lowered Lake Chicago to the Toleston stage, at 605 feet, and because the Huron and Erie basins were drained down to the same level by reason of their discharge into Lake Chicago. While these events were occurring, the Valders ice margin undoubtedly was receding northward in the Ontario basin, but it must have remained at an elevation sufficiently high on the south slope to prevent discharge to the east from the Erie basin. Finally, after the Early Algonquin stage was attained, the ice retreated far enough to permit drainage to the east. The earliest discharge in this direction is recorded by several shallow spillways and gravel deposits, which are described in the "Niagara Folio" (Kindle and Taylor, 1913, pp. 13–14). The outflow soon became concentrated, however, and began cutting the Niagara gorge. By this time the water surface in the Erie basin was drained below the Algonquin level, and when this occurred the water in the Huron basin began to discharge southward through the channels in the Port Huron moraine (Fig. 68). This was the first southward discharge in that area, which Leverett and Taylor had dated so confidently as having occurred at a slightly earlier time.

The Early Algonquin stage of the Erie basin thus was brought to a close, probably soon after it was formed, but the lake in the Huron basin remained at the Algonquin level for a longer time, and its connection with the Toleston stage in the Michigan basin was broadened as the ice margin retreated. The Early Algonquin–Toleston lake, which may be identified simply as the Early Algonquin stage, now had two outlets—the original one at Chicago and the Port Huron outlet, which previously had been an inlet.

The discussion of later events in the Erie and Huron basins is postponed until after the early stages in the other Great Lakes have been reviewed.

chapter nine

EARLY LAKE STAGES IN THE MICHIGAN BASIN

INTRODUCTION

The early stages of the Great Lakes in the Michigan basin were formed by recession of the margin of the glacial ice northward from the divide which encircles the south end of the lake. This divide lies along the Valparaiso morainic system, which was deposited during the Cary substage of the Wisconsin glacial stage. Two low places in this morainic system became simultaneous dischargeways, converging to become a single channel a few miles downstream.

Several of the early lake stages are referred to as Lake Chicago, the name generally being applied to any of the lakes which were dammed on the north by glacial ice and which discharged southwestward through the Chicago outlet (Figs. 56–58, 60–61, 63–67).[1] Several other stages occurred, some of which were below the present level of Lake Michigan. All of these discharged northeastward, either through the lowland between Little Traverse Bay and Lake Huron, or through the Straits of Mackinac. Lake Chicago received inflowing water not only from the ice in the Michigan basin but also at times from glacial lakes to the east, in the Huron and Erie basins. The route of this discharge into the Michigan basin was via the Grand River valley across the state of Michigan during some of the earlier stages. Later, the Huron and Michigan basins were connected through the lowland

---

[1] Figures 53–75, lake stage maps, follow Chapter 16.

east of Little Traverse Bay, and still later through the Straits of Mackinac.

South of a line between Frankfort, Michigan, and Sturgeon Bay, Wisconsin, all of the beaches are horizontal. North of that line all beaches above the present lake beach rise to the north.

Table 11 gives a list of the various lake stages in the Michigan basin, with their altitudes and outlets. The stages are listed in the order of their elevations above sea level, insofar as the elevations are known. This is not the order in which they occurred; it is known that several of the lower stages intervened between higher stages. Figure 37 shows the lake stages of the Michigan basin in the order in which they occurred, according to the author. The exact order of occurrence and the time of occurrence of some of the lake stages in relation to readvances of glacial ice are still controversial, and the principal alternative interpretations will be given in the following paragraphs.

EARLY LAKE CHICAGO

An early stage of Lake Chicago (Fig. 54), occurring prior to the extensive Glenwood stage, has been inferred from evidence in the outlet area near Chicago. This has been discussed in detail by Bretz (1951b, p. 404), and his conclusions may be summarized as follows. The Valparaiso morainic system was deposited on a pre-existing topography in a manner which left two important transmorainic low areas which coincided with pre-Valparaiso channels. When the glacial ice front receded northward from the Valparaiso system, a lake was ponded between the ice and the divide and its water discharged through these two low areas. One of these channels is occupied at present by a portion of the Des Plaines River, the Illinois and Michigan canal, and the sanitary and ship canal, which connects the south branch of the Chicago River with the Illinois River system. The other channel is the site of the Calumet-Sag canal, which connects Lake Michigan, through the Calumet River, to the Illinois and Michigan canal near Sag Bridge, Illinois.

The elevation of the Early Lake Chicago surface is not known, as no beaches of that lake have been observed. Bretz (1951b, p.

Table 11. *Lake stages of the Michigan basin.*

Name	Altitude above sea level (feet)	Outlet
Glenwood stage	640	Chicago
Calumet stage	620	Chicago
Toleston stage	605	Chicago
Algonquin stage	605	Chicago and northeastern
Nipissing stage	605	Chicago and Straits of Mackinac
Algoma stage	595	Straits of Mackinac [a]
Lake Michigan	580	Straits of Mackinac
Two Creeks–interval low stage	580–	Northeastern
Cary–Port Huron–interval low stage	580–	Northeastern
Kirkfield	580–	Northeastern
Post-Algonquin low stages—inferred	580–	Straits of Mackinac
Chippewa stage	230	Straits of Mackinac

[a] High-water phases also discharged through the Chicago outlet.

405) suggests that the outlet channel may have been cut down sufficiently to bring the lake surface as low as 605 feet.

THE TINLEY GLACIAL ADVANCE

A readvance of the ice front, to the inner margin of the Valparaiso morainic belt, destroyed Early Lake Chicago by occupying its entire basin (Fig. 55). At this time "The Tinley ice barely more than reached the outlet-valley heads but it built a dam across each and its outwash built a valley train in each. . . . The surface of the deposit is close to 630 feet A.T., 10 feet lower than the Glenwood level of Lake Chicago" (Bretz, 1951b, p. 405).

THE FIRST GLENWOOD STAGE

Retreat of the ice front from the Tinley moraine again allowed water to pond between the ice and the divide, and this initiated the Glenwood stage of Lake Chicago (Figs. 37 and 56).

The crest of a surviving buttress of the Tinley moraine beside the Sag channel is at an elevation of 685 feet A.T., but it is likely

166　GEOLOGY OF THE GREAT LAKES

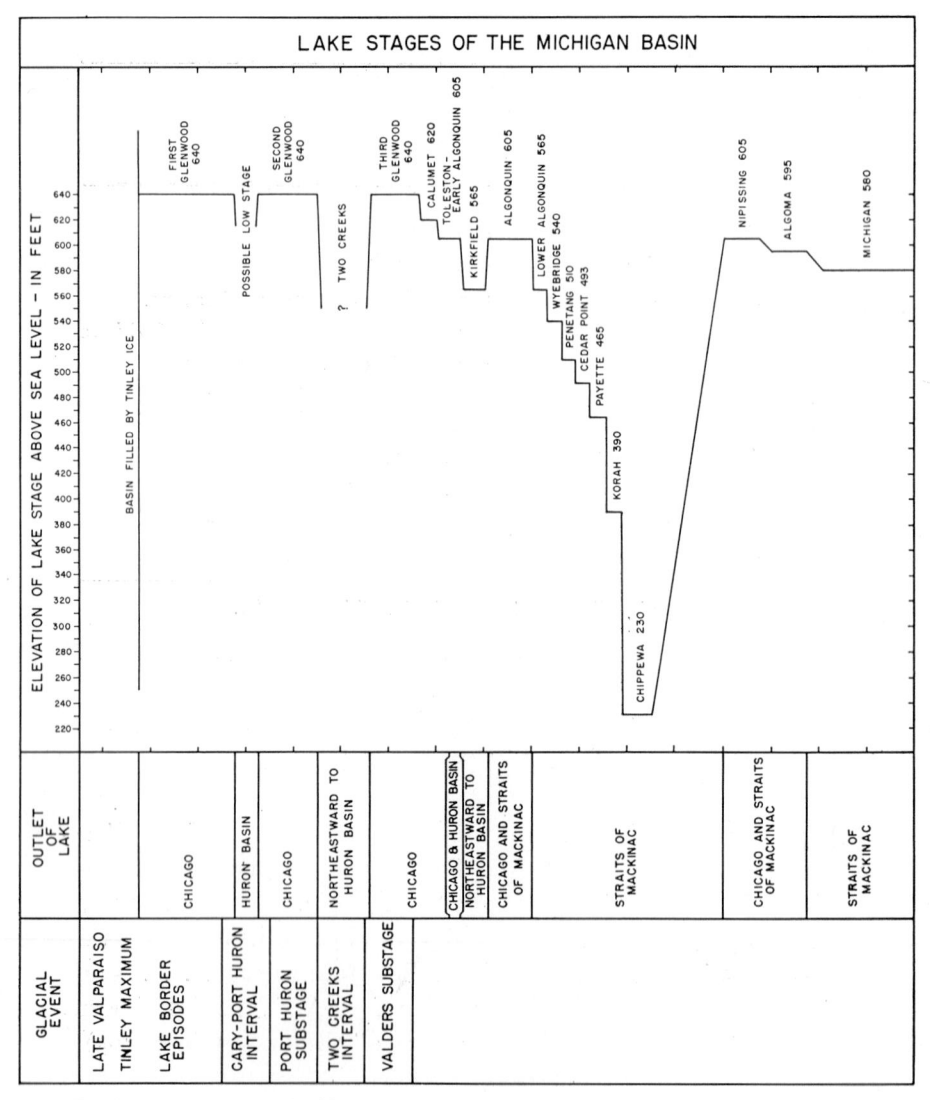

Fig. 37. Lake stages of the Michigan basin. Lake levels are plotted against an unspecified time scale in the upper part of the figure, and lake outlets and correlative glacial events are shown in the lower part.

that the first discharge from the new lake must have rapidly eroded a channel down to the surface of the valley train, which probably extended several miles downvalley. This great mass of outwash in the outlet-channel system, with a gentle gradient and a surface which rose to about 630 feet A.T. near the head of the channel, was the prime factor in determining the level of the Glenwood stage, as further deepening of the channel would require the removal of channel-bed material for a distance of many miles downstream (Bretz, 1951b, pp. 405–6). Another cause of the stabilization of the outlet channel to hold the lake at the Glenwood level, cited by Bretz (1951b, p. 407), was the accumulation of a lag concentrate of large boulders from the till, forming an armor on the channel bed, so that only a marked increase in erosive ability of the outlet river could clear the channel bed and allow further deepening.

The Glenwood stage of Lake Chicago is recorded in many places in the southern part of the Michigan basin by well-developed shoreline features, including sand and gravel beaches, spits and bars, and wave-cut terraces and cliffs. These are described in detail by Goldthwait (1907, pp. 43–52), Leverett and Taylor (1915, pp. 350–54), and Bretz (1955, pp. 108, 113–17).

During an early part of the Glenwood lake stage there was a readvance of the glacial ice front, but this failed to reach the previously formed southwestern rim of the Michigan basin, and the lake continued to exist as a narrow crescent (Fig. 57). Four closely spaced moraines were built by the Lake Border ice in the area north of Chicago where the ice did advance to the shore, and a Lake Border morainic system was deposited elsewhere around the Michigan basin without leaving a record of the number of distinct fluctuations of the ice front. The amount of retreat of the ice margin between the Tinley and Lake Border advances is unknown (Fig. 56), because the latest advance probably destroyed all traces of the Glenwood beach which may have existed in the area it overrode.

During the retreat of the ice front from the Lake Border moraines the Glenwood stage expanded northward (Fig. 58). Shore features at the Glenwood level are easily recognizable at intervals along the eastern side of the basin as far north as Muskegon, Michigan, and along the western shore as far north as Racine, Wisconsin. From Racine northward to Milwaukee, Wisconsin, shore erosion by several later lake stages has destroyed all evidence of the higher beaches. The maximum northward extent of the Glenwood stage, along with the question of its duration in time, is a controversial subject which is discussed in greater detail in subsequent paragraphs. Bretz (1951b, pp. 421 and 425, and 1955, p. 114) has stated that the Glenwood stage was brought to a close during the Whittlesey stage of the Erie and Huron basins, which existed during the Port Huron glacial advance. The present author believes that the Glenwood stage persisted at least that long. Because of this degree of agreement, it appears reasonable to suppose that the lake expanded northward in the Michigan basin following the ice front as it retreated from the Lake Border (latest Cary) maximum, to the position of maximum retreat before Port Huron time (Fig. 59), and that the lake was then compressed in size as the ice front advanced to the position of the Port Huron maximum (Fig. 60). The Port Huron ice overrode any Glenwood shore features which may have been formed within the area it subsequently covered.

INTRA-GLENWOOD LOW-WATER STAGE

The extent of retreat of the ice front during the Cary–Port Huron interval is not known. If the retreat were great enough to uncover a lower outlet to the east for the waters in the Erie and Huron basins, and an outlet through which the Michigan basin could discharge northeastward to the Huron basin, the water of the Glenwood stage would have been drained to a lower level (Fig. 59). This possible low stage would be correlative with a similar low stage in the Erie and Huron basins for which there is some evidence (Chap. 8). The "possible low stage" in the Michigan basin is shown in Fig. 37.

Evidence for an intra-Glenwood low-level stage was found, and recorded in an unpublished thesis by Workman (1925), south of Chicago near the intersection of Halstead Street and Ridge Road (S.E. ¼ sec. 32, T. 36 N., R. 14 E). The evidence indicates (1) that swamp and driftwood debris was deposited on earlier Glenwood lagoonal sediments, (2) that this deposit then was trenched several feet deep by a stream draining to a lake level somewhere below 615 feet A.T., and (3) that still later the Glenwood level of the lake was restored, the small valley was filled, and much sand and gravel (the Franklin bar) was deposited on the eroded early Glenwood materials. Workman concluded that "this period of peat formation and erosion was an interruption of the lake during the Glenwood stage."

An intra-Glenwood organic lagoon deposit near Dyer, Indiana, has been described by Bretz (1955, p. 129); this consists of a deposit of peat, driftwood, and a few ice-rafted boulders underlying a sand-and-gravel spit which is at the Glenwood level of Lake Chicago. The deposit does not necessarily record a lower stage of the lake, but simply indicates a change of local shoreline conditions. It has, however, provided material for dating by the radiocarbon method. The first two ages obtained (Libby, 1954a, p. 135) were 10,661 ± 460 years and 11,284 ± 600 years. These range from 116 to 739 years younger than the Two Creeks forest bed, which is a post–Port Huron, pre-Valders deposit. The second pair of ages obtained for this material by the same laboratory (Libby, 1954b, p. 735) were 18,500 ± 500 years and a value greater than 21,000 years. A third pair of ages for organic material from the same deposit (Flint and Rubin, 1955, p. 650) is 12,650 ± 350 years and 12,200 ± 350 years. The average of these last two dates is an age of 12,425 years, approximately 1000 years older than the Two Creeks material.

If the lagoon deposit near Dyer is a correlative of the lagoon deposit described by Workman, the date of the Dyer deposit may be used to date the intra-Glenwood low-water stage. There is no assurance, however, that the two deposits are correlative.

The intra-Glenwood low stage described by Workman may have occurred during the Cary–Port Huron interval, and thus be a representative of the "possible low stage" shown in Fig. 37. On the other hand, it may be a correlative of the Two Creeks low stage, which occurred during the Port Huron–Valders or Two Creeks interval. If this is correct, the Cary–Port Huron "possible low stage" is left unsupported by field evidence in the Michigan basin but it still remains as a possible stage, with some support from evidence in the Erie basin.

THE SECOND GLENWOOD STAGE

Assuming that a low stage occurred during the Cary–Port Huron interval, it would have left the Chicago outlet dry at its Glenwood level (Fig. 59). When the Port Huron ice advanced and closed the low-level outlet at the northeastern corner of the Michigan basin, the lake water rose until it spilled through the Chicago outlet and thus formed a Second Glenwood stage (Fig. 60). This, according to the present interpretation of events, then persisted throughout the Port Huron glacial substage, while the Whittlesey (Fig. 60) and the Highest and Middle Warren stages (Fig. 61) of the Erie and Huron basins were discharging their water down the Grand River valley to the Michigan basin.

Lake Chicago then was brought to a lower level when retreat of the Port Huron ice again uncovered a low outlet at the northeastern corner of the Michigan basin.

THE TWO CREEKS LOW-WATER STAGE

The retreat of the ice margin during the interval between the Port Huron and Valders glacial substages, known as the Two Creeks interval (Fig. 62), has been described in detail in a foregoing review of the Wisconsin glacial stage. It was noted that the section near Two Creeks, Wisconsin (illustrated in Fig. 31), recorded the following events: deposition of gray till, presumably of Port Huron age; a recession of the ice front which allowed the water of Lake Chicago to flood the area and deposit lake sediments; a further recession of the ice sufficient to uncover a new, low outlet (which could only have been located to the northeast—probably through the lowland between Little Traverse Bay and the Lake Huron basin) and to permit the lake to be drained down at least as low as the present level of Lake Michigan; growth of a forest on the emerged lake bottom; a readvance of the ice margin to cover the low outlet, resulting in the flooding of the forest and deposition of lake sediments over the bed and around the trunks of the submerged trees; further advance of the ice, which rode over the area, breaking off the trees and burying them in red till. The low-water stage recorded in this section is here named the "Two Creeks stage."

Radiocarbon dates have been obtained by analysis of several samples of wood from the buried forest at Two Creeks, by several different laboratories, so that the forest bed is the most thoroughly dated reference point in Great Lakes history. The dates have a narrow range and average about 11,400 years before the present (Arnold and Libby, 1951; Crane, 1956; Preston, Person, and Deevey, 1955). The most significant aspect of the Two Creeks section is that it represents a low lake stage in the Michigan basin and that this in turn requires a low stage throughout the other Great Lakes to the east. Another point of interest is that the Valders till which overlies the interstadial forest bed has a high content of red clay, which must have been acquired from extensive lake clay deposits formed during the Port Huron–Valders interval. Evidences of the direction of ice movement during the Valders advance indicate that the ice came from the north and northwest. It has been concluded that the Superior basin was largely if not completely free of ice during the Two Creeks interval and was occupied by Lake Keweenaw (Fig. 62), which is described in a subsequent chapter.

Deposition of red lake clay in the Michigan basin occurred during the retreat of the Port Huron ice, probably after discharge into that basin from the Superior basin had begun. During the advance of the Valders ice, there again was deposition of red clay, both in the main Michigan basin and in the Green Bay re-entrant, where an extensive ice-margin lake occurred and covered several hundred square miles.

The Two Creeks stage may possibly be correlative with either or both of the intra-Glenwood peat deposits near Dyer and those described by Workman (discussed in a foregoing paragraph). As has been noted, the radiocarbon dates of the Dyer material range from 10,661 years to more than 21,000 years, or from 739 years younger than the Two Creeks to more than 9600 years older. If the average of the dates obtained in the latest analyses, 12,425 years, is accepted, the Dyer material is 1000 years older than the Two Creeks stage.

THE THIRD GLENWOOD STAGE

Because the Two Creeks low-water stage was produced by drainage through a different outlet, the Chicago outlet may be assumed to have remained unaltered during the Two Creeks interval. When the lake rose to flood the forest, it returned to the same level it had maintained during its previous period of discharge through the Chicago outlet (Fig. 63). According to Bretz, Lake Chicago had already been lowered to the Calumet stage, at an elevation of 620 feet A.T., before the Two Creeks interval began; according to the present author, Lake Chicago was still at the Glenwood level when the Two Creeks interval began, and the water was returned to that level at the close of the Two Creeks interval. The reasons for this conclusion are stated in subsequent paragraphs. Meanwhile, however, the possible relationship of the Two Creeks stage to a hypothetical "Bowmanville low-water stage" is discussed.

THE BOWMANVILLE LOW-WATER STAGE

The Bowmanville deposits, which were described in detail by Baker (1920, pp. 65–67), occur below the level of the Calumet beach. These include a succession of layers which has been summarized by Bretz (1955, p. 92) in the following generalized column:

7. Sandy silt and peat beds. Average thickness 20 inches. Shallow-water mollusks in lower layers.
6. Silt and peat beds, 15 to 52 inches thick. Oxidized zone and crayfish burrows in upper part.
5. Silt, sand, peat, and marl; 19 to 59 inches thick. Lower layers generally without fossils; upper layers filled with shells and fish remains, representative of shallow-water types, oxidized zone and crayfish burrows in upper part.
4. "Heavy" bed of shells, mostly *Unios*.
3. Gravel and sand, 2 to 19 inches thick. No record of life except in upper part where *Unios* begin to appear.
2. Silty, carbonaceous, and peaty material; 1 to 40 inches thick. Shells of shallow-water mollusks, spruce cones, oak leaves.
1. Sand or sand with pebbles. From a fraction of an inch to 12 inches thick. No evidences of life.

Layer 1 was considered Glenwood in age, layer 3 Calumet, and layers 4 and 5 Toleston. Following Andrews (1870), layer 2 was interpreted as a record of a low-water stage, the Bowmanville, presumably occurring between the Glenwood and Calumet stages.

Several writers have not accepted the Bowmanville as a low-water stage of Lake Chicago; Goldthwait (1908a) stated, "It has long been supposed (following Dr. Andrews) that the Glenwood and Calumet stages were separated by a stage of low water when the lake fell to a level at least as low as the present and probably much lower. The evidence cited is a peat bed which lies beneath the Calumet ridge at Grosse Point. But recent study of this locality strongly suggests the peat is merely a lacustrine deposit, formed in quiet water behind the barrier during the Calumet stage, and buried by shoreward advance of the reef" (p. 61). Leverett and Taylor (1915) questioned the Bowmanville low-stage interpretation as follows: "However, although the presence of peat under the gravel suggests a lower stage of water it can scarcely be said to prove it conclusively, for a bar might be extended out over a peaty deposit standing at the same level as the lake and might press it down and thus give it a lower level than it had while in process of growth. . . . If a lower lake level preceded the development of the Calumet beach other evidence than that from the buried peat deposits should be found. For instance, the valleys which entered the lake at this lower stage should have been cut to a level below the Calumet beach and then the beach should have been built across the beds of these channels" (p. 356). Alden (1918, pp. 332–33) mentioned the "Bowmanville," but he also rejected it as a bona fide low-water stage.

This hypothetical Bowmanville low-water stage has been correlated with the well-established Two Creeks low-water stage by Bretz (1951b, p. 410). Because the name "Bowmanville" had priority, the Two Creeks lake stage was referred to as the "Bowmanville" stage (Bretz, 1951b, p. 410; Flint and Deevey, 1951, p. 261; Flint, 1953, p. 905), the "Bowmanville or Two Creeks" stage (Bretz, 1953, p. 378), and the "Bowmanville–Two Creeks–Wayne" low water (Bretz, 1953, p. 380, and 1955, p. 111).

The present author prefers the term "Two Creeks" stage as a designation of the low-level stage which occurred during the Two Creeks interval, and he suggests that the term "Bowmanville" be discontinued except as a designation for the deposits of the original Bowmanville locality. It has been noted in the foregoing discussion of the "Bowmanville low-water stage" that the deposits at Bowmanville do not contain definite evidence of a low-water stage. Even if the lowest peat in those deposits does represent a low water, there is no assurance that it was contemporaneous with the Two Creeks stage; it may have been later or earlier than the Two Creeks. As M. M. Leighton has pointed out,[2] the greater abundance of remains of animal and plant life in the Bowmanville deposits argue for a younger age than that of the Two Creeks stage. If the Two Creeks stage occurred as an intra-Glenwood event, which seems probable to the present author, and if the Bowmanville peat was formed between the Glenwood and Calumet stages as supposed by Andrews, Baker, and Bretz, the Bowmanville should be later than the Two Creeks. However, according to the present author's interpretation, Lake Chicago was lowered from the last Glenwood stage to the Calumet, and from the Calumet to the Toleston, without any intervening lower-water stages. According to this interpretation, if the Bowmanville peat is post-Glenwood it cannot represent a low-water stage. The alternative explanations of Goldthwait (1908a), and Leverett and Taylor (1915), that the Bowmanville peat was simply formed in a shallow-water embayment and covered by a Calumet beach without any change in water level, thus are more acceptable.

TERMINATION OF THE GLENWOOD STAGE

The last Glenwood stage was brought to a close by downcutting of the Chicago outlet. This evidently was an abrupt event, because there are no shore features in the Michigan basin between the Glenwood level and the Calumet level, 20 feet lower. The causes of the static levels of Lake Chicago in the Glenwood, Calumet, and Toleston stages, and of the two lowerings, from Glenwood to Calumet and from Calumet to Toleston, have been discussed in detail by Bretz (1951b). He has indicated clearly (1) that the Glenwood and Calumet stages were maintained while the bottom of the outlet channel was in unconsolidated material (till, at the head of the outlet, and valley train material farther

---

[2] Personal communication, 1956.

downstream), and that during these two static stages the erosive power of the discharge was not sufficient to deepen the channel, and (2) that only a significant increase in the volume of discharge could cut the outlet channels deeper.

Bretz (1951b, p. 408) concluded that the minimum age of the Glenwood shore was no younger than Port Huron, because he could find no Glenwood-level shore features north of Muskegon, Michigan, and Milwaukee, Wisconsin. This conclusion is questioned, however, because of the following considerations. If Glenwood shores had been formed on the inner margin of the Port Huron morainic system, after retreat of the Port Huron ice, they would have been overridden and destroyed by Valders ice in the main basin of Lake Michigan. Glenwood features might be expected to have survived destruction in the embayments of the major river valleys in western Michigan, because the Valders ice did not completely fill those areas. Martin (1955), presumably following Leverett, shows Glenwood beaches in the valleys of Stony Creek and the Pentwater, Pere Marquette, Lincoln, and Big Sable rivers, while Bretz (1951b, p. 421), in a thousand-mile trip in the coastal belt between Grand River and Grand Traverse Bay, failed to find any Glenwood features. The different opinions regarding the presence of a Glenwood shore on Port Huron drift in the area discussed here cannot be resolved without a field examination by all of the parties concerned. It should be pointed out that if Glenwood shore features were formed on the Port Huron drift in the embayments beyond the limits later reached by Valders ice, those features would not have been so strongly developed as shores on the open coasts, and further that when the Valders ice reached its maximum it would have dammed all of the embayments in question, flooding the areas to levels above the Glenwood level. Some modification of the Glenwood features, if they were present, might be expected to have occurred during the transgression and regression of these ice-margin lakes. The net result of these two factors is that the Glenwood shore would be very obscure.

If a post–Port Huron existence of the Glenwood stage is not precluded, the question remains as to whether the Glenwood stage may have existed in post-Valders time. As stated previously,

Bretz searched for Glenwood features in the coastal belt of western Michigan from Grand River to Grand Traverse Bay, and he also searched the western shore of Lake Michigan north of Milwaukee. Because he found no Glenwood-level features on the Valders drift, he concluded that the Chicago outlet had been cut down, bringing the lake below the Glenwood level, before the beginning of retreat of the Valders ice (Fig. 63). The failure to find a Glenwood beach on the Valders drift is not, however, proof that one was not there. Shore erosion during several later lake stages (the Algonquin, Nipissing, and Algoma), and especially by the present Lake Michigan, has removed practically all traces of any earlier beaches which may have existed in the area in question. Even the Algonquin beach is missing from a considerable part of the stretches of shore which were examined by Bretz.

On the western side of Lake Michigan the margin of the Valders drift crosses the shore at Milwaukee. It is impossible to determine whether a Glenwood beach was formed on the Valders drift early in the retreat of the Valders ice, because "Between the termination of the Glenwood beach near Wind Point [18 miles south of Milwaukee] and Port Washington [25 miles north of Milwaukee] the active retreat of the clay bluffs has removed almost all the old shore records" (Goldthwait, 1907, p. 51).

On the eastern shore of Lake Michigan, the margin of the Valders drift, according to Bretz, crosses the shore in southern Oceana County, a few miles north of Muskegon. Again, it is impossible to determine whether a Glenwood beach was formed on the Valders drift, early in the retreat of the Valders ice, along the shore of the main basin of the lake, because "From near the line of Muskegon and Oceana Counties northward for several miles the bluff of the modern lake rises above the level of the Glenwood beach, some points being fully 200 feet above Lake Michigan" (Leverett and Taylor, 1915, p. 353).

It appears obvious, from the two foregoing paragraphs, that an early post-Valders age of the Glenwood stage is not impossible. Evidence supporting such an age will be cited in a subsequent discussion.

Another reason why Bretz believed that the Glenwood stage was terminated by Port Huron time was that he had recognized two periods of discharge from the eastern lakes, through the Grand River valley, which added to the total discharge of Lake Chicago, and he correlated these with the two lowerings of the Chicago outlet, which produced the Calumet and the Toleston stages. The first period of discharge from the eastern lakes to Lake Chicago embraced the outflow from Lakes Maumee, Arkona, and Whittlesey; then, after a time of discharge to the east, the second period of discharge from the eastern lakes to Lake Chicago embraced the outflow from a single stage, Lake Warren. The mechanism of downcutting of the Chicago outlet by erosion during periods of increased discharge is not questioned here. What is questioned is the assignment of the long history of fluctuating discharge, from some of the Maumee stages, all three of the Arkona stages, and the Whittlesey stage, to a single period of increased discharge which caused the lowering of Lake Chicago from the Glenwood to the Calumet level. The rapidity of the downcutting of the outlet, noted on a foregoing page, is recognized by Bretz (1951b) as follows: "without leaving any surviving shorelines, Lake Chicago rather abruptly dropped to the Calumet level as the large volume of water poured into it via Grand River . . ." (p. 408). It appears more reasonable to assign the increased discharge of a single stage, such as the Whittlesey, to the brief period of downcutting of the Chicago outlet. If the Whittlesey stage is correlated with the downcutting, then the previous fluctuations in discharge from the Maumee and Arkona stages occurred without affecting the level of the outlet channel. On the other hand, if the channel could survive the early increases in discharge it also could survive the Whittlesey-stage increase, and even that of the later Warren stage, only to be cut down by a still larger increase in discharge which may have occurred at a later date. It may be noted that the evidence from the Michigan basin indicates that the Glenwood stage persisted at least through the retreat of the ice front from the Tinley moraine, through the various fluctuations of the Lake Border episodes, and through the retreat which preceded the Port Huron advance.

In summary, it may be stated that Bretz's ingenious explanation of the history of Lake Chicago, with its detailed and specific correlations with events in the basins to the east, is not necessarily the correct one. There are some other details of the lake history, involving later lake stages in the eastern basins, which seem to require that the Glenwood stage persisted in Lake Chicago until after the Valders ice had begun to retreat (Fig. 65A).

In the discussion of the early lakes in the Erie and Huron basins it was pointed out that the Early Algonquin stage most probably came into existence by lowering from the 620-foot Lundy level to the 605-foot Algonquin level while water was discharging to the Michigan basin. This conclusion was based on the need for an adequate explanation of the stabilization of the Port Huron channel at the Algonquin-Toleston level. It was further pointed out that three successive stages of the Erie and Huron basins, the Grassmere, at 640 feet, the Lundy, at 620 feet, and the Early Algonquin, at 605 feet, had correlatives in the Michigan basin in the Glenwood, at 640 feet, the Calumet, at 620 feet, and the Toleston, at 605 feet.

The outlets of the Grassmere and Lundy stages were unknown, according to Leverett and Taylor (1915, pp. 400, 406–8), but were assumed to lie somewhere to the east of Lake Erie (Figs. 65B and 66B). Because it appears reasonable to presume that the Glenwood and Grassmere stages, both at an elevation of 640 feet, the Calumet and Lundy stages, both at 620 feet, and the Toleston and Algonquin stages, both at 605 feet, were connected, and that their elevations were controlled by the sill elevations of a single outlet (the Chicago outlet), a search has been made for a route by which the Huron and Michigan basins could have been joined.

The margin of the Valders ice separated the lakes in the Huron and Michigan basins prior to the Grassmere stage (Fig. 64), while Lowest Lake Warren was discharging down the Grand River valley to Lake Chicago (which still stood at the Glenwood stage, according to the present interpretation). This ice margin, which has been mapped by Melhorn (1954), extended along the northern slopes of the inner Port Huron moraine in the northern part of the Lower Peninsula of Michigan, and did not lie very far above the elevation of the Lowest Warren stage. The Valders ice, as shown by the distribution of the red Valders till, was an elongate mass which filled the northern two-thirds of the Michigan basin without rising very high on its flanks. Only a slight retreat of the ice margin would permit discharge of lake water between the ice and the side of the basin (Fig. 65A).

In the search for evidences of such discharge, the author found a channel in the area east of Grand Traverse Bay which is quite adequate to have carried the flow from the Huron basin to the Michigan basin during Glenwood-Grassmere time. This is shown in Fig. 38, with the margin of the retreating Valders ice in a position permitting the water from the Huron basin to enter the channel at a point a few miles west of Petoskey, Michigan. The Algonquin beach, formed at an elevation of 605 feet, and the Nipissing and Algoma beaches, formed at 605 feet and 595 feet respectively, and since raised by upwarp of the land, extend through this valley as easily recognizable features except in a segment just west of Chestonia. In that segment the valley maintains its broad, smoothly curving shape, and its floor lies only a few feet above the Algonquin beach and well below the warped Glenwood plane. From the lower end of the segment just described, the channel extends northward to Ellsworth and then southward, passing through the basins of Lake Bellaire and of Torch, Round, and Elk lakes, then to the southern end of Grand Traverse Bay. With the ice still presumably occupying the northern part of Grand Traverse Bay, the Grassmere discharge would have passed from the southern part of the bay to the margin of the main Michigan basin through a broad lowland in Leenanau County, and from there it presumably went southward along the margin of the ice lobe. No traces of this ice-margin flow have been found, because the shore of Lake Michigan from the Grand Traverse area southward for many miles has been extensively eroded by wave action and much of the present beach lies against steep bluffs which extend up to elevations above the Glenwood level.

In the Jordan valley system (Fig. 38), above the Algonquin beach, there appear to be several alternate possible routes for discharge from the Huron to the Michigan basin. There is a sill almost exactly at the level of the upwarped Grassmere-Glenwood plane in the valley of Minnehaha Creek, six miles east of Petoskey, which might have been the site of the first discharge in Grassmere-Glenwood time. Another possible route is through the Bear Creek

180   GEOLOGY OF THE GREAT LAKES

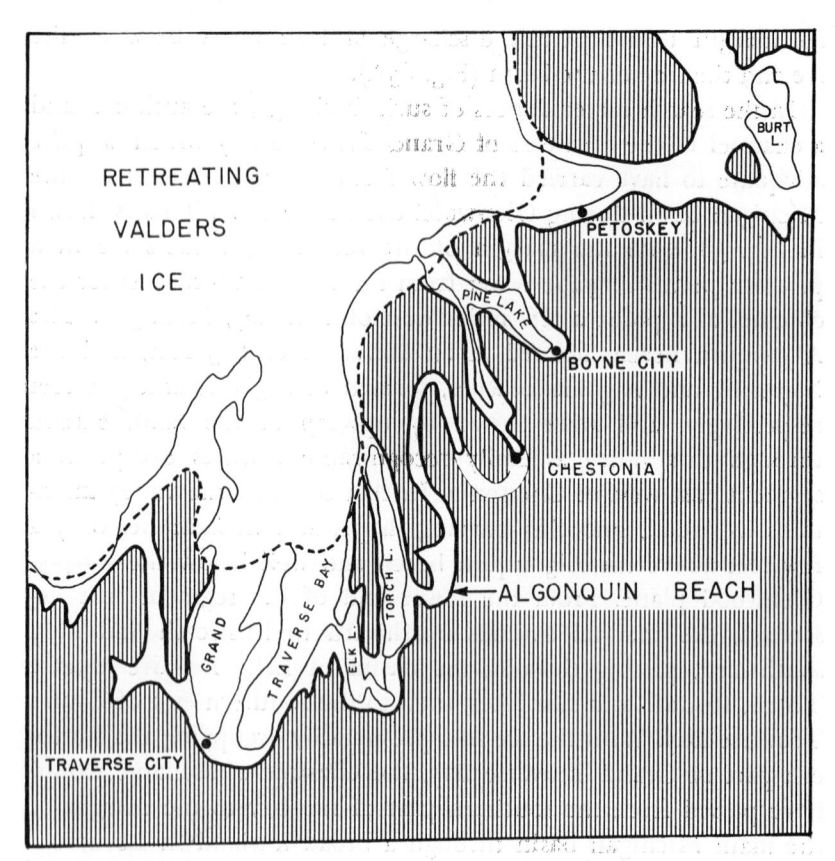

Fig. 38. The Jordan valley in the Grand Traverse Bay region, Michigan: a possible route of discharge from the Huron to the Michigan basin during the Grassmere and Glenwood III stages.

valley, extending southward from Petoskey; still another is the Walloon Lake valley, five miles west of Petoskey. Still farther west, at Bayshore, there is a short valley extending through the divide between Lake Michigan and Pine Lake and carrying the Algonquin beach on its sides. At Charlevoix the main Pine Lake valley joins the Michigan basin. These valleys, from Minnehaha Creek westward to Charlevoix, are progressively deeper, and it appears likely that the discharge shifted from one to another as the Valders ice margin retreated westward.

When detailed topographic maps of the area, which are now in preparation, become available, it may be possible to work out the precise sequence of events involved in the Huron-Michigan discharge in Glenwood to Toleston time. Altimeter traverses, tied into present lake level and to local beach marks, have shown the presence of Glenwood-level deposits as far east as Boyne Falls, five miles southeast of Boyne City.

Probably by Calumet time, and almost certainly by Toleston time, the discharge from the east would have had to pass around the high ground west of Charlevoix, and it is presumed that the margin of the Valders ice retreated far enough to permit this discharge.

According to the author, the final Glenwood stage of Lake Chicago persisted until early in the retreat of the Valders ice (Fig. 65A), then was terminated by downcutting of the Chicago outlet. The discharge which accomplished the downcutting at this time must have been of significantly greater volume than at any previous time in the history of Lake Chicago. It is presumed that rapid melting of the Valders ice, along a front extending from the western part of the Ontario basin to the western side of the Michigan basin, provided the necessary volume.

THE CALUMET STAGE

The Calumet stage is recorded by several well-developed beaches found at intervals around the southern part of the Michigan basin, at elevations approximately 620 feet above sea level. These are described in detail elsewhere (Bretz, 1955, and in the references cited by Bretz). The strand line can be traced into the Chicago outlet, where the lake obviously discharged.

The Calumet stage appears to have been coexistent with, and receiving the discharge of, the Lundy stage, which stood at the same elevation in the Huron and Erie basins (Fig. 66A). The justification of this view has been stated in the discussion of the Grassmere, Lundy, and Early Algonquin stages of the Erie and Huron basins. An additional point supporting this view is found in Bretz's study of the Glacial Grand River. Bretz (1953, p. 379) states that there is no terrace in the entire length of the Glacial Grand River valley which can be correlated with the Calumet stage of Lake Chicago, and further, he states that "There seems to have been no recognizable increment to the Allendale delta [at the mouth of the Glacial Grand River] at the Calumet level . . ." and that "were there no Calumet beach elsewhere around the margin of Lake Chicago, our interpretation of the Allendale delta might not include this intermediate level" (p. 380). This absence of any indication of a grading of the Glacial Grand River channel and delta to the Calumet level is easily accounted for by the present author's hypothesis that discharge from the Huron basin to the Michigan basin was shifted from the Grand River valley to a more northerly route while the Glenwood stage of Lake Chicago was still in existence (Fig. 65A). If the Glacial Grand River had ceased to exist before the lake surface was lowered to the Calumet level, no evidence of grading of its valley to the Calumet level would be expected.

It was presumably the Chicago outlet sill that maintained both the Calumet and the Lundy stages long enough for shore features to be formed (Fig. 66A). The Calumet shoreline is a very distinct one, but unlike the earlier Glenwood and the later Toleston-Algonquin shores, it is represented only by beach sands and gravels and not by wave-cut terraces or shore cliffs. For this reason, it is permissible to assume that the Calumet stage may have had a relatively brief existence.

The bottom of the Chicago outlet channel in Calumet time must have been on unconsolidated material, and the most likely reason for a cessation of the downcutting which produced the Calumet stage is that the volume of the discharge was reduced, probably because of a diminution in the runoff from the wasting Valders ice sheet. After a period of stability, downcutting was resumed and the Calumet stage came to an end.

THE TOLESTON STAGE

Next in elevation below the Calumet beach, in the southern part of the Michigan basin, are the shore features at 605 feet above sea level which were first designated as representatives of the Toleston stage of Lake Chicago (Fig. 67A). These were described in some detail by Leverett and Taylor and earlier writers. Leverett and Taylor (1915, p. 357) indicated the possibility that the Toleston-level shoreline may have been only partly formed by Lake Chicago, because it has a level that was "closely approximated if not reached by Lake Algonquin."

The Toleston-level shore is traceable into the Chicago outlet channel. By Toleston time this had been cut down to a bedrock sill which prevented further deepening, and the bed of the now-abandoned channel is still preserved, except where it has been cut away in the construction of the various canals which connect Lake Michigan with the Illinois River system.

Leverett and Taylor (1915, p. 409) believed that Lake Toleston existed in the Michigan basin, and discharged to the Illinois River, while Early Lake Algonquin was formed at essentially the same level in the Huron basin and first discharged southward to the Erie basin (Fig. 67B). They suggested that the ice barrier on the high ground northwest of Alpena, Michigan, then retreated and allowed the separate lakes to join. The present author has proposed in another chapter that the waters of the Huron and Erie basins were already discharging to the Michigan basin in Grassmere and Lundy time, and that the stabilization of the Chicago outlet to form Lake Toleston also determined the Early Algonquin level in the Huron and Erie basins. The Port Huron outlet at the south end of the Huron basin was cut at this time by northward-flowing water from the Erie basin (Fig. 67A). A short time later, recession of ice in the Ontario basin opened an outlet to the east, the Erie basin was drained below the level of the Port Huron channel, and the Toleston–Early Algonquin lakes of the Michigan and Huron basins then discharged through both the Chicago and the Port Huron outlets (Fig. 68).

According to this interpretation of events, the Toleston and Early Algonquin stages came into existence simultaneously or nearly so. At first the connection between them may have been a narrow channel, but further recession of the ice front soon allowed the connection to widen, and at this stage of development the combined lakes may be referred to as Early Lake Algonquin.

Discussion of succeeding events in the Michigan basin is deferred to subsequent chapters dealing with the later stages of the Great Lakes.

chapter ten

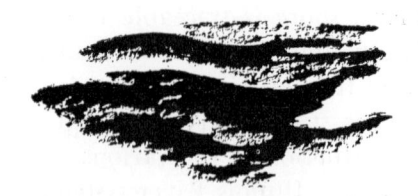

EARLY LAKE STAGES IN THE SUPERIOR BASIN

INTRODUCTION

The history of lake stages in the Superior basin (Table 12) is less perfectly known than that of any of the other Great Lakes, primarily because of the primitive condition of the area but also because none of the older shorelines are horizontal. The "hinge line," or line of no upwarp, for each of the older beaches apparently lies outside the basin, to the south, and it is therefore impossible to know the true elevation of the water planes at the times when the beaches were formed. A further difficulty in the development of a history of the lake stages in this basin is that much of the basic data observed by the earlier investigators has not been published in a readily usable form. In the summaries by Leverett and Taylor (1915), and by Leverett (1929a), the distribution of shore features is described in terms of preconceived identifications and correlations. The beaches assigned to a single lake stage are described in their distribution throughout the basin, but rarely is the complete vertical sequence of beaches in any one locality described. One source of difficulty in understanding the history of the Superior basin is the concept that the main Algonquin stage of the lower lakes extended into the Superior basin. This concept was based on very meager evidence, but it can be disproved by data published by Leverett (1929a) as well as by information observed by the present author.

EARLY LAKE STAGES IN THE SUPERIOR BASIN   185

This subject is reviewed in Chapter 12, in a discussion of the extent of Lake Algonquin.

Table 12. *Lake stages of the Superior basin.*[a]

Named stages, listed in order of decreasing age	Outlet
Keweenaw	Unknown; to Michigan basin?
Various ice-margin lakes	(see text)
Lake Duluth I	St. Croix River; to Mississippi River
Lake Duluth II	St. Croix River; to Mississippi River
Lake Duluth III	St. Croix River; to Mississippi River
Lake Duluth IV	St. Croix River; to Mississippi River
Sub-Duluth	Ice-margin channel, Huron Mountains; to Green Bay in Michigan basin?
Algonquin [b]	
Lutsen	St. Marys River valley?
Marais	St. Marys River valley?
Tofte	St. Marys River valley?
Kodonce	St. Marys River valley?
Deronda	St. Marys River valley?
Minong	St. Marys River valley?
Nipissing	St. Marys River valley
Sub-Nipissing	St. Marys River valley
Superior	St. Marys River

[a] Altitudes are not given because the original altitudes are unknown and all beaches are tilted, with present altitudes increasing toward the northeast. The Nipissing and sub-Nipissing beaches apparently cross the planes of the earlier stages.
[b] The beach represented by this entry is incorrectly identified (see text).

CARY–PORT HURON–INTERVAL INFERRED LAKE

The earliest known lakes of the Superior basin came into existence much later than those in the basins farther south. Figure 39 indicates that the presence of a lake in the western part of the Superior basin during the Cary–Port Huron interval is inferred. This is based on studies of the glacial drift in Minnesota,[1] west of the basin, which suggest that the ice front probably retreated into

---

[1] Most of this recent work, by H. E. Wright and A. F. Schneider, is as yet unpublished except for summaries contained in the guidebook for Field Trip No. 3, Ann. Meeting, Geol. Soc. of America, 1956.

186   GEOLOGY OF THE GREAT LAKES

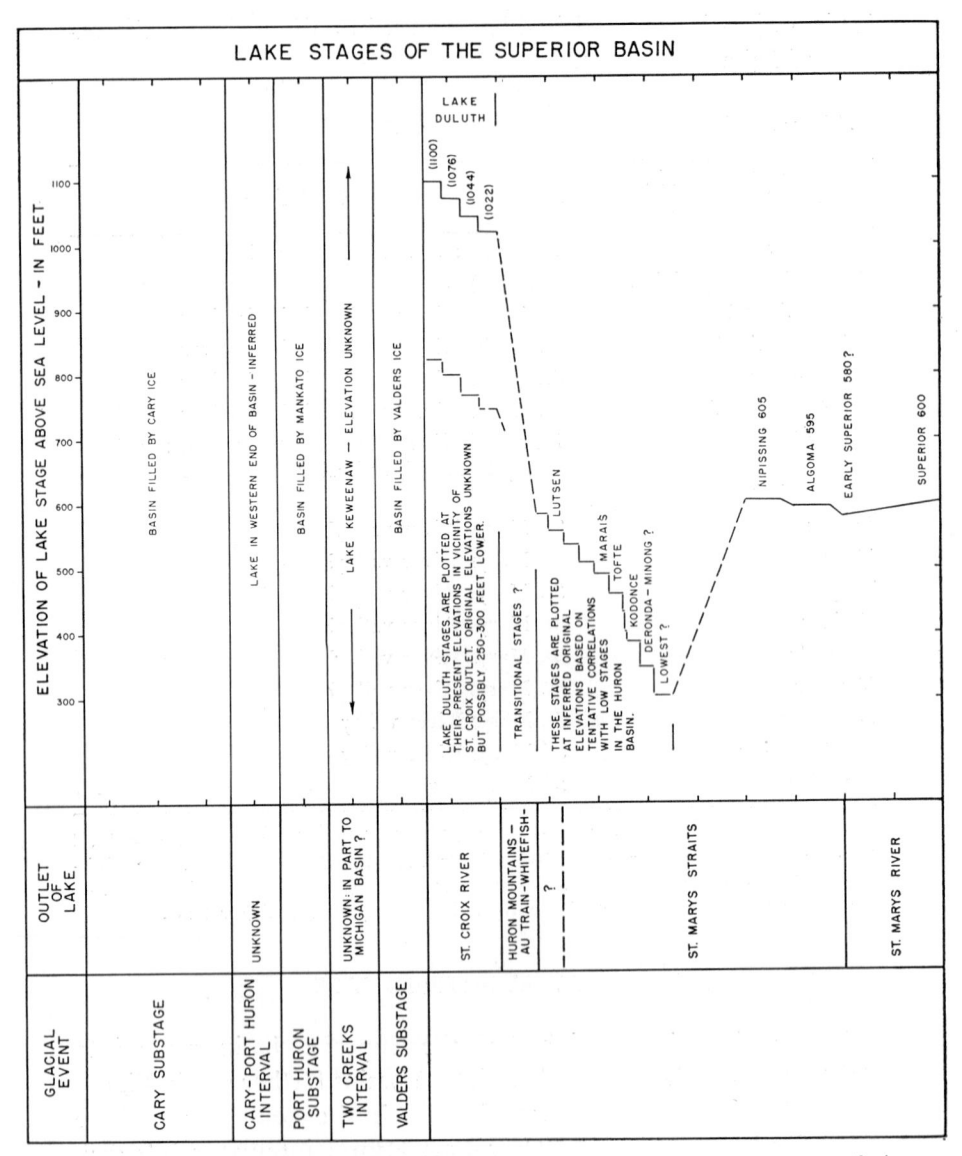

Fig. 39. Lake stages of the Superior basin. Lake levels are plotted against an unspecified time scale in the upper part of the figure, and lake outlets and correlative glacial events are shown in the lower part.

the Superior basin during this interval (Fig. 59).[2] No direct evidences of such a lake have been found.

---

[2] Figures 53–75, lake stage maps, follow Chapter 16.

The Port Huron substage is here considered a correlative of the Mankato substage of the Des Moines lobe. It should not be confused with the Valders substage, which occurred at a later time, after the Two Creeks interval.

LAKE KEWEENAW OF THE PORT HURON–VALDERS INTERVAL

Lake Keweenaw (Fig. 62) is not represented by any known beaches or other deposits related to shoreline features. It has been inferred from the obvious requirement that the red clay component of the Valders glacial drift must have had an origin in interglacial lake beds. This was suggested as early as 1918 by Alden, but the inferred lake was named by Murray (1953, p. 153).

Murray's evidence for Lake Keweenaw is as follows: the mineral suites, including the heavy minerals of the sand and coarse silt fractions of the red Valders till and of the underlying gray till, are identical. Because the Valders ice followed a course decidedly to the west of that of the earlier ice (Thwaites, 1934), this indicates that the Valders ice sheet accomplished little or no bedrock erosion, but removed and redeposited a portion of the older glacial deposits. This implies a relatively thin Valders ice sheet incapable of accomplishing extensive bedrock erosion, which is in agreement with evidence presented by Thwaites (1943). The red color of the Valders till is seen under the microscope to result from fragments of hematite in the clay size range and to a fine stain on the larger grains. The most obvious source of the hematite is in the red rocks of the Lake Superior basin. Because only the finer fractions of the Valders till have been enriched in this material, it is concluded that the fine silt and clay were deposited in lake beds and were then picked up by the advancing Valders ice and mixed with the older till (Murray, 1953, pp. 148–53).

Murray has concluded that at the time of maximum retreat, during the Two Creeks interval, the ice front had receded far enough to permit Lake Keweenaw to occupy about two-thirds of the Superior basin (Fig. 62). The discharge from this lake probably went down the St. Croix River at first but later it poured into the Michigan basin, contributing the red clays which are found there. When the ice front advanced during the Valders substage it completely filled the Superior basin, destroying Lake Keweenaw (Fig. 63).

POST-VALDERS ICE-MARGIN LAKES

The margin of the shrinking Valders ice sheet retreated within the drainage divide of the Superior basin at various points, giving rise to local ponding of water. At first these ice-margin lakes were isolated bodies of water standing at various elevations in separate valleys (Fig. 65A). Later they were joined by further recession of the ice front, and came to a common level, thus initiating Lake Duluth. Only the larger of these ice-margin lakes are described in the following paragraphs.

Glacial Lake Nemadji

Water was ponded between the retreating ice front and a divide southwest of the southwestern corner of Lake Superior. This lake, named Nemadji, drained westward through an outlet (located near Moose Lake, Minnesota) which stands at a present elevation of 1070 feet above sea level. The discharge was to the Kettle River, a tributary of the St. Croix River. Lake Nemadji was later joined with Lake Brule to form Lake Duluth, which was the first major lake stage in the Superior basin in late Wisconsin time (Leverett, 1929a, p. 55).

Glacial Lake Brule

While the ice front stood across the lower part of the Brule River valley, located on the south side of Lake Superior a few miles east of Duluth, water was ponded in that valley and its shores are marked by beaches which stand at a present elevation of 1125 feet. This lake drained southward to the St. Croix River, and thence to the Mississippi. It received the discharge from other ice-margin lakes and streams to the east, probably from as far as Baraga County, Michigan, which lies south of Keweenaw Bay. When the ice margin receded farther in the Superior basin, Lake Brule was merged with Lake Nemadji, forming Lake Duluth (Fig. 65A), and this lake continued to discharge down the St. Croix River (Leverett, 1929a, p. 56).

Glacial Lake Ashland

Lake Ashland shore features occur at a present elevation of 1125 feet, and the lake covered an area of several townships in Ashland County, Wisconsin, and extended westward in Bayfield County to the eastern slope of the Bayfield Peninsula (Fig. 65A). It was bordered on the north by glacial ice, and it discharged westward across the peninsula to Glacial Lake Brule. Lake Ashland was drained down to, and merged with, Lake Duluth (Fig. 66A) when the ice front retreated from the Bayfield Peninsula (Leverett, 1929a, p. 56).

Glacial Lake Ontonagon

This lake occupied much of the Ontonagon River drainage basin, in parts of Ontonagon, Gogebic, and Houghton counties, Michigan, while the ice front stood along the copper range in northeastern Ontonagon County. The shorelines of the lake are at a present elevation of 1320 feet above sea level at the west end, and they rise to 1340 feet at the east end, the greater elevation there being due to postglacial uplift of the land. Lake Ontonagon received drainage from the east as well as from its ice border on the north, and it discharged westward overland to Glacial Lake Ashland (Figs. 65A and 66A). It was succeeded by Lake Duluth, at a lower elevation (Fig. 67A), when the ice barrier receded (Leverett, 1929a, p. 57).

LAKE DULUTH

Glacial Lake Duluth came into existence when recession of the ice front in the area between the western end of Lake Superior and the Brule River valley allowed Glacial Lake Nemadji to join with Glacial Lake Brule (Fig. 65A). Further recession of the ice drained Glacial Lakes Ashland (Fig. 66A) and Ontonagon (Fig. 67A) and added the lower portions of their basins to Lake Duluth (Leverett, 1929a, p. 57). Because this lake continued to grow in size with continued retreat of the ice front, it became a major stage of the Superior basin.

The outlet of Lake Duluth was the old outlet of Lake Brule,

into the St. Croix River valley. Leverett (1929a) does not explain why the Lake Nemadji outlet, which stands 1070 feet above sea level, was abandoned, and why Lake Duluth drained through the old Lake Brule outlet, if Lake Brule stood at the stage marked by its 1125-foot beaches at the time it was joined with Lake Nemadji. The explanation must be that the Brule level was lower at the time and that upwarp of the land has since raised the Brule–St. Croix outlet area.

Beaches in the western part of the Superior basin near the outlet area, which are assigned to Lake Duluth stages by Leverett (1929a, p. 58), include one at a present elevation of 1100 feet above sea level which is "rather indefinite," and three others, which are strongly developed, at present elevations of 1076, 1044, and 1022 feet above sea level. These are shown in Fig. 39.

The floor of the now-abandoned outlet channel is at an elevation of 1022 feet, but it is underlain by five feet of muck and peat which apparently have accumulated since the outlet was last occupied by discharge from the Superior basin. The final bottom surface of the actual outlet channel thus stands at 1017 feet. It is apparent that the lake stage marked by beaches at the 1022-foot level is the last one that could have drained through the St. Croix outlet, and that all lower stages in the Superior basin must have drained elsewhere, to the east.

The St. Croix outlet apparently was cut down and then stabilized three or four times, in order to produce the stages of Lake Duluth which are recorded by the beaches mentioned in the foregoing paragraph.

The Lake Duluth beaches of the outlet area have been correlated with other beaches occurring at intervals along both the south and the northwestern sides of the Superior basin, as is described in detail by Leverett (1929a, pp. 59–61). These beaches rise to the eastward and northward, with the line of maximum tilting bearing about N. 20° E. Along this line the lowest Lake Duluth beach rises about 25 inches per mile. The full series of Lake Duluth features apparently does not extend eastward on the south shore beyond the Bayfield Peninsula, nor does it extend as far as the east edge of Lake County, Minnesota, on the north shore (Leverett, 1929a, p. 61). The upper beach was abandoned, apparently, because of downcutting of the outlet, by the time the ice front had retreated to the north end of the Bayfield Peninsula. It is suggested here that the outlet was cut down at this time because of the drainage of Lake Ashland down to the Duluth level. This would have resulted in an increased volume of discharge at the outlet.

The drainage of Lake Ontonagon, a little later, may be the cause of another one of the periods of downcutting at the St. Croix outlet. These episodes of downcutting may, however, have been unrelated to drainage of the early marginal lakes and may have been caused, instead, by periods of rapid melting of the glacial ice which swelled the discharge through the outlet. One is tempted to correlate these early post-Valders maximum episodes of ice retreat and of downcutting of the St. Croix outlet with similar episodes of downcutting of the Chicago outlet, which produced the Calumet and Toleston stages in the Michigan basin.

Because the Lake Duluth beaches are not horizontal in any part of the basin, but have a distinct southward slope even in the vicinity of the outlet, it is obvious that the hinge line must lie to the south of the outlet. The amount of uplift at the outlet has been calculated by Leverett (1929a, pp. 62–63) as follows: assuming that a beach which occurs at a present elevation of 850 feet near the outlet is the Algonquin beach, and knowing that the Algonquin stage in the Huron and Michigan basins occurred at an elevation of 605 feet, this 850-foot beach must have been raised 245 feet. The outlet area, therefore, has been raised at least 245 feet. Leverett gave no reason, however, for the Algonquin correlation of the 850-foot beach, and the present author believes that the main or highest Algonquin stage never existed in the Superior basin. The validity of the calculated amount of uplift of the outlet is, therefore, open to question. The 850-foot beach near the outlet of Lake Duluth may correspond to a lower Algonquin level in the Huron basin, perhaps to one which occurred at an elevation of 565 feet. The Lake Duluth stages are plotted in Fig. 39 at the present elevations of their beaches. It must be understood that they actually occurred at lower levels, the exact elevations of which are unknown.

On the Keweenaw Peninsula, beaches identified by Leverett as lower members of the Duluth group have a rate of tilt of about 2.7 feet per mile, and the trend of the tilt line seems to be about N. 30° E. (Leverett, 1929a, p. 63). Identification of the beaches of any particular stage of the lake in this area must be considered very tentative, however, because Leverett (1929a) has noted that "in exposed situations, as on the northwest slope of the Keweenaw Peninsula, there is a beach for about every 20-foot interval, and in places the interval is but 10 or 15 feet" (p. 60).

The lower Lake Duluth beaches have been mapped in detail in Cook County, Minnesota, on the northwestern shore of the lake, by Sharp (1953). The lowest of this group recognized by Sharp occurs at an elevation of 1255 feet near Grand Marais, Minnesota, and this presumably is the equivalent of the lowest Duluth beach, occurring at an elevation of 1022 feet near the outlet.

TRANSITION FROM LAKE DULUTH TO A POST-ALGONQUIN STAGE

During the episodes of Lake Duluth recorded by the group of beaches which can be correlated with discharge through the St. Croix outlet, the glacial ice front was retreating northeastward in the basin. When the ice margin retreated far enough down the slope of the Huron Mountains, between Keweenaw Bay and Marquette, Michigan, water began to discharge southeastward toward the Michigan basin (Fig. 71). The details of this part of the history have not been deciphered, but Leverett (1929a) states, "Lower and lower passages for border drainage were opened from time to time as the ice melted back. But we may infer from the strength of the beaches formed in the western part of the Lake Superior basin at levels between the Lake Duluth and Lake Algonquin shores and controlled by the level of the heads of border drainage channels that the ice margin may have held a given position for considerable time" (p. 63). A "sub-Duluth" beach, at elevations ranging from 1160 to 1205 feet (going northeastward) in Cook County, Minnesota, has been correlated with this period of discharge along the slope of the Huron Mountains by Sharp (1953, pp. 117–18).

Further retreat of the ice later opened a dischargeway leading directly from the Superior basin to the head of Little Bay de Noc in Green Bay (Michigan basin), but by this time the main Algonquin stage of the Michigan and Huron basins had come to a close. Further discussion of events in the Superior basin is deferred to Chapter 13.

chapter eleven

LAKE STAGES IN THE ONTARIO BASIN

STAGES DESCRIBED BY FAIRCHILD

The lake history of the Ontario basin as described by Fairchild (1909) was worked out before several events of the late Wisconsin glacial and lake history in the western basins were known. Some reinterpretation of Fairchild's history is necessary if the events he described are to be placed in their most probably correct relationship to the events which occurred elsewhere in the Great Lakes region. The summary given in Table 13 is modified from Fairchild (1909). It has been designed to include all of the known stages represented by shoreline features and associated outlet channels.

All of the early stages, before the Iroquois, were rather narrow ice-margin lakes. One "episode of free eastward drainage and no lakes" is noted, and this may be assumed to be correlative with one of the late Wisconsin interstadial low-water events of the other lake basins. There is, in addition, the possibility that another low stage occurred within the period of time represented by the events of Table 13 and that it left no record in the Ontario basin. The recorded "episode of free eastward drainage and no lakes" is here correlated with the Cary–Port Huron interval, simply because the number of lake stages in the list can be distributed most conveniently in relationship to events of the other lake basins on the basis of that assumption.

LATE CARY EVENTS

The Valley Heads moraine, which forms the divide at the south ends of several of the Finger Lakes,[1] has been identified as late Cary in age by MacClintock and Apfel (1944), and Holmes (1952) apparently accepts this identification. Figure 40 accordingly indicates that the Ontario basin was filled by glacial ice in Cary time.

A POSSIBLE LOW-WATER STAGE DURING THE CARY–PORT HURON INTERVAL

The Cary–Port Huron interval, as described in connection with the early lake stages in the Erie and Huron basins, was a time when retreat of the ice from the southern edge of the Ontario basin apparently allowed the lakes in the western basins to drain eastward. This may have left no record in the Ontario basin, or it may be represented by Fairchild's "episode of free eastward drainage and no lakes." The latter possibility is assumed to be the correct one, in the present discussion (Fig. 59).[2] If this is correct, Fairchild's Newberry, Hall, and Vanuxem I stages probably occurred soon after the retreat of the ice front from the Valley Heads moraine and during the Lake Border episodes, and were coexistent with early Arkona stages (Fig. 58).

CARY–PORT HURON–INTERVAL EVENTS

The extent of retreat of the ice front during the Cary–Port Huron interval is unknown, except that it apparently must have been great enough to permit drainage of the lake in the Erie basin down to a level below the Lowest Arkona stage (Chap. 8). As stated in a foregoing paragraph, this is here assumed to have occurred between the Vanuxem I and Vanuxem II stages (Table 13), but an alternative time of occurrence, sometime before the Vanuxem I stage, cannot be certainly ruled out.

---

[1] Canandaigua, Keuka, Seneca, Cayuga, Owasco, and Skaneateles lakes.
[2] Figures 53–75, lake stage maps, follow Chapter 16.

Table 13. *Lake stages of the Ontario basin.*[a]

Lake stage	Present altitude, vicinity of outlet (feet)	Outlet	Correlative stage in Erie basin after Fairchild	Correlative stage in Erie basin present revision (Hough)
Separate ice-margin lakes	Various	Various; southward to Susquehanna River		
Newberry (ice at Batavia blocking westward drainage)	1000	Horseheads; southward to Susquehanna River	Whittlesey	Late Maumee
Hall	1000–900	Westward to Erie basin	Warren	Early Arkona stages
Vanuxem I	900	Syracuse; eastward to Mohawk and Hudson rivers		Late Arkona stages
Episode of free eastward drainage and no lakes				Cary–Port Huron interval
Vanuxem II and possibly Hall II	900 1000–900	Syracuse; eastward to Mohawk and Hudson rivers		Whittlesey and early Warren
				Two Creeks–interval low stage, and Wayne

			Warren	Late Warren
Warren	880	Westward to Erie basin		
Hyper-Iroquois, Lake Dana	700	Marcellus-Cedarvale; eastward to Mohawk and Hudson rivers		Grassmere Lundy Early Algonquin
Hyper-Iroquois, Lake Dawson in west	480	Fairport-Palmyra; eastward to early Iroquois		
Early Iroquois in east	440	Rome; eastward to Mohawk and Hudson rivers		Early Lake Erie
Iroquois	440	Rome; eastward to Mohawk and Hudson rivers		
Frontenac		Covey Hill ice-margin channel		
St. Lawrence marine embayment		St. Lawrence valley		
Ontario	246	St. Lawrence River	Lake Erie	Lake Erie

[a] Modified from Fairchild (1909, pp. 54–59, pls. 34–42).

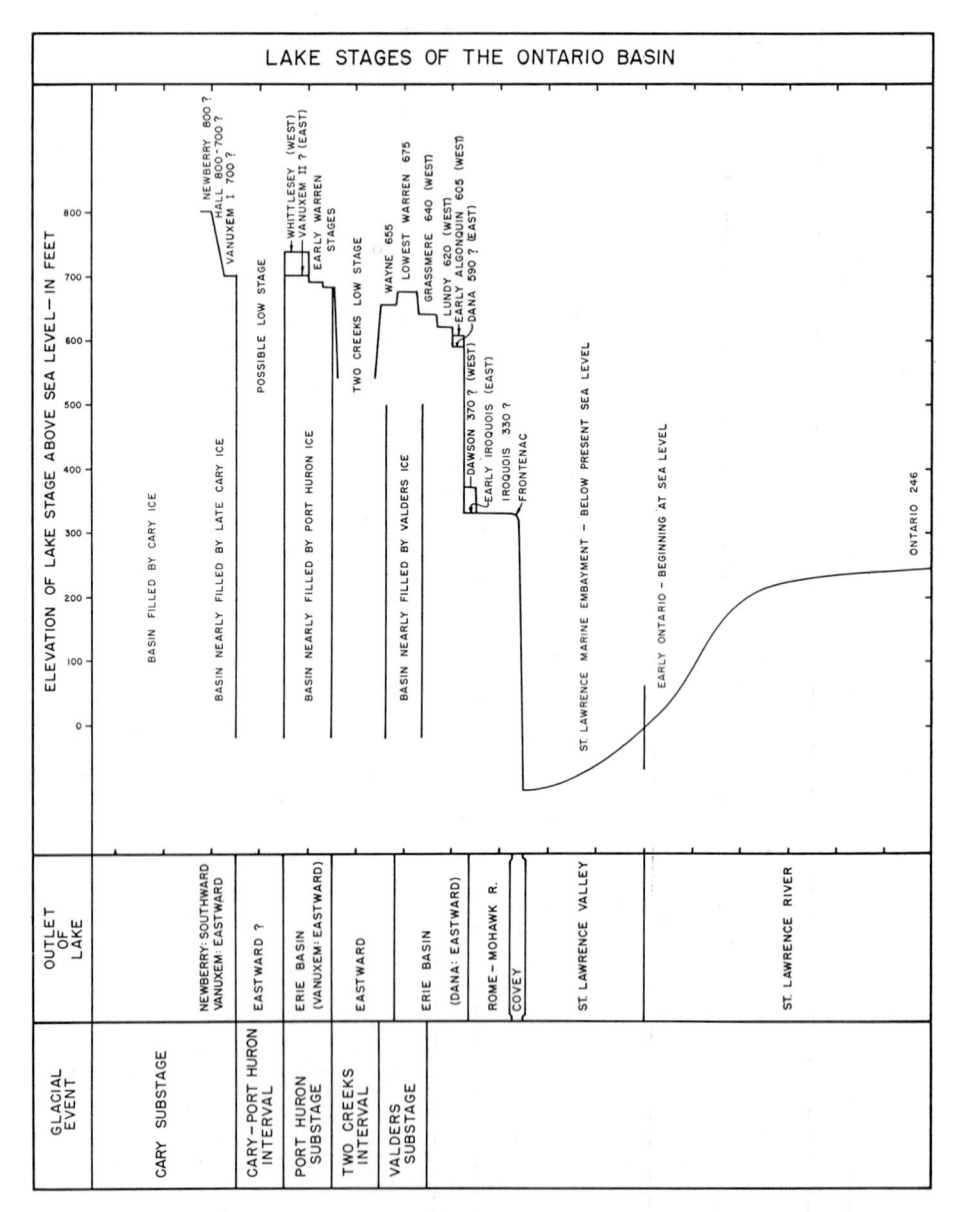

Fig. 40. Lake stages of the Ontario basin. Lake levels are plotted against an unspecified time scale in the upper part of the figure, and lake outlets and correlative glacial events are shown in the lower part.

PORT HURON–SUBSTAGE EVENTS

When the lake stages of the Ontario basin were mapped by Fairchild (1909), the Port Huron (Mankato) and Valders glacial substages had not been recognized. At the present time the exact correlatives of the Port Huron and Valders moraines in the Ontario basin are not known. It is assumed here, meanwhile, that the Port Huron advance coincided with (and produced) Fairchild's Vanuxem II stage (Fig. 60). Fairchild indicated that in addition to the Vanuxem II, a second Hall stage possibly existed after his "episode of free eastern drainage and no lakes."

There is a possibility that a considerable amount of upwarp of the land occurred during the Cary–Port Huron interval. The Vanuxem I–stage outlet, the Syracuse channels, if used in late Cary time, may have been uplifted during the Cary–Port Huron interval. If so, the lake stage formed during the Port Huron advance and again using the Syracuse outlet would stand at a relatively higher position; the resulting shorelines may be represented by Fairchild's second Hall stage, which is 100 feet higher than the present elevation of the Vanuxem shore features.

The Vanuxem II, and the Hall II stage if it existed, were approximately correlative with the Whittlesey and early Warren stages of the Erie basin, under the present tentative interpretation.

TWO CREEKS–INTERVAL EVENTS

If Fairchild's "episode of free eastward drainage and no lakes" is assigned to the Cary–Port Huron interval, another low stage is required in the Two Creeks interval (Fig. 62). This is arbitrarily assumed to have occurred after the Vanuxem II stage (and after the Hall II stage, if it existed), and before the Warren stage of Fairchild (Table 13). The Wayne stage of the Erie basin, which is correlated with the Two Creeks interval, also is assumed to have occurred at this stage of the events in the Ontario basin (Fig. 63).

VALDERS-SUBSTAGE EVENTS

The Ontario basin Warren stage (Table 13, first column) was correlated with a late Warren stage of the Erie basin by Fairchild (as indicated in Table 13, fourth column). This correlation is retained in the present interpretation of lake history, as is shown in the fifth column of Table 13 and in Fig. 64. At this time the ice front blocked all possible eastern outlets and the ice-margin lake along the southern edge of the Ontario basin discharged freely to the Erie basin (Fairchild, 1909, pl. 39).

The Warren beaches stand at a present elevation of 880 feet above sea level on the Batavia-Syracuse parallel (Fairchild, 1909, p. 54). The Lowest (latest) Warren beach in the western part of the Erie basin, where it has not been warped, is at an elevation of 675 feet above sea level. If the Warren beaches in the two areas are correlative, there has been an upwarp of 205 feet along the Batavia-Syracuse parallel since Warren time. This appears to be a reasonable amount, in view of the information that more than 100 feet of uplift has occurred in that area since the time of Lake Iroquois (refer to a subsequent discussion of Lake Iroquois).

TRANSITION FROM LOWEST LAKE WARREN TO LAKE IROQUOIS

Early in the retreat of the ice front from its Valders maximum position, an outlet to the east was opened through the Marcellus-Cedarvale channel southwest of Syracuse, New York. This permitted the water along the ice margin in the southern part of the Ontario basin to drain down to the "Hyper-Iroquois, Lake Dana" level (Fairchild, 1909, p. 53). The shore features of this stage have a present elevation of 700 feet above sea level. The amount of uplift which has occurred in the area since late Warren time apparently is approximately 200 feet (as deduced in a foregoing paragraph), and the amount of uplift since Iroquois time is approximately 100 feet (as deduced from the warping of the Iroquois beach, described in a subsequent paragraph). The Dana level therefore originally stood somewhere between 500 and 600 feet above sea level. This was below the level of the Lundy stage in the Erie and Huron basins (620 feet), and the suggestion made by Leverett and Taylor (1915, p. 399) that the Lundy and Dana stages were correlative (as represented in Fig. 66B) cannot be accepted. It appears likely that the lakes in the Erie and Huron basins had passed through the Grassmere and Lundy stages before the Dana stage occurred in the Ontario basin.

With further retreat of the ice margin, discharge at the eastern end of the Ontario basin was shifted to the Rome outlet, and the earliest stage of Lake Iroquois came into existence as a small body of water in the southeastern part of the basin. For a time the ice remained as a barrier in the central part of the southern margin of the Ontario basin and ponded water from there westward as the "Hyper-Iroquois, Lake Dawson" stage (Fairchild, 1909, p. 58). This discharged eastward through the Fairport-Palmyra channel to early Lake Iroquois. The Lake Dawson outlet channel is at a present elevation of 480 feet above sea level, and allowing for at least 100 feet of uplift in the area its original elevation must have been no higher than 380 feet. It was obviously much lower than the Lundy or Early Algonquin stages of the Erie basin; it must, therefore, have occurred during the lowering of the level in the Erie basin from the Early Algonquin to an early Lake Erie level.

EARLY LAKE IROQUOIS

When the ice front in the Ontario basin retreated far enough to permit the lakes along the southern margin to come to a single level, while discharging eastward through the Rome outlet, the main stage of Lake Iroquois came into existence (Fig. 68). The present elevation of the Rome outlet is 440 feet above sea level, but its elevation in early Iroquois time was approximately 330 feet (this is derived in a subsequent paragraph). The discharge through the Rome outlet went down the Mohawk and Hudson valleys to the Atlantic Ocean.

Lake Iroquois expanded as the glacial ice wasted away and ultimately it spread over the entire basin (Fig. 69). This lake has left a clear record in shore features, which have been mapped and described by several writers. The broad outlines of Lake Iroquois were described by Spencer (1890, p. 446), who drew on the work of G. K. Gilbert for the part of the shore in New York. Spencer suggested the name "Iroquois beach," and showed that the shore was not horizontal but was tilted up toward the northeast. In 1907 Spencer published a paper on "The Falls of Niagara" which gave his final views on the tilting of the beach and references to the literature up to that time. Fairchild published the first large-scale map of the beach in 1900, but the Ontario portion of it was enlarged from the small outline given by Spencer. Coleman produced a somewhat more detailed map and description of the Iroquois beach in 1904 and then, more than 30 years later, when modern maps had become available, he published an extremely detailed map and account of Lake Iroquois (Coleman, 1936).

The precise original elevation of the Iroquois beach is not known because the beach is not horizontal in any part of the Ontario basin. It descends from an elevation of 1030 feet at its most northeasterly point down to an elevation of 362 feet above sea level at the southwestern corner of the basin. A curve constructed from the known elevations at various localities shows that the amount of uplift which has occurred decreases southward (actually, the line of maximum tilt bears N. 20° E.). By projection of the curve, the points of no uplift may be inferred to exist along a line which passes through the northeastern part of the Erie basin a few miles southwest of Buffalo, New York. By inspection of this projected curve, it is judged that the original elevation of Lake Iroquois was approximately 330 feet above sea level. A part of the uplift which deformed the Iroquois beach apparently occurred while the lake was in existence. As the outlet at Rome was raised, the lake level rose and water flooded beyond the original Iroquois shores at all points south of an isobase passing through the Rome outlet; north of this isobase the land was raised above lake level and the shores receded somewhat.

LAKE FRONTENAC AND THE TERMINATION OF LAKE IROQUOIS

The Lake Iroquois beach has been traced northeastward to Covey Hill, at the northern end of the Adirondack Mountains about one mile north of the international boundary, in Ontario. North of Covey Hill the land slopes steeply down to the broad, low valley of the St. Lawrence River. When the edge of the ice sheet retreated northward down this slope, Lake Iroquois was drained (Mather, 1917, p. 542). The name "Lake Frontenac" has been applied to the last episode of Lake Iroquois history, when the lake drained through an ice-margin channel at Covey Hill (Leverett and Taylor, 1915, p. 445). This episode was brief, and no separate shorelines are referable to the event.

ST. LAWRENCE MARINE EMBAYMENT

When the ice margin retreated from Covey Hill, the water in the Ontario basin was drained down to the level of a marine embayment which apparently was already in existence in the St. Lawrence valley. The Drummondville moraine, which lies in the St. Lawrence valley south of the St. Lawrence River in the area between Montreal and Quebec, has been correlated tentatively with the Valders substage. The till of the ground moraine north of the Drummondville terminal moraine is overlain by marine sediments, and the absence of weathering and leaching of the till indicates that there was no long interval between recession of the ice and the marine invasion (Gadd, 1955, pp. 171–73).

The marine waters of the St. Lawrence valley became confluent with the waters of the Ontario basin when Lake Iroquois was drained. The extension of the marine embayment into the Ontario basin has been referred to as the Gilbert Gulf (Fairchild, 1907, p. 112, and Leverett and Taylor, 1915, p. 445), but this name was not continued in use (Mather [1917, p. 542] refers to it as the "Champlain Sea"). The entire embayment in the St. Lawrence valley has been referred to as the "Champlain Sea." Chapman (1937, pp. 113–15) has shown that the marine invasion of the Champlain valley was a brief event of limited extent. The present author therefore considers the name "Champlain" inappropriate as a designation for an extensive body of water which occupied the St. Lawrence valley for a long time, and is therefore using the terms "St. Lawrence marine embayment" and "St. Lawrence Sea" for this body of water.

When the St. Lawrence marine embayment extended into the Ontario basin, the level of the sea was below present sea level because there were still extensive ice sheets on the land, withholding a considerable volume of water from the oceans of the world. Data presented by Shepard and Suess (1956) indicate that sea level was about 100 feet below the present level in Valders time. The present author estimates that Lake Iroquois was drained about 8000 years ago, and the Shepard-Suess curve indicates that sea level then stood about 55 feet below its present level. The floor of the St. Lawrence valley and the northeastern rim of the Ontario basin were still lower. Mather (1917, p. 543) has concluded that rise of sea level (due to melting of glacial ice) went on at a faster rate than the land was uplifted in this region, and that the shore of the marine embayment therefore encroached on the land for a considerable period of time.

The highest shoreline of the marine embayment in the Ontario basin occurs at a point 11 miles north of Kingston, Ontario, at a present elevation of about 505 feet above sea level (Mather, 1917, p. 545). The uplift of this locality since the shoreline was formed is, therefore, 505 feet plus whatever rise has occurred in sea level between the time of the marine embayment and the present: a total uplift of perhaps 560 feet. A preglacial stream in Nepanee valley, 20 miles west of Kingston, Ontario, discharged into the marine embayment of the Ontario basin at a point which is now 325 feet above sea level (Mather, 1917, p. 548).

The St. Lawrence marine embayment also extended up the Ottawa River valley, probably beginning the invasion sometime after it was joined with the waters of the Ontario basin (Fig. 71). Near Ottawa the upper limit of marine submergence has been determined to be 690 feet above present sea level (Johnston, 1916b, p. 5).

EARLY LAKE ONTARIO

The St. Lawrence marine embayment stage of the Ontario basin was brought to a close when upwarp of the land raised the northeastern rim of the basin to sea level. It is not known what the elevation of sea level was at the time this occurred. Sea level had been rising, because of return of water to the oceans by the wasting of continental ice sheets, and the rate of rise of sea level probably was diminishing before the rate of uplift of the land was at its maximum.

The subsequent history of Lake Ontario is one of continued upwarp of the area, with the sill at the northeastern end of the basin rising faster than the remainder of the basin and causing the lake to rise progressively higher on its shores.

CORRELATION OF EVENTS IN THE ONTARIO BASIN WITH ALGONQUIN STAGES IN THE HURON BASIN

It has been inferred that glacial ice still stood high enough on the southern edge of the Ontario basin to serve as a barrier to the lake waters when the Early Algonquin stage of the Michigan, Huron, and Erie basins came into existence and the three basins were all discharging through the Chicago outlet (Chap. 8). Soon after this, however, the ice in the Ontario basin retreated far enough to allow the Erie basin to drain down to an early Lake Erie stage and then Early Lake Algonquin of the Michigan and Huron basins discharged through both the Chicago and Port Huron outlets. After pausing at the Dana and Dawson stages, the water in the Ontario basin soon came down to the Iroquois stage.

The discharge of Lake Iroquois continued to pass through the Rome outlet until long after the Ontario basin was free of glacial ice. During the expansion of Lake Iroquois to fill the entire basin there were several changes in the routes of discharge from the upper lakes to the Ontario basin. The early draining of the Erie basin, and the discharge of the waters of Early Lake Algonquin from the Huron to the Erie basin and from there to the Ontario basin via the Niagara River have been mentioned. Later, during its Kirkfield stage, the Huron basin discharged through the Kirkfield outlet and the Trent valley, across southern Ontario directly to Lake Iroquois (Fig. 69). This discharge built a large delta at the Iroquois level at Peterborough, Ontario (Coleman, 1936, p. 24). This delta, and the absence of other deltas above it, show that Lake Iroquois was in existence at the time of the first discharge from Lake Algonquin down the Trent valley (Leverett and Taylor, 1915, p. 444). Evidence in the Huron basin (Chap. 12) indicates that the Kirkfield outlet discharge was discontinued, probably for a long period of time, while the main Algonquin stage was in existence (Fig. 70), and that discharge through the Kirkfield outlet was resumed at the close of the Algonquin stage.

The channel of the Trent valley glacial dischargeway continues beyond the Peterborough Iroquois-stage delta, down to the present shore of Lake Ontario, and it probably extends an appreciable distance below present lake level (Leverett and Taylor, 1915, pp. 444-45, and Mather, 1917, p. 549). This portion of the channel, from Rice Lake down, apparently was formed by the later discharge through the Kirkfield outlet, at the close of Algonquin time, and it must have coincided with the St. Lawrence marine embayment stage of the Ontario basin (Fig. 71).

chapter twelve

LAKE ALGONQUIN

INTRODUCTION

The name "Algonquin" has been applied to a number of events which make up a rather long and complicated part of the Great Lakes history. The central event, the main Algonquin stage (Fig. 70),[1] is well represented by extensive beaches occurring at many places throughout the Huron and Michigan basins. These have been described in detail by Leverett and Taylor (1915, pp. 414–33). In the unwarped southern portions of the Huron and Michigan basins the Algonquin beaches occur at an average elevation of 605 feet above sea level. North of a hinge line extending through Grand Bend, Ontario (on the southeastern shore of Lake Huron), and Frankfort, Michigan (on the eastern shore of Lake Michigan), the beach rises to the north (Fig. 42). It generally is the highest beach feature to be found in the warped area in the Huron and Michigan basins, and it reaches a present elevation of 1015 feet at a point on the Root River six miles north of Sault Ste. Marie, Ontario.

Glacial ice formed some part of the shores of the lakes during all of the principal Algonquin stages, and the fluctuation of the ice front is an important part of the Algonquin history.

The lakes of the Huron and Michigan basins have stood at the 605-foot level three times, with intervening low-water stages. The discharge of the lakes has shifted between various outlets, and the lake levels have been changed, in part because of fluctuation of

---

[1] Figures 53–75, lake stage maps, follow Chapter 16.

the ice front and in part because of upwarp of parts of the region. Lake level has been lowered by downcutting of an outlet.

The Algonquin history generally is understood to include the following:

1. An Early Algonquin stage (designated Algonquin I by Flint [1947, p. 260]) which existed at an elevation of 605 feet in the southern part of the Huron basin. Contemporaneous with this was the Toleston stage of Lake Chicago, which existed in the Michigan basin at an elevation of 605 feet. Recession of the glacial ice margin allowed a connection between the waters of the Huron and Michigan basins to widen into a broad strait, and both the Early Algonquin stage of the Huron basin and the Toleston stage of the Michigan basin may then be referred to as "Early Lake Algonquin."

2. The Kirkfield stage (Algonquin II [Flint, 1947, p. 260]), which was caused by diversion of discharge when retreat of the ice front uncovered the Kirkfield outlet to a route down the Trent valley leading to the Ontario basin. Water level stood approximately 40 feet below the main Algonquin level at one time during the Kirkfield stage, and it may have been 50 or 60 feet below the main Algonquin level during another part of the Kirkfield. This stage ended when the Kirkfield outlet was rendered inoperative. Whether this occurred by readvance of the ice or uplift of the land is a question which will be discussed in this chapter.

3. The main Algonquin stage (Algonquin III [Flint, 1947, p. 260]), which began when water level was returned from the Kirkfield level to the original Algonquin level, 605 feet. This stage existed in both the Huron and Michigan basins, which were connected, and it discharged through both the Port Huron and the Chicago outlets. At first, the lake was bounded on the north and northeast by glacial ice which probably stood well out into the Michigan and Huron basins. As the ice front receded, the lake expanded until it filled the Michigan and Huron basins. According to Leverett and Taylor (1915, pl. 21), it also extended throughout the Superior basin. This chapter cites evidence indicating that glacial ice still occupied the eastern part of the Superior basin during Algonquin time, excluding the lake from that area.

Water level was lowered from the main Algonquin level by the opening of a lower outlet somewhere to the east.

4. *Later lake stages at lower elevations.* Leverett and Taylor (1915) considered that several later stands of the lake should be included in the Algonquin group, because they believed that several of the sub-Algonquin beaches in the north converged with the main Algonquin beach to the south. Stanley (1937, p. 1680) showed that four distinct beaches at levels well below the upwarped Algonquin beach were, in fact, made at lower elevations and undoubtedly discharged through new outlets. Flint (1947, p. 261), while recognizing Stanley's work, still classed these later events as Lake Algonquin IV. The present author proposes that the portion of lake history designated as Algonquin should be terminated with the Algonquin III stage. The lowest levels reached, named Lake Chippewa in the Michigan basin and Lake Stanley in the Huron basin (Hough, 1955, p. 965), are not considered part of the Algonquin sequence.

The main Algonquin beach has been elevated 410 feet near Sault Ste. Marie, Ontario, in the northern part of the Huron basin. The greater part of this uplift occurred after the low-water stage. The uplift ultimately raised the low-stage outlet, at North Bay, Ontario, to the level of the old Algonquin outlets at Port Huron and Chicago. Thus the Nipissing stage was initiated.

For an adequate understanding of the Algonquin stage it is necessary to include a brief discussion of the Nipissing. This is not considered part of the Algonquin in any of the references cited, but the author (Hough, 1953c, pp. 137-38) has shown that it stood at the Algonquin level, 605 feet, and therefore occupied the same beaches in the unwarped southern parts of the lake basins.

EARLY LAKE ALGONQUIN

The manner in which Early Lake Algonquin came into existence has been discussed in detail in Chapter 8. A brief summary of the author's concept of the origin of the Early Algonquin stage is given in the following paragraph.

The Erie basin was discharging northward to the Huron basin, and the Huron basin was discharging northwestward to the Michigan basin, possibly during both of two stages which preceded the Early Algonquin and which occurred at elevations of 640 and 620 feet (Figs. 65A and 66A). When a further drop in lake level in the Michigan basin occurred, due to downcutting of the Chicago outlet, the Toleston stage of Lake Chicago came into existence at an elevation of 605 feet. This allowed the water in the Huron basin to be lowered to the same level, and this in turn allowed the northward-discharging Erie basin to drain down to the 605-foot level also (Fig. 67A). During this lowering of the Erie basin level a channel was incised in the Port Huron moraine at Port Huron, Michigan, and this was cut only deeply enough to permit the adjustment to a common level. This manner of formation of the Erie-Huron connecting channel is of importance because it explains the delicate adjustment in elevation between the Port Huron and the Chicago outlets.

The Early Algonquin stage in the Erie basin, discharging northward to the Huron basin, was brief. Aside from the remnants of a probable Algonquin surface extending from St. Clair 18 miles southward, no trace of a stand at the Algonquin level in the Erie basin is known. It is concluded, therefore, that soon after the Port Huron channel was cut to carry the Erie basin discharge northward, a retreat of the ice front in the Ontario basin opened a lower outlet there. When water level in the Erie basin was lowered, the direction of flow through the Port Huron outlet was reversed (Fig. 68). The lakes in the Huron and Michigan basins remained at an elevation of 605 feet above sea level, and they therefore continued to discharge through the original outlet at Chicago as well as through the Port Huron outlet. The Chicago outlet was floored in part by bedrock and could not be cut lower by scour of the outflowing water. The Port Huron outlet was floored by glacial deposits which were susceptible to scour, but it was not cut lower until a much later time, after the Nipissing stage. The reason for the stability of the Port Huron outlet during the Early Algonquin, main Algonquin, and Nipissing stages apparently is that the original channel was cut by the northward-flowing discharge from the Erie basin and thus had a much smaller size and possibly a slightly higher bed elevation than that of the

Chicago outlet, which was cut to the Toleston level by the discharge from the Michigan, Huron, and Erie basins. The Early Algonquin stage persisted in the Huron and Michigan basins until retreat of the ice front had permitted Lake Iroquois to expand over a considerable part of the Ontario basin, and until it uncovered an outlet east of the southeastern corner of Georgian Bay.

THE KIRKFIELD STAGE

The concept of the Kirkfield low-water stage, as presented by Leverett and Taylor (1915, pp. 409–13), was based mainly on the apparent necessity of incorporating the Kirkfield outlet channel into the Algonquin history. Leverett and Taylor gave no proof of the existence of the Kirkfield stage, and stated that no beaches of that stage had been observed.

An unconformity in the sandy lake deposits at Jackson Point in the Simcoe basin, Ontario, has been interpreted as the product of an early low stage of Lake Algonquin at least 50 feet below the main Algonquin level. The sands just below the unconformity are barren, while the sands above it contain shells identical with those in the highest Algonquin beach deposits nearby (Johnston, 1916a, pp. 15–17). Evidence for a pre–main Algonquin low-level stage also was published by Stanley (1938a), and the Kirkfield stage is now a credible event (Fig. 69). Stanley's section at Sucker Creek, Grey County, Ontario, shows fragmentary spit beaches of a low lake stage which are impounded behind a strong Algonquin bar which truncates and partly covers them. The crests of the impounded beaches are 36 feet below the Algonquin crest, suggesting a stage 36 feet below the Algonquin. The bottoms of the same impounded beach deposits are 60 feet below the Algonquin crest, and this may indicate that water surface in the Huron basin had been as much as 60 feet below the Algonquin level.

The Kirkfield stage discharged through the Trent valley of Ontario, which extends from the vicinity of Kirkfield (12 miles northeast of Lake Simcoe and 35 miles east of the easternmost point of Georgian Bay) to Trenton, on the Bay of Quinte, in the northeastern part of the Lake Ontario basin. The actual sill which determined the elevation of the Kirkfield stage was at Fenelon

Falls, 12 miles east of Kirkfield. The name "Kirkfield" is used as a designation for the outlet, however, because the lake narrowed in the vicinity of Kirkfield to pass into the outlet channel. Detailed descriptions of the Trent valley dischargeway have been given by Johnston (1916a, pp. 1–22), by Deane (1950, pp. 87–90), and by Chapman and Putnam (1951, pp. 120–22).

While the Trent valley outlet channel is recognizable as far east as Trenton, it is apparent that the Kirkfield-stage discharge reached the shore of Lake Iroquois at a point much farther west. This was at Peterborough, Ontario, where an extensive delta was built. The more easterly portion of the channel apparently was submerged at this time, and was not used as a dischargeway until after the main Algonquin stage, when Lake Iroquois had been drained.

During the Kirkfield stage the water surface in the Michigan basin probably also was lowered to the same level as that in the Huron basin, and both the Chicago and Port Huron outlets were temporarily abandoned. Some of the trenching and filling below the "Toleston" (Algonquin) level in the Allendale delta, Grand River (Bretz, 1953, p. 380), probably occurred at this time. An event named "pre-Algonquin lake" by Spurr and Zumberge (1956, pp. 100–101) is recorded in Cheboygan and Emmet counties, Michigan, north and east of Little Traverse Bay. The shore features of this lake lie 18 to 25 feet below the level of the highest Algonquin stage and they are distinctly older because they are covered in places by wave-worked sands of the higher stage. This "pre-Algonquin lake" apparently is a phase of the Kirkfield stage.

THE KIRKFIELD–MAIN ALGONQUIN TRANSITION

It is generally agreed that the Kirkfield stage came into existence as a result of retreat of the glacial ice front to uncover the Kirkfield outlet. The cause of the rise of water surface back to the 605-foot level, to initiate the main Algonquin stage, has not been established unequivocably.

Two possible causes of the Kirkfield–main Algonquin transition are considered: (1) uplift of the Kirkfield outlet, and (2) readvance of glacial ice to close the Kirkfield outlet. Taylor favored the first, uplift; Johnston (1916a) favored the second, readvance of ice. The unconformity in the Algonquin sandy lake deposits at Jackson Point in the Simcoe basin was interpreted by Johnston as a consequence of the Kirkfield–main Algonquin relationship. Because the locality is so close to the Kirkfield outlet, he supposed that any uplifting movement which affected the outlet would affect the Jackson Point locality also and therefore that uplift alone could not have produced the required amount of submergence. Stanley (1938a), however, in his discussion of the Sucker Creek locality, considers uplift as a more probable cause.

Johnston's Jackson Point locality and Stanley's Sucker Creek locality are on approximately the same isobase, which passes only 20 miles south of the outlet. The hinge line, marking the southern limit of post-Algonquin uplift, lies many miles to the south. It must be assumed that the Sucker Creek and Jackson Point localities would be involved in any warping which affected the Kirkfield outlet area. At Sucker Creek the main Algonquin beach is a large gravel bar with a present elevation of 800 feet. This overlies the Kirkfield-stage beach deposit, whose crest is 36 feet lower, at 764 feet. At the Kirkfield outlet both the Kirkfield and the main Algonquin beaches should join, if the uplift hypothesis is correct. The main Algonquin beach at Sucker Creek has been raised from 605 feet (its original elevation) to 800 feet (its present elevation), or 195 feet, in post-Algonquin time. The main Algonquin beach at the Kirkfield outlet has been raised from 605 feet (its original elevation) to 870 feet (its present elevation), or 265 feet, in post-Algonquin time. The ratio of the observed uplift at Sucker Creek to the observed uplift at the Kirkfield outlet is 195:265, or 100:136. In other words, the Sucker Creek locality was raised 100 feet for every 136 feet of rise of the Kirkfield outlet. Because the Kirkfield beach happens to be 36 feet below the Algonquin beach at Sucker Creek, we may apply the above ratio directly and postulate, under the uplift hypothesis, that the level of the Kirkfield stage was 136 feet below the Algonquin-stage level, and that in the Kirkfield-Algonquin transition the Kirkfield outlet was raised 136 feet, to the Algonquin level, while the Kirkfield beach at Sucker Creek was raised only 100 feet, leaving it 36 feet below the Algonquin level. This amount of uplift during the

Kirkfield-Algonquin time interval appears to be unreasonably large. It is equal to half of the total uplift which occurred in the region in all of post-Algonquin time. Incidentally, Stanley has indicated that only about 19 per cent of the post-Algonquin uplift had occurred by Payette time. In view of the foregoing discussion the present author rejects uplift as the cause of the closing of the Kirkfield stage. A small amount of uplift may have occurred at this time, but apparently it was insufficient to raise the Kirkfield outlet to the main Algonquin level.

The only other cause of the Kirkfield–main Algonquin transition which appears reasonably probable is a readvance of the glacial ice front to block the outlet. Deane (1950, p. 89 and fig. 9-D) describes such a readvance and correlates it with the building of the Lake Simcoe moraine. Assuming that no appreciable uplift occurred at this time, the Kirkfield stage as represented by the tops of the impounded bars at Sucker Creek would have been only 36 feet below the main Algonquin level; the possible lower Kirkfield level represented by the bottoms of the gravel bars would have been 60 feet below the main Algonquin level. If a small amount of uplift occurred in Kirkfield–main Algonquin time, the Kirkfield levels would have been slightly lower.

The occurrence of a second period of discharge down the Trent valley, which is correlated with the extension of the Algonquin-age channel from the Iroquois delta at Peterborough, Ontario, down to the present level of Lake Ontario at Trenton, is an indication that the Kirkfield outlet was not raised appreciably at the end of Kirkfield time but was used again, at the end of main Algonquin time.

Closing of the Kirkfield outlet caused the water in the Huron and Michigan basins to rise back to the original Algonquin level, 605 feet above sea level, and thus the main Algonquin stage was initiated.

THE MAIN ALGONQUIN STAGE

Evidence for the Main Algonquin Stage

The Algonquin beach was first recognized in the southern part of the Georgian Bay region, where it is uplifted and tilted to the south. Spencer (1891b) made determinations of its elevation at various localities, but did not trace it continuously across country. He believed that in its extension southwestward it descended below the level of Lake Huron. Taylor, who correlated this beach with the highest-level beach in the warped areas of the western shore of Lake Huron, was of the same opinion until Goldthwait, in a personal communication,[2] presented evidence indicating that the highest, presumably Algonquin, beach of the northern Wisconsin shore of Lake Michigan probably had been continuous with the horizontal Toleston-level beach of the southern shore. The evidence later was published by Goldthwait (1907, pp. 101-3). Publications by various writers, as listed in the bibliography of the Leverett and Taylor monograph (1915, pp. 34-54), and especially some of the letters written by Taylor to J. W. Goldthwait in 1905-7 (personal file of R. P. Goldthwait), reveal the extreme difficulties and halting progress in development of the concept of the main Algonquin stage.

The final concept of the highest or main Algonquin beach as presented by Leverett and Taylor (1915) is as follows. It is a feature which lies at an elevation approximating 605 feet in the unwarped southern parts of the Huron and Michigan basins; north of a hinge line which extends through Grand Bend, Ontario (on the southeastern shore of Lake Huron), and Frankfort, Michigan (on the eastern shore of Lake Michigan), the beach rises to the north. It generally is the highest beach feature to be found in the warped area, and it reaches a present elevation of 1015 feet at a point on the Root River six miles north of Sault Ste. Marie, Ontario. Taylor gives detailed descriptions of the highest Algonquin beach for the many localities where it is preserved (Leverett and Taylor, 1915, pp. 414-33), and these descriptions generally may be accepted as a report of the field evidence for the Algonquin stage. However, in his summaries of the highest Algonquin stage he implies that the beach is an essentially continuous feature which is easily identifiable throughout the Huron and Michigan basins. This, actually, is not true; the highest Algonquin stage is

---

[2] Goldthwait, J. W., personal communication to F. B. Taylor, as acknowledged by letter from Taylor dated Sept. 9, 1905, now in the files of R. P. Goldthwait, Ohio State Univ., Columbus.

a concept based on the correlation of a number of separate features.

The principal difficulty in correlating the beaches of the northern, warped, areas with the horizontal beaches in the southern parts of the Huron and Michigan basins is that all beaches above the present lake shore have been removed by erosion in all of the critical stretches of shore where the tilted beaches presumably come to horizontality. This absence of the Algonquin beach in the critical areas on both the eastern and western shores of both the Huron and Michigan basins becomes apparent only by a perusal of the detailed descriptions of the beach (Leverett and Taylor, 1915, pp. 416–33). The absence of the Algonquin beach in a stretch of the southeastern shore of Lake Huron, from the supposed location of the hinge line marking the southern limit of uplift, northward for nearly 60 miles, is mentioned only in a note on Taylor's plate 25 (Leverett and Taylor, 1915).

The author believes that the Leverett and Taylor synthesis of the main Algonquin stage is essentially correct for the Huron and Michigan basins, excluding their northernmost shores (Fig. 70). This synthesis includes the interpretation that the main Algonquin-stage lake, at 605 feet above sea level, discharged through both the Chicago and the Port Huron outlets, as had its predecessor at this level, the Early Algonquin stage. A major revision of the Leverett and Taylor history is necessary, however, in regard to the northward extent of the lake.

The Maximum Extent of the Main Algonquin Stage

During the main Algonquin stage the glacial ice front retreated northward, allowing the lake to expand until it filled the Michigan basin and nearly filled the Huron basin. The position of the ice front in the Huron basin at the close of main Algonquin time is not known in detail. The lake shore lay along the edge of a crystalline-rock upland, a few miles north of Sault Ste. Marie, Ontario. The lake surrounded a hill on Manitoulin Island, three miles south of Little Current, Ontario (Greenman and Stanley, 1943, p. 524). Apparently glacial ice covered most of the north shore of the North Channel of Lake Huron, because no Algonquin-level beaches have been found there (Stanley, 1934, pp. 409–10, and Greenman and Stanley, 1943, pp. 525–26). East of Georgian Bay, the elevations of the highest beach found north of Kirkfield show that it lies in a plane below that of the Algonquin (Goldthwait, 1910, p. 31, and Chapman, 1954, fig. 2). The ice blocked all possible outlets to the east of Georgian Bay, but whether it extended into the bay is not known.

Leverett and Taylor (1915, pl. 21) show Lake Algonquin as extending throughout the Superior basin, with glacial ice having retreated entirely from all of the Great Lakes basins. The author will cite evidence indicating that this was impossible. First, however, it is of interest to examine the basis for the interpretation made by Leverett and Taylor.

The conclusion that Lake Algonquin existed in any part of the Superior basin was founded on very questionable evidence. Gravelly ridges east of Lake Superior in the area north of Michipicoten Bay, reported by Coleman (1900), occur at elevations up to 1400 feet. On McKay Mountain, south of Fort William, Ontario (on the northwest shore of Lake Superior), indications of lake action were found up to 1350 feet, and high beaches were found on the Keweenaw Peninsula (Leverett, 1929a). All of these observations were plotted on an isobase map of the Great Lakes region and they appeared to represent points on a warped Algonquin plane. That hypothetical plane passes above almost the entire area of the peninsula of Michigan which separates the Michigan and Superior basins; the boundary of Lake Algonquin was therefore drawn to include the Superior basin. With this concept in mind, the early investigators could select additional features to support it.

Taylor's best evidence for the existence of the Algonquin stage in the Superior basin is quoted:

Coleman has reported deposits which he regards as beaches at high levels in the region north of the Great Lakes. The highest is 30 miles northwest of Michipicoten Harbor and 70 miles west of Missinaibi. It is described as a "very distinct terrace, of coarse but well-rounded gravel and stones, 1,445 feet above the sea." Near the same locality is another sandy terrace at 1,380 feet. Similar deposits are at Gondreau Lake, 20 miles southwest of Missinaibi, altitude near 1,500 feet; near Meteor Lake, 1,420 feet; near Monabasing Lake, 1,335, and 1,316 feet; near Geneva, north of Cartier, 1,400 feet; and 12 miles northwest of Wahnapitae Lake, 1,047 feet.

It may be doubted whether any of these deposits are true beaches due to wave action in large bodies of water. The descriptions seem to reveal forms which were much more probably formed by glaciofluvial action or were associated with relatively small and local bodies of ponded waters. *Nevertheless, some of them may be beaches and may belong to the Algonquin group.* It is at least a very singular *coincidence* that if the plane of the highest Algonquin beach, from Hessel, Michigan to Root River, north of Sault Ste. Marie, be produced northward, it passes the locality at Gondreau Lake at about 1,470 feet.

In the writer's reconnaissance on the north coast of Lake Superior and in Lawson's more accurate work *the highest Algonquin was not certainly identified at any place* [Leverett and Taylor, 1915, p. 433].

The italics in the foregoing and following quotations are those of the present author.

The best evidence for the existence of Lake Algonquin in the Superior basin which was presented by Leverett in his "Moraines and Shore Lines of the Lake Superior Region" (1929a) is quoted in the following paragraphs:

There are so few data available on the Lake Algonquin shore lines in the Canadian part of the Lake Superior Basin that it cannot be adequately treated at this time. . . . In the northern part of the northern peninsula there are half a dozen or more distinct shore lines, all referable to Lake Algonquin.

At the top of the series are usually two or three ridges that are especially strong and continuous which are separated by intervals of but 5 to 10 feet. . . . The lowest member of this upper series *may be* the one that should be correlated with the single strong beach that leads to the Port Huron outlet and to the Chicago outlet. It *seems probable,* also, that the western part of the Lake Superior Basin had become connected with Lake Algonquin by the time this strong series was completed. On the whole, the Lake Algonquin beaches in the western part of the Lake Superior Basin are weak and widely spaced, as if they might have been formed during the time of rapid uplift.

The limits of Lake Algonquin are *less definitely known* in the uninhabited districts of northeastern Delta County and adjacent parts of Alger and Schoolcraft counties, Michigan, but *it is thought* that an aggregate area of about 200 square miles in this region may have stood above the Algonquin *level* [p. 66].

This morainic system [in the eastern part of the Upper Peninsula of Michigan] is in part above and in part below the *level* reached by the waters of Lake Algonquin. The part below that level has a much stronger morainic expression than is commonly exhibited by moraines laid down in water, such as those of the Saginaw and Erie Basins. The effect of the [supposed] lake has been remarkably slight in toning down the morainic features. The basins are only partly filled and the knolls bear only slight notches *cut by the lake waves* [?]. The moraines are bordered in places on their *south* sides by plains of sandy gravel, which by their situation as well as the character of their material appear to be outwash aprons, yet most of them are considerably lower than the highest level of Lake Algonquin [p. 49].

It appears obvious, from the foregoing quotations from Leverett and Taylor and from Leverett, that there is absolutely no evidence requiring the existence of the main or highest Lake Algonquin stage in the Superior basin. The Algonquin beach has not been traced into the basin. The warped Algonquin plane has been projected into the basin, and beaches of various degrees of strength of development and other types of deposits found there at seemingly appropriate levels have been assigned to the Algonquin stage. Some of these beaches may be more appropriately assigned to the Lake Duluth stages, existing west of a mass of glacial ice. If glacial ice filled the eastern part of the Superior basin while Lake Algonquin existed in the lower lakes, there could be no connection between Lake Algonquin and the Superior basin.

The author has observed evidence that glacial ice did exist in the eastern part of the Superior basin in Algonquin time. North of Sault Ste. Marie there are sand and gravel deltas at the level of the Algonquin beach (1015 feet). These are located at the southern ends of valleys which traverse the crystalline-rock upland separating Whitefish Bay and Goulais Bay of Lake Superior from the Sault Ste. Marie area (Fig. 41). The valley bottoms, floored with sand and gravel, are graded to the level of the Algonquin beach, and their profiles rise gently from the Algonquin beach northward or northwestward toward Lake Superior, then drop more abruptly into the Superior basin in youthful-appearing sections. These relationships may be seen on sheet 41 K/9 of the National Topographic Series, Canadian Department of National Defence (1939). A mass of till located in the Goulais River valley near Bellevue Station, 14 miles north of Sault Ste. Marie (Fig. 41), has an outwash plain on its eastern upvalley side, and a thick de-

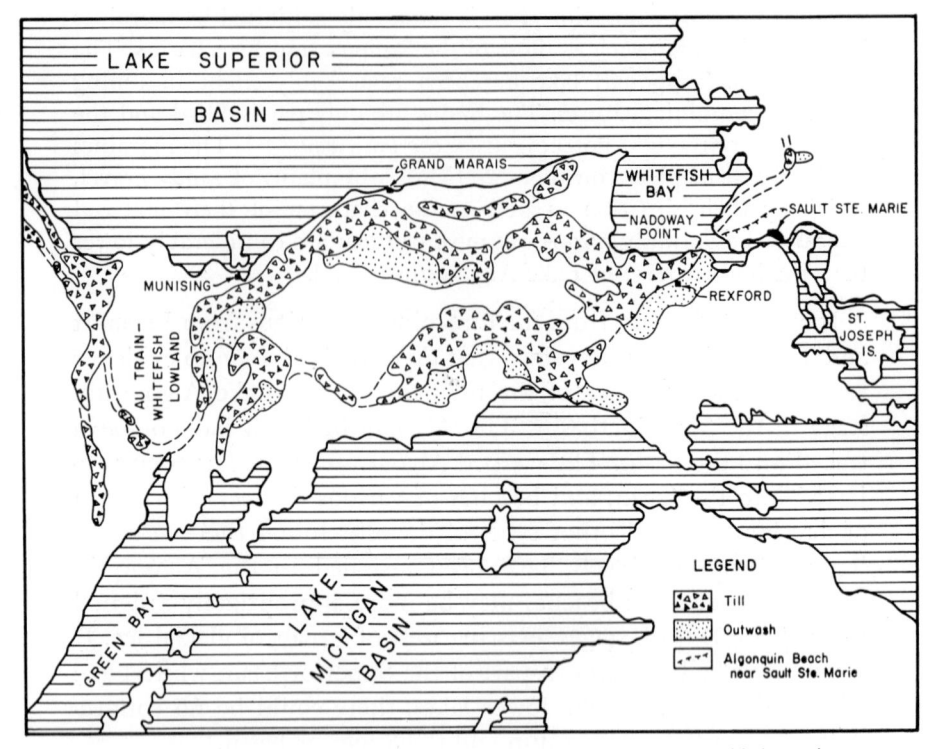

Fig. 41. The northern shore of the main stage of Lake Algonquin in the Michigan basin, shown by the southern margin of a glacial moraine extending from the Sault Ste. Marie area southwestward and westward across the Upper Peninsula of Michigan.

posit of varved clay extends from there eastward up the Goulais valley for several miles. A line drawn from this mass of till southwestward across the upland valleys, through the points in them which mark the upper limits of their gradation to the Algonquin beach, apparently marks the position of the front of a glacier. If this line is continued southwestward across the entrance of St. Marys River, to Nadoway Point, Michigan, it joins a moraine which has been mapped by Leverett (1929a, pl. 1). This relationship may be seen in Fig. 41.

Leverett's map and data given in his text (1929a) provide proof that glacial ice excluded Lake Algonquin from the Superior basin. The Algonquin plane descends from an elevation of 1015 feet, six miles north of Sault Ste. Marie, to an elevation of 934 feet on St. Joseph Island, 25 miles to the south. The surface of a glacial outwash plain at Rexford, Michigan, 35 miles west of St. Joseph Island, is at an elevation of 930 feet. This feature has been cited as a record of the Algonquin stage (Leverett, 1929a, p. 66). An isobase drawn from Rexford to St. Joseph Island bears about 15° south of east, which is the bearing of the Algonquin isobases in the Huron basin. The moraine, on Leverett's map (1929a, pl. 1, and reproduced in part in Fig. 41), occurs at Nadoway Point (near the entrance to the St. Marys River) and extends southwestward along the northwestern edge of this outwash plain. The moraine extends westward across the south side of the Upper Peninsula of Michigan to the head of Green Bay. The existence of this moraine with an Algonquin-level outwash plain on its south side indicates that glacial ice was standing in the Superior basin while the lake to the south was at the Algonquin level.

A second, more northerly, moraine swings away from the southerly one at a point 18 miles southwest of Nadoway Point, extends northwesterly and westerly between Taquamenaw Swamp and Lake Superior, and parallels the shore of Lake Superior from Grand Marais to Munising (Fig. 41). A few miles southwest of Munising this moraine swings southward, indicating that a lobe of ice occupied the Au Train–Whitefish River lowland between the Superior basin and the head of Green Bay to the south. This more northerly of the two moraines is the one with outwash plains on its south side which are "considerably lower than the highest level of Lake Algonquin" (Leverett, 1929a, p. 49). It is apparent that between the time the ice front stood at the southern moraine and the time of its stand at the northern moraine the level of Lake Algonquin had been lowered below its highest elevation. It is suggested here that this probably occurred when the ice front in the Kirkfield outlet region retreated sufficiently to open the first of the eastern outlets which drained Lake Algonquin and initiated the post-Algonquin sequence of events. It would thus correspond with the second period of discharge through the Kirkfield outlet.

Further evidence bearing on this point is given by Bergquist (1936):

The highest shore work recorded in the Taquamenon area of Luce County is to be found in the high moraine immediately south of Mc-

Millen. The slopes of the moraine in this area show the effects of wave washing up to an altitude of 870 feet. This places the highest Algonquin level about 30 feet below the plain at Rexton, southeast, in Mackinac County, as determined by Leverett. The cause for this difference in shore levels is not determined but may have been due to the lodgement, in the McMillen straits, of a large block of stagnant ice. . . . The higher water, in the basin to the east, then must have had its shore pressed against the ice mass, and no records of wave action in this stage were preserved. However, when the water plane had been lowered to the level of 870 feet, the ice mass had either disappeared entirely or had left the slopes of the moraine to allow shore work to become effective there [p. 95].

Instead of "a large block of stagnant ice" the present author believes that the margin of the main glacier existed in this locality while "the higher water . . . had its shore pressed against the ice mass."

Leverett cited no evidence for an Algonquin-level beach on the north side of the peninsula which separates the Superior and Michigan basins, but he does give evidence that the highest lake stage on the north side, in the eastern end of the basin, was considerably lower: "Along the inner slope, on the south shore of Whitefish Bay, the morainic contours extend down within 50 feet of the Lake Superior level, or 650 feet above the sea . . ." (Leverett, 1929a, p. 52). The elevation of the Nipissing beach in the Sault Ste. Marie area 30 miles to the east is 650 feet.

The conclusion of the present author is that the northernmost shore of Lake Algonquin extended along the front of a glacier from the Sault Ste. Marie area southwestward and westward across the head of the Lake Michigan basin to the northern end of Green Bay, and that the main or highest Lake Algonquin did not occupy any part of the Superior basin (Fig. 70).

In the Au Train–Whitefish River lowland, lying between Lake Superior and the head of Green Bay, there is a record of discharge of a large stream of water. This discharge undoubtedly was from the Superior basin to Green Bay, and occurred at a time when glacial ice had receded from the highlands to the west but still stood in the eastern part of the Superior basin, blocking lower dischargeways to the east (Leverett, 1929a, p. 63). The Superior basin discharge through the Au Train–Whitefish lowland must have occurred after main Algonquin time, because it has been shown in a foregoing paragraph that glacial ice extended across the eastern part of the Upper Peninsula of Michigan and into the lowland, after the main Algonquin level was abandoned.

As has been noted in Chapter 10, little is known of events in the Superior basin during the transition from the time of the Lake Duluth stages to the time of the main Algonquin stage of the Michigan and Huron basins. A discussion of later events in the Superior basin will be given in a subsequent chapter.

Warping of the Algonquin Beach

The beaches of the Algonquin stage, as recognized in this chapter, have been warped as shown by Fig. 42. The hinge line crosses the Huron basin near its southern end and the Michigan basin near its middle. South of this line the Algonquin beaches are horizontal at an altitude of 605 feet above sea level. From the hinge line northward the beaches rise, reaching their highest altitude at a point six miles north of Sault Ste. Marie, Ontario. This altitude, 1015 feet above sea level, is 410 feet above the original Algonquin surface. The Algonquin beach occurs at an altitude of 1013 feet on Manitoulin Island, three miles south of Little Current, Ontario (Stanley, 1934, p. 410), 407 feet above the original surface.

The straight lines of Fig. 42 are isobases, each representing equal amounts of elevation above the original altitude of the beach. The northernmost isobase (representing 410 feet of uplift) is drawn from the locality just north of Sault Ste. Marie to a point just north of the Manitoulin Island locality. The next isobase to the south (representing 250 feet of uplift) extends eastward to a point just south of the Kirkfield outlet, which is the northernmost locality of the Algonquin beach on the eastern side of the Huron basin.

The warped Algonquin plane is represented in vertical sections by the uppermost lines of Figs. 43 and 44 (Chap. 13).

If the warped plane represented by the isobases of Fig. 42 is projected northeastward, it is found to extend above the highest beaches which occur in the northeastern part of the Huron basin.

224   GEOLOGY OF THE GREAT LAKES

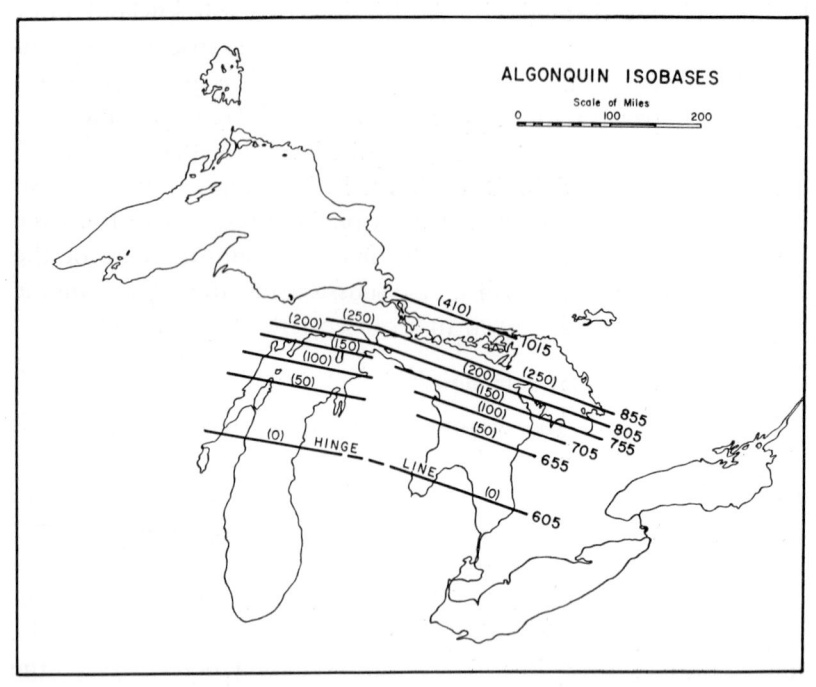

Fig. 42. Algonquin beach isobases. Lines of equal uplift of the beaches, with present altitudes above sea level in feet given by figures at ends of lines, and amount of uplift above original altitude in feet given by figures in parentheses. The line of zero uplift is the hinge line, and south of it the beaches are horizontal. (Modified from Leverett and Taylor, 1915, pl. 8.)

This is taken as evidence that glacial ice covered the area in Algonquin time. The projected Algonquin plane lies about 800 feet above North Bay, Ontario, where the present elevation of a post-Algonquin outlet channel is 700 feet. In Algonquin time that ice-covered locality must have been below sea level. The total post-Algonquin uplift at North Bay apparently amounts to nearly 900 feet.

Post-Algonquin uplift of the region is discussed in greater detail in the following chapters.

chapter thirteen

POST-ALGONQUIN LOW STAGES

CLOSE OF THE MAIN ALGONQUIN STAGE

There are several beaches below the Algonquin in the upwarped northern parts of the Huron and Michigan basins. Leverett and Taylor (1915, pp. 415–16) regarded most of these as "lower" or "later" Algonquin beaches. This association with the Algonquin was made because they believed that most of these beaches converged southward to join the Algonquin in its area of horizontality, as shown in Fig. 43. If this were true, these beaches would have been formed at the Algonquin level, 605 feet, during upwarp of the land (or during various halts in a period of upwarp). The convergence to the Algonquin beach is not supported by field evidence, however.

As has been noted in Chapter 12 on the Algonquin stage, all beaches above the present lake shores are missing from all of the critical areas where the Algonquin beach presumably approaches horizontality. From the hinge line, or line through the northernmost points of no uplift of the Algonquin beach, northward for several miles on both shores of Lakes Huron and Michigan, wave erosion has removed all evidence of former lake shore development. In Fig. 43 these areas are included in the section where the beaches are represented by dashed lines. F. B. Taylor once considered the interpretation that several of the sub-Algonquin beaches descended below the Algonquin beach in their southward projections, and were crossed by the Nipissing beach,

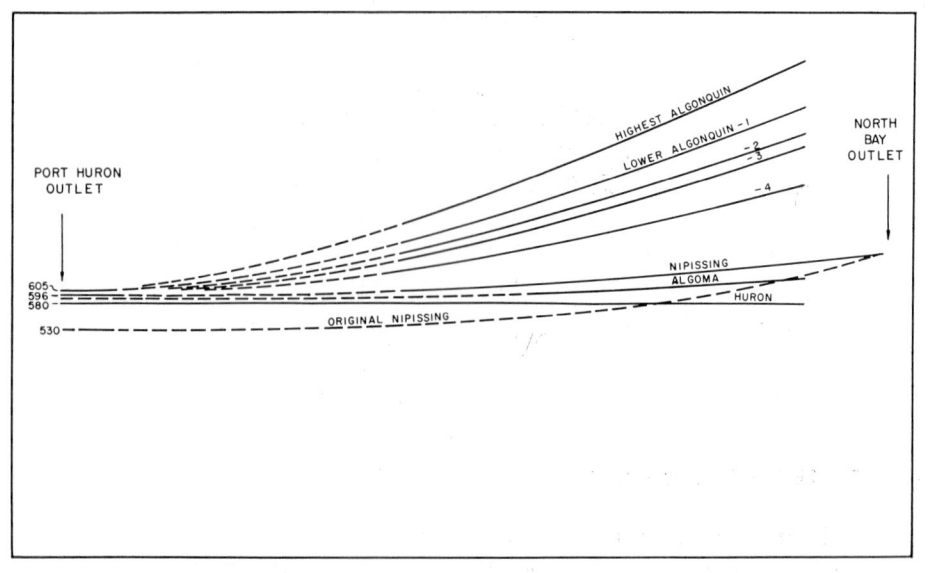

Fig. 43. Relations of the sub-Algonquin beaches, as interpreted by Leverett and Taylor (1915). The convergence of "lower" Algonquin beaches with the highest Algonquin beach supposedly occurred in areas where all beaches above the present lake shore have been destroyed by erosion (shown by dashed lines above the Nipissing).

on the western side of the Huron basin. This representation was made in an aforementioned letter to J. W. Goldthwait.[1] Taylor later rejected this interpretation in favor of the one published in 1915 (Leverett and Taylor, 1915) and represented in Fig. 43.

Stanley (1936, p. 1958, and 1937, p. 1685) has demonstrated convincingly that the major beaches lying below the Algonquin, but not including the Nipissing, are nearly parallel as shown in Fig. 44. This means that they were formed at successively lower levels and must have had successively lower outlets. It further indicates that uplift of the northern parts of the region must have been largely delayed until after the lowest beach was formed.

The conclusions reached by Deane (1950) in his study of the Lake Simcoe district, Ontario, confirm the work of Stanley except in one particular point which is italicized in the following quotation: "The following conclusions may be drawn from a study of the profile map of the water planes: the first is the stability of

---

[1] See n. 2, Chap. 12.

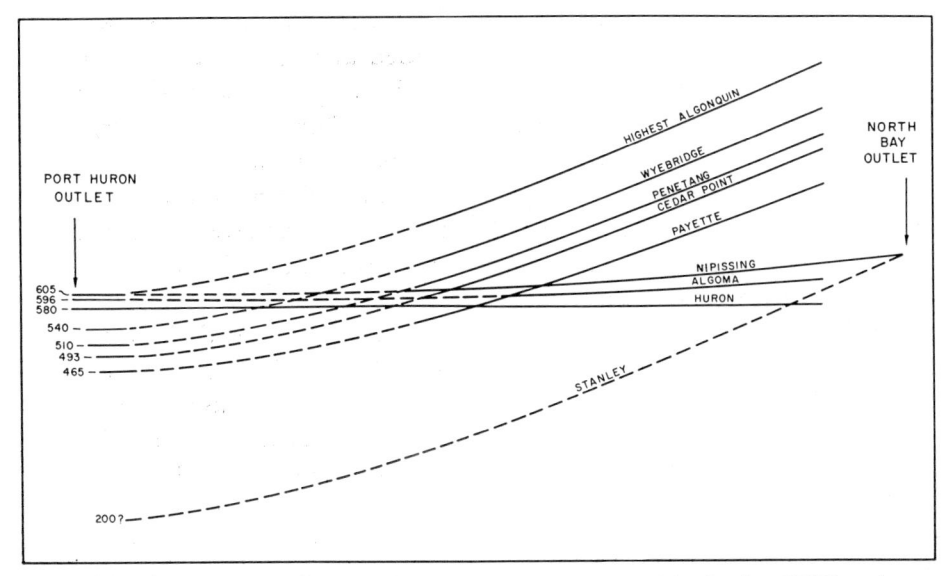

Fig. 44. Relations of the sub-Algonquin beaches, as interpreted by Stanley (1937) and by Hough (1953c). The near parallelism of the four beaches below the Algonquin and the crossing of those beaches by the Nipissing were observed in eastern Georgian Bay.

most of the area during the formation of the main Algonquin beach; the second is the *convergence of the uppermost beaches* Algonquin and Ardtrea, *indicating differential uplift;* and the third is the general parallelism of the lower Algonquin beaches, again indicating stability" (Deane, 1950, p. 77).

The author is not convinced that Deane's (1950, fig. 7) profiles prove convergence of the Algonquin and Ardtrea beaches. These are shown on Deane's profile A-B as being essentially parallel in the northern 26 miles covered by the figure, and as converging in the southernmost 12 miles; this convergence is based, however, on the projection of the Ardtrea trace to a single beach elevation symbol six miles from the south end of the profile. If the trace had been projected instead to another symbol one mile from the south end of the profile, no appreciable convergence would have been shown. On Deane's profile C-D the convergence depends on projection of the Ardtrea trace to a single beach elevation symbol plotted 10 miles from the south end of the profile. If the trace had been projected, instead, through two other beach symbols at 9 and 11 miles from the south end, no appreciable convergence would have been shown.

Stanley's (1936) conclusion is regarded as valid: "Lake Algonquin was not spilled out southward by a great, spasmodic uplifting movement, but, after the episode of Port Huron discharge and construction of the Algonquin beach, it was drained to successively lower levels by the opening of new outlets" (p. 1956). Because of this, the present author believes that the Algonquin group of stages should be terminated with the close of the main Algonquin stage.

The outlet through which drainage occurred to end the main Algonquin stage was not specified by Stanley. The present author suggests that it was the old Kirkfield outlet, used during the Kirkfield stage and then blocked by a readvance of the ice front. If no appreciable upwarp had occurred since Kirkfield time (and none would be expected, if there had been a slight advance of the ice to close the outlet), the retreating ice front again would uncover the outlet and it would still be low enough to drain the lake below the Algonquin level. Deane (1950, p. 89) states that the Fenelon Falls (Kirkfield-Trent) outlet was in operation a second time.

As was noted in the discussion of Lake Iroquois (Chap. 11), the first period of discharge down the Trent valley, in Kirkfield time, correlated with the building of the Lake Iroquois–stage delta at Peterborough, Ontario. The second period of discharge, at the end of the main Algonquin stage, is correlated with the extension of the "Algonquin" channel from that delta down to and beyond the present shore of Lake Ontario at Trenton, Ontario, and thus with the St. Lawrence marine embayment stage of the Ontario basin.

THE "UPPER GROUP" OF BEACHES

At Mackinac Island Leverett and Taylor (1915, p. 416) found an "upper group" of beaches occurring at frequent intervals in the first 47 to 50 feet below the highest or main Algonquin beach. Below this upper group is an interval of 25 or 30 feet apparently without beaches. At Sucker Creek, Grey County, Ontario, Stanley (1938a, p. 480 and fig. 3) found an "upper group" of four beaches constructed on the outer face of the main Algonquin bar and extending below the highest Algonquin level for about 40 feet. There is an absence of distinct beaches for the next 44 feet down the slope. Mackinac Island and Sucker Creek are on nearly the same isobase (the highest Algonquin beach on Mackinac is at 809 feet, while it is at 800 feet at Sucker Creek). These observations, indicating periodic pauses in the first lowering of the lake surface through a range of approximately 50 feet, then a more abrupt lowering of 40 or 50 feet, suggest that the lakes represented by the "upper group" drained through one periodically deepened outlet, and that possibly discharge was shifted to a new outlet when the next lowering occurred.

In the Lake Simcoe area the Ardtrea beach is 20 feet below the highest Algonquin and the Upper Orillia beach is 50 feet below the highest Algonquin (Deane, 1950, p. 77). These may be correlative with the "upper group" of beaches described in the foregoing paragraph.

The author has stated that the first lowering from the main Algonquin stage may have been caused by discharge through the old Kirkfield outlet, opened a second time by retreat of glacial ice. The "upper group" of post-main Algonquin beaches at Sucker Creek extends down to approximately the same elevation (762 feet) on the outside of the main Algonquin bar as is occupied by the Kirkfield beach (764 feet) upon which the main Algonquin bar rests. This indicates that the same outlet, the Kirkfield, may have controlled the elevations of both the Kirkfield stage (Fig. 69)[2] and the lowest of the "upper group" of post–main Algonquin beaches (Fig. 71).

THE WYEBRIDGE, PENETANG, CEDAR POINT, AND PAYETTE STAGES OF THE HURON BASIN

Four prominent beaches below the main Algonquin, and below the "upper group" of beaches, occur at several localities in the Georgian Bay area. These have been studied in detail by Stanley (1936 and 1937). From their elevations and nearly parallel

---

[2] Figures 53–75, lake stage maps, follow Chapter 16.

relationships to each other and to the main Algonquin beach (Fig. 44), it is concluded that they represent successively lower lake levels. These four beaches were named the Wyebridge, Penetang, Cedar Point, and Payette. Only the last of these is shown on a lake stage map (Fig. 72). Table 14 gives their present elevations on Giant's Tomb Island, near the eastern edge of Georgian Bay, and their present slope, as observed by Stanley (1936, table 3, p. 1951); their estimated original elevations are by the present author but are based on Stanley's estimate that the original Payette water plane was about 115 feet below the present Lake Huron surface near Grand Bend, Ontario (Stanley, 1936, pp. 1957–58). The slight convergence of these beaches indicates that some uplift occurred during the period of their formation. The difference in slope between the Algonquin and the Payette beaches (Table 14) indicates that approximately 19 per cent of the total observed uplift in Algonquin to present time occurred in the Algonquin to Payette time $(3.38 - 2.75/3.38 = 18.7\%)$.[3]

The evidence reviewed in the foregoing paragraphs indicates that the lake surface in the Huron basin was lowered from the Algonquin level by diversion to a new, lower outlet (presumably the Kirkfield), leaving the old Chicago and Port Huron outlets dry; and that further lowerings of level occurred, presumably because of further diversions through other outlets which were progressively lower. An outlet at North Bay, Ontario, has long been recognized as a final dischargeway from the northeastern part of Georgian Bay to the Mattawa and Ottawa rivers and thence to the St. Lawrence valley (Leverett and Taylor, 1915, p. 410). Stanley postulated other, intermediate, outlets through the largely unsurveyed upland east of Georgian Bay, which were made available because of ice recession across the area. He visualized the various lowerings of lake level, represented by the "upper group" and the Wyebridge, Penetang, Cedar Point, and Payette stages as resulting from drainage through various unknown outlets to the east. He stated that after the Payette stage, lowering occurred at least once again and probably several times.

Whether the Minong stage in Lake Superior is the equivalent of the Payette stage or whether it is a much later one and represents an

---

[3] Stanley, personal communication, 1954.

Table 14. *Data concerning Algonquin and lower beaches in the Georgian Bay area.*[a]

Beach	Present altitude at Giant's Tomb Island (feet)	Slope of beach (feet per mile)	Original altitude, estimated (feet)
Algonquin	875	3.38	605
Wyebridge	785	?	540
Penetang	748	3.00	510
Cedar Point	724	2.875	493
Payette	686	2.75	465

[a] After Stanley (1936, p. 1951).

additional drop in lake level is not yet certain. Discharge past North Bay and down the Mattawa ultimately brought an end to these events. Although some slight tilting accompanied these changes in level, the greater part of the remarkable deformation of the Algonquin beach was accomplished after Payette time. It may even have been antedated by . . . the first use of North Bay–Mattawa as an outlet. . . . It remains an important point to discover just how far below the Algonquin plane the lake level stood at that time. At North Bay, it may have been some 600 feet below the Algonquin plane, or nearly at sea level [Stanley, 1936, pp. 1956–58].

The Battlefield beach of Mackinac Island, which lies 90 feet below the Algonquin (Leverett and Taylor, 1915, p. 433), is an obvious correlative of the Wyebridge beach which Stanley found to be 90 feet below the Algonquin at Cape Rich, Grey County, Ontario, which lies on the same isobase. Lower beaches at Mackinac are distributed among a large number of levels so that specific correlations with other Georgian Bay beaches cannot be made with any assurance.

Probable correlatives of the Wyebridge, Penetang, Cedar Point, and Payette beaches of the Georgian Bay area may be identified on the slopes north and northwest of Sault Ste. Marie, Ontario. The probable elevation of the Payette beach in this area may be calculated as follows. The Algonquin beach, at an elevation of 1015 feet, has been raised 410 feet; assuming that 19 per cent of the post-Algonquin uplift had occurred by Payette time, as calculated by Stanley (refer to a previous paragraph), the Algonquin beach had been raised 78 feet by Payette time, to an elevation of 683 feet. Taking the original elevation of the Payette stage as 465 feet (refer to Table 14), the Payette beach would be found at 683 minus 465, or 218 feet below the Algonquin. The present elevation of the Payette beach therefore should be at 1015 minus 218, or at an elevation of 797 feet. There is a definite terrace on the slope northwest of Sault Ste. Marie, between the elevations of 780 and 790 feet; this is presumed to be a representative of the Payette beach. Leverett and Taylor (1915, pl. 23) recorded 15 more or less distinct terraces between this presumed Payette beach and the Algonquin beach, in the Sault Ste. Marie area, from Sugar Island to Root River. The present author has found, by altimeter surveys in the area, that the most distinctive of these terraces occur at elevations of 860, 880, and 935 feet, and he suggests that these may represent the Cedar Point, Penetang, and Wyebridge beaches. More precise determinations of the levels of the most distinctive terraces are needed before positive identification of these beaches can be made.

The main Algonquin beach and members of the "upper group" have been identified along the brow of an escarpment on Manitoulin Island, three miles south of Little Current, Ontario, by Stanley (1934, p. 410). These occur at elevations ranging from 1013 down to 940 feet. The Little Current locality is on approximately the same isobase as Sault Ste. Marie, and the Algonquin and lower beaches should occur at nearly the same elevations in both localities. The present author has made an instrumental survey (by plane table and alidade) of the slope on the eastern side of Manitoulin Island, a few miles southeast of Stanley's Little Current locality and approximately three miles north of the village of Sheguiandah, and has found a number of terraces which are tentatively correlated as follows:

Elevations	Possible Correlatives
962	Lowest of "upper group"
935	
918	Wyebridge
900	
848	Cedar Point

POST-ALGONQUIN LOW STAGES 233

Elevations	Possible Correlatives
782	Payette
748	
695	
650	Nipissing
628	Algoma
610	
590	

The terrace at 782 feet is one of the more prominent ones, and it is correlated with the Payette stage, with a fair degree of confidence. Assuming this correlation to be correct, the next higher terrace at 848 feet is at approximately the correct elevation for the Cedar Point beach. The Penetang beach would be expected at approximately 870 feet; because there is no terrace there, it is presumed to have been destroyed during the cutting of the Cedar Point terrace, which locally is on the steepest part of the eastern slope of the island. A terrace does appear at approximately 870 feet, as determined by hand level, at another locality in the same vicinity. Because the terrace at 900 feet is poorly developed, and because the Wyebridge beach would be expected to occur at about 913 feet, the more strongly developed terrace at 918 feet is tentatively correlated with the Wyebridge. The terraces below 782 feet are discussed in the following section.

None of the post-Algonquin stages, from the "upper group" through the Payette, has been recognized in the Michigan basin, but it seems certain that they must have occurred there.

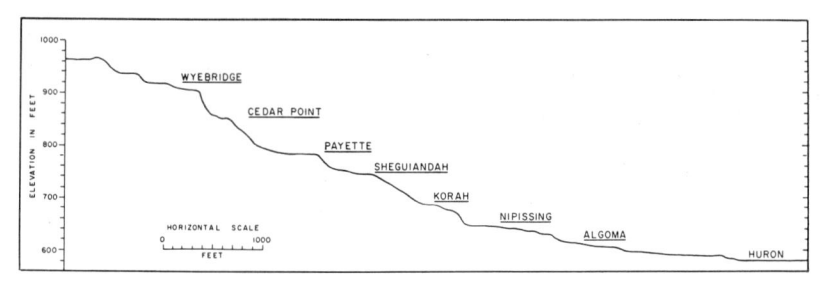

Fig. 45. Profile of the eastern slope of Manitoulin Island, three miles north of Sheguiandah, Ontario, surveyed with plane table and alidade. Two new lake stages, the Sheguiandah and Korah, are recognized on this profile.

THE SHEGUIANDAH AND KORAH STAGES
OF THE HURON BASIN

Stanley (1937, p. 1681) has suggested that following the Payette stage there may have been several later and lower stages before the North Bay outlet was in full use. The present author has found beach forms between the Payette and the Nipissing beaches at two localities in Ontario, near Sheguiandah on Manitoulin Island, and in Korah Township west of Sault Ste. Marie, which represent additional low-water stages.

Figure 45 shows a profile of the eastern slope of Manitoulin Island, surveyed by plane table and alidade along an east-west line about three miles north of Sheguiandah. This profile shows terraces which can be correlated with several of the known post-Algonquin lake stages. Between the Payette and Nipissing are two additional terraces representing lake stages which have not been recognized heretofore. The upper of the two, at an altitude of 748 feet, is here named the Sheguiandah [4] and is considered a representative of a Sheguiandah stage of the Lake Huron basin.

The Sheguiandah stage apparently is represented by a relatively narrow terrace in the area west of Sault Ste. Marie, 130 miles west of its type locality. This terrace, lying just above 750 feet, crosses a north-south road 2.7 miles north of the southern boundary of Korah Township and 2 miles west of the eastern boundary of the township (Fig. 46). An extensive terrace at the same elevation occurs just east of Sault Ste. Marie, in Tarentorus Township (see sheet 41 K/9 of the National Topographic Series, Canadian Department of National Defence, 1939).

The lower of the two beaches which are intermediate between

---

[4] The Sheguiandah archeological site (Lee, 1954) is located east of the village on a hill whose crest, at an elevation of 736 feet, is just below the Sheguiandah-stage plane. The Indian stone quarries and most of the artifacts of the site occur from near the crest of the hill down to an elevation of about 690 feet, and no archeological material has been found below the Nipissing beach at 650 feet. It appears that the site was occupied in pre-Nipissing time, and it is possible that it was occupied shortly after the lake level fell below the Sheguiandah-stage level.

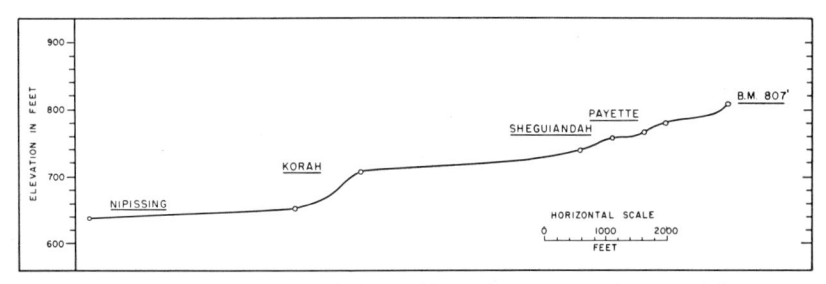

Fig. 46. Profile of the slope west of Sault Ste. Marie, Ontario, two miles west of the eastern boundary of Korah Township. Based on repeated altimeter and odometer readings, tied to a bench mark at the upper end of the profile. The newly recognized Sheguiandah and Korah beaches appear on this profile.

the Payette and the Nipissing beaches on Manitoulin Island is less well developed than the upper one, and if there were no supporting evidence the author would hesitate to recognize it as a representative of a major lake stage. Supporting evidence is found in abundance, however, in the Sault Ste. Marie area in the form of a broad and extensive terrace which extends across the full width of Korah Township, where it is bounded on the south by the 700-foot contour (see sheet 41 K/9 of the National Topographic Series, Canadian Department of National Defence, 1939). The village of Korah, in Korah Township, is located on this terrace. The terrace is here named the Korah beach and is considered a representative of the Korah stage of the Lake Huron basin. The Korah terrace is shown in Fig. 46. The lower of the two terraces between the Payette and Nipissing beaches on Manitoulin Island (Fig. 45) is correlated with the Korah.

The present slope of the Sheguiandah and Korah beaches has not been determined because the two localities where they have been recognized have undergone nearly the same amount of uplift. It is assumed, in the present discussion, that they probably are nearly parallel with the next higher beach, the Payette, and that their original levels were determined by successive lowerings of the lake surface in the Huron basin. It is further assumed that the Sheguiandah and Korah stages occurred before the final low-water stages, the Chippewa stage of the Michigan basin and the Stanley stage of the Huron basin, which discharged through the

North Bay outlet. The Stanley beach should lie below the level of Lake Huron in the vicinities of the type localities of the Sheguiandah and Korah beaches.

THE KORAH–FORT BRADY RELATIONSHIPS

The Korah beach, in the vicinity of Sault Ste. Marie, is actually the same feature as the original Fort Brady beach recognized and named by Leverett (Leverett and Taylor, 1915, p. 436) for a cut beach and cliff at Fort Brady in the southwest part of Sault Ste. Marie. The present author believes that the original name should be discarded because other beaches at several localities, some of them far to the south, have been almost certainly erroneously correlated with the original Fort Brady beach. In particular, certain beaches between the elevations of 635 and 685 feet on Mackinac Island have been rather widely cited as Fort Brady beaches (Goldthwait, 1908b, p. 471; Leverett and Taylor, 1915, pp. 436–37; Stanley, 1936, p. 1954). If the interpretation given by the author in the foregoing paragraph is correct, that is, that the Sheguiandah- and Korah-stage beaches are nearly parallel with the next older Payette beach, the Korah beach would lie below present lake level in the vicinity of Mackinac Island. Stanley (1936, p. 1955) has suggested that both the Fort Brady and the Battlefield terms be suspended.

THE CHIPPEWA AND STANLEY LOW-WATER STAGES

Introduction

The existence of an extremely low-level stage in the Huron basin, occurring in post-Algonquin and pre-Nipissing time, was inferred by Stanley (1936, p. 1958, and 1937, p. 1681) from his interpretation of the relationships between the Nipissing beach and the Wyebridge–Penetang–Cedar Point–Payette series of beaches. Because all of these post-Algonquin beaches are nearly parallel, and they rise to the northward as a result of upwarp of the land, they may be projected to elevations well above that of the North Bay outlet (Fig. 44).

North Bay Outlet

The North Bay outlet is an abandoned channel which extends from North Bay, Ontario (northeast of Georgian Bay), down the Mattawa River to the Ottawa River and thus is connected to the St. Lawrence valley. Detailed study of the outlet channel indicates that it once was occupied by a river having a discharge comparable to that of the St. Clair River, the present outlet of the upper Great Lakes (Leverett and Taylor, 1915, p. 448). The present elevation of the sill of the North Bay outlet is not recorded in the literature, but the present elevation of the beach, which records the water plane of the lake which discharged through the outlet, is given as 700 feet (Leverett and Taylor, 1915, n. 4, p. 448).

Because the Algonquin beach may be projected to an elevation of approximately 1500 feet at North Bay (Chapman, 1954, fig. 2), the total post-Algonquin uplift of that locality apparently is 1500 minus 605 feet, or approximately 895 feet. The North Bay outlet, now at an elevation of 700 feet, apparently was below sea level during Algonquin time, when the area was covered by glacial ice. The amount of uplift which occurred from Algonquin time to the time of the first use of the North Bay outlet is unknown; it is therefore impossible to determine the elevation of the first lake stage which drained through it. Stanley (1936) has stated, "The lake must have . . . reached a minimum level at the inception of Mattawa discharge. It remains an important point to discover just how far below the Algonquin plane the lake level stood at that time. At North Bay, it may have been some 600 feet below the Algonquin plane, or nearly at sea level" (p. 1958).

The Submerged Mackinac River Valley

Another feature which is related to the low-level lakes is the submerged valley incised in the Straits of Mackinac (Fig. 47), which has been described in detail by Stanley (1938b, p. 966). The head portion of this valley in Lake Michigan has a present depth of 150 feet, and in general the valley deepens in its 70-mile course

238   GEOLOGY OF THE GREAT LAKES

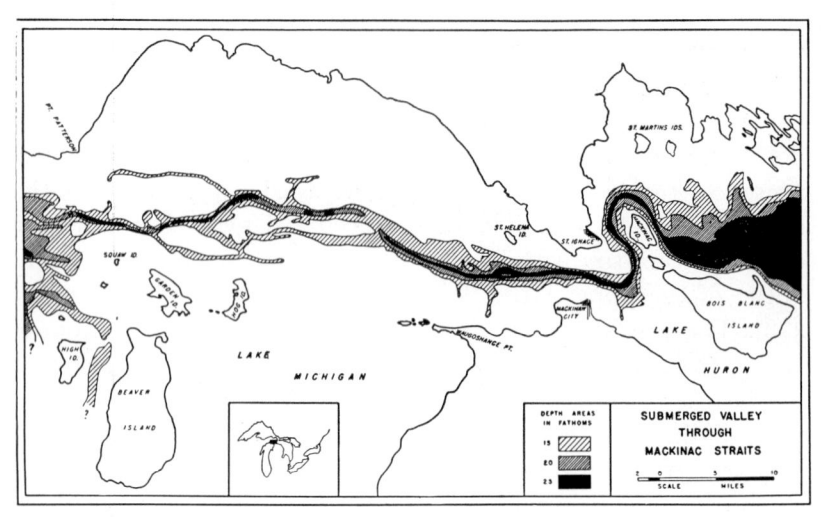

Fig. 47. The submerged Mackinac River channel. (Reproduced from fig. 1 of Stanley, 1938b, Univ. of Chicago Press, by permission.)

eastward, through the straits, into Lake Huron, where channel depths in excess of 200 feet occur. Uplift of slightly more than 200 feet has occurred at Mackinac Island, where the highest Algonquin beach is now 232 feet above lake level. By adding approximately 200 feet of depth lost, due to uplift, to a present controlling depth of 150 feet, it is calculated that the Michigan basin waters would have been lowered approximately 350 feet below present lake level by drainage through the straits channel into the Huron basin. This inferred low stage of Lake Michigan has been amply confirmed by detailed studies of the lake-bottom sediments, which are summarized in the following paragraphs.

Lake Chippewa in the Michigan Basin

Evidence for an extremely low stage in the Michigan basin, described elsewhere in detail by the author (Hough, 1955, pp. 957–68), may be summarized as follows. A 35-foot core sample from the deepest point in the lake, 923 feet, contained a complete sequence of deposits from the Valders substage to the present, as described in Chapter 1 and in Fig. 26. Several other cores from various depths greater than 350 feet established the widespread

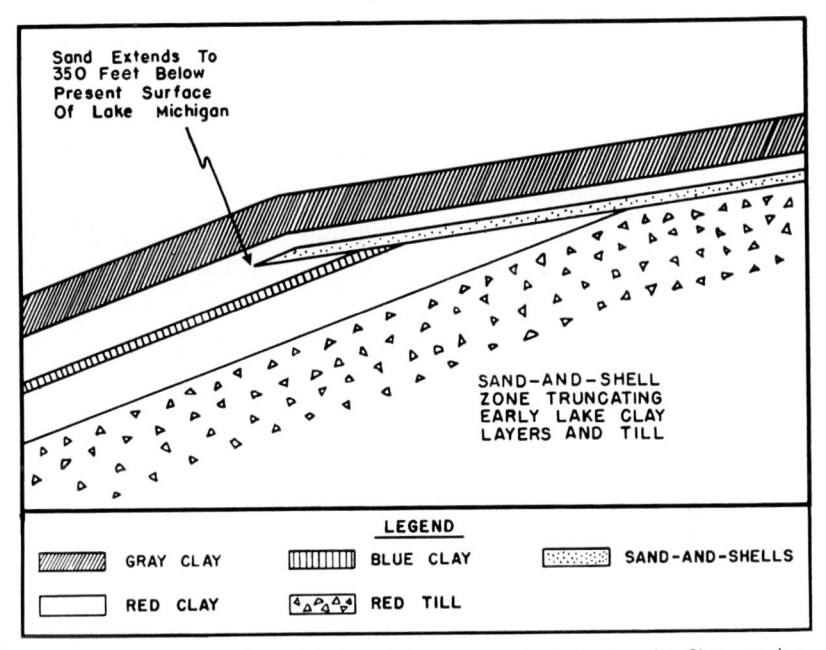

Fig. 48. Unconformity in Lake Michigan bottom sediments, indicating the Chippewa low-water stage 350 feet below present lake level. (Reproduced from fig. 6 of Hough, 1955, by permission.)

occurrence of this normal deep-water sequence, which is tabulated below:

> gray lake clay, grading to
> red lake clay
> bluish-gray lake clay
> red lake clay
> red varved clay
> red till

Core samples taken from present depths of less than 350 feet show that portions of the normal deep-water sequence are missing and are replaced by a thin zone of sand or sand and shells. The sand is overlain by the upper portion of the normal deep-water sequence. Figure 48 shows these relationships as an unconformity in the lake-bottom deposits. The shells contained in the sandy zone have been identified [5] as those of gastropods and pelecypods which live in extremely shallow water, and many of them were broken or considerably abraded, indicating wave action. Shells are not found in the normal lake clay zones in the core samples, but are restricted to the zone of the sandy layer and the immediately superjacent sandy clays. All of this information may confidently be interpreted as evidence that the lake in the Michigan basin was lowered to an elevation of approximately 350 feet below the present surface, or to 230 feet above sea level. It is thus seen that the evidence from the bottom sediments confirms the inference drawn from the present sill depth of the incised channel in the Straits of Mackinac and the observed uplift of the outlet area, that a lake stage existed 350 feet below the present lake level.

This stage, which has been named the Chippewa [6] (Hough, 1955, p. 965), is represented in Fig. 49. In the preparation of this map, it was found that the 350-foot depth contour closes off the southern third of the lake slightly north of Grand Haven, Michigan. Farther south is a deep basin with a maximum present depth of 564 feet. The deepest point on the divide between the southern and the northern basins is 336 feet, and the bottom in this area is composed of red till; therefore, there must have been a separate lake in the southern basin at an altitude of not more than 336 feet below present lake level, during the Chippewa stage. This is designated as Southern Lake Chippewa. A channel which connected Southern Lake Chippewa with the main Lake Chippewa to the north can be traced on a detailed topographic map of the divide area (prepared by the author from unpublished soundings made by the U.S. Lake Survey), and it is revealed by fathogram

---

[5] By M. R. Matteson, Dept. of Zoology, Univ. of Illinois, Urbana.

[6] The name "Chippewa" was chosen in order to use an Indian name to associate with "Algonquin," which has been used to designate an earlier, higher lake stage. The Chippewa Indians lived in this region and were related to the Algonquins in many aspects of their culture. After the name was applied to the lowest stage in the Michigan basin (Hough, 1955), the author found that Taylor (1895b) had used "Chippewa" in connection with a high beach in the northern Huron and Michigan basins. The beach which was so named is not closely related to that described here, and Taylor later abandoned the term. It does not appear in his monograph (Leverett and Taylor, 1915).

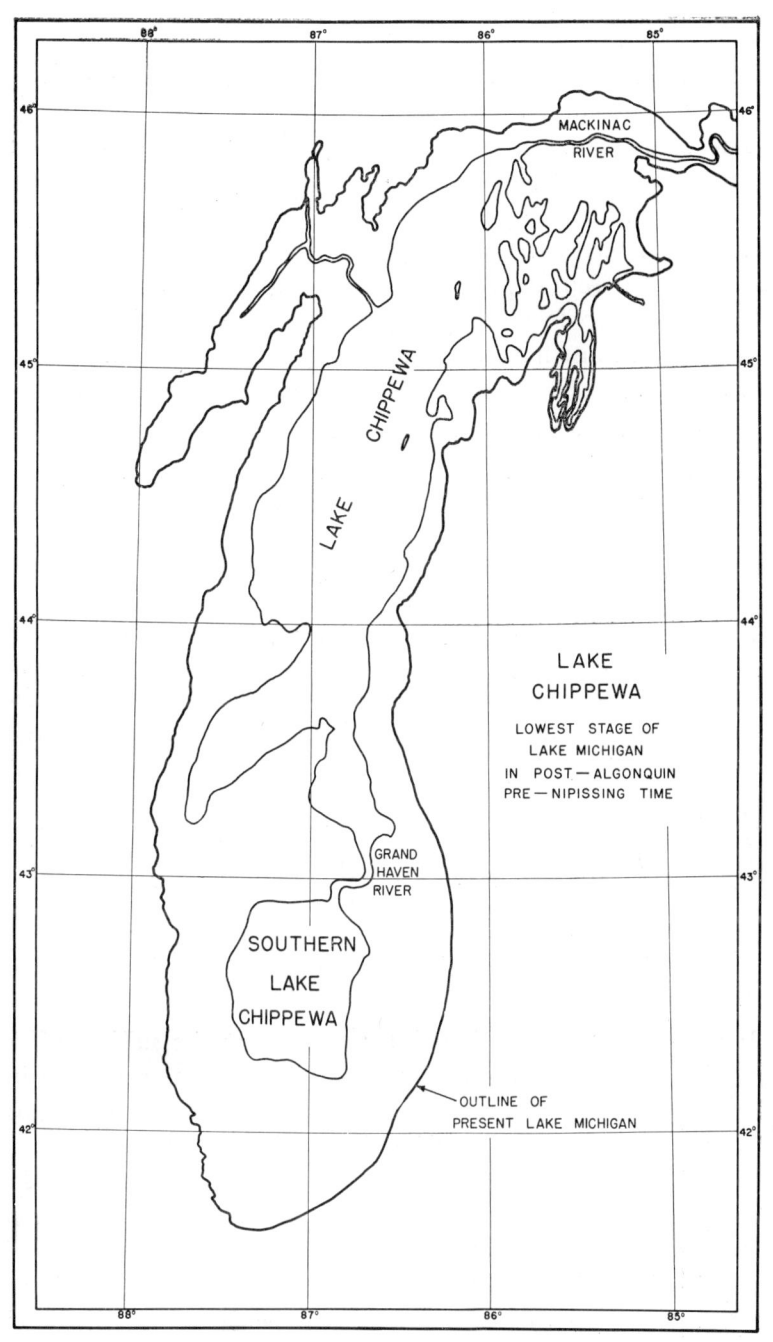

Fig. 49. The Lake Chippewa stage of the Michigan basin. (Reproduced from fig. 7 of Hough, 1955, by permission.)

cross sections of the area. This channel has been named the Grand Haven River channel.

The discharge from the main body of Lake Chippewa passed down the incised valley leading through the Straits of Mackinac, as a stream which may be named the Mackinac River. Because this river channel extends into the Huron basin as a recognizable feature to a present depth of 200 feet, and because there has been a little over 200 feet of upwarp in the area, lake level in the Huron basin must have been at least 400 feet below the present lake surface.

Lake Stanley in the Huron Basin

The low-level lake in the Huron basin, whose existence was inferred by Stanley (1936 and 1937), is now required as a cause for the draining of the Michigan basin down to the Chippewa level. The name "Lake Stanley" has been applied to the lowest stage of the lake in the Huron basin (Hough, 1955, p. 965). The exact level of this stage has not been determined, but it was at least 400 feet below present lake level or not more than 180 feet above sea level. This lowest stage of the Huron basin discharged through the North Bay outlet, as described in a foregoing paragraph (Fig. 73).

The lake surface was raised from its Stanley-stage level by uplift of the North Bay outlet, and was ultimately returned to the level of the old southern outlets to initiate the Nipissing stage. This part of the history is discussed in a subsequent chapter.

Outlets

It is appropriate, at this point, to return to the subject of the possible outlets of several of the post-Algonquin, pre-Nipissing lake stages. The author has stated that the immediately post-Algonquin "upper group" of beaches probably were formed by a lake which discharged down the Kirkfield–Trent valley outlet. Stanley recognized four subsequent major stages (the Wyebridge, Penetang, Cedar Point, and Payette) which existed at successively lower levels, and he postulated that they drained through unknown outlets to the east, somewhere between the Kirkfield–Trent valley route and the North Bay–Mattawa River route. The present author has added two additional stages, the Sheguiandah and Korah, which require appropriately located outlets.

Two new outlets have been described by Chapman (1954, p. 67). The first of these, named the Fossmill, is a channel several miles in length extending from Fossmill, Ontario (about 23 miles southeast of North Bay), eastward along the line of the Canadian National Railway and the Petawawa River to Petawawa, where it opens out into the Ottawa valley. The second is a passage beginning farther northeast and extending southeastward up the Amable Du Fond valley to join the Fossmill channel at Kiosk, about 60 feet below its sill elevation. The sill of the Fossmill outlet apparently is well below the plane of the Payette stage, but it is above the North Bay outlet. It appears possible, therefore, that the Sheguiandah and Korah stages may be related to the Fossmill and the Amable Du Fond channels. Further study of these stages is needed before definite correlations can be made.

The outlets for the stages from the Wyebridge to the Payette are still unknown. Two possibilities should be considered. One is that additional outlets may yet be found between the Kirkfield and the Fossmill channels, and the other is that variations in the extent of ice damming in either or both the Kirkfield and Fossmill channels may have produced these lake stages.

POST-ALGONQUIN, PRE-NIPISSING STAGES OF THE SUPERIOR BASIN

It has been stated (see Chap. 12) that the main Algonquin stage, which existed at an elevation of 605 feet above sea level, was restricted to the basins of Lakes Huron and Michigan and that its northern shore lay along a glacial ice front in that part of the Upper Peninsula of Michigan which extends from Sault Ste. Marie westward to the Au Train–Whitefish lowland (Fig. 70). The author believes that the water surface in the Huron and Michigan basins was lowered below the Algonquin level before the glacial ice in the eastern part of the Superior basin retreated sufficiently to permit the water surface of the Superior basin to be drained down to the level of the lakes in the Huron and Michigan basins.

At a much later time, during the Nipissing Stage, the Superior, Huron, and Michigan basins contained confluent lakes at a single elevation (Fig. 74). This is abundantly shown by the Nipissing beach, which is an easily recognizable feature and which has been traced into the Superior basin.

While several beaches which are almost certainly post-Algonquin and pre-Nipissing in age are known in the Superior basin, no positive correlations have been made between them and any of the stages of the Huron and Michigan basins.

The pre-Algonquin stages of Lake Duluth, which existed in the western part of the Superior basin, are known in some detail (see Chap. 10). Post–Lake Duluth beaches have been recorded in greatest detail for the area of Cook County, Minnesota, on the northwestern shore of Lake Superior (Sharp, 1953). This area has been uplifted several hundred feet since the retreat of the glacial ice. In the vicinity of Grand Marais, Minnesota, there are shoreline features identified as representative of the lowest stage of the Lake Duluth series, occurring between the elevations of 1240 and 1270 feet. Below these are traces of a sub-Duluth shore, at an average elevation of 1180 feet, which are assigned to a stage when the waters of the western part of the Superior basin discharged along the ice margin on the flanks of the Huron Mountains in Marquette County, Michigan, and thence to the Lake Michigan basin (Sharp, 1953, p. 117). The next lower distinct shoreline features occur as a group between the elevations of 1021 and 1055 feet. These were designated as "highest Algonquin" and "higher Algonquin" shorelines by Sharp (1953, pp. 121 and 135), who followed the then generally accepted notion that Lake Algonquin had existed in the Superior basin. In view of the conclusions of the present author regarding the exclusion of Lake Algonquin from the Superior basin, these shoreline features in Cook County between 1020 and 1055 feet must represent something other than the Algonquin stage.

Other shore features, at successively lower levels in Cook County, Minnesota, have been named the Lutsen, Marais, Tofte, Kodonce, Deronda, and Nipissing. The average elevations of these features in the vicinity of Grand Marais are given in Table 15, which has been derived from Sharp's (1953) figure 12. There

Table 15. *Shoreline features in the vicinity of Grand Marais, Cook County, Minnesota.*[a]

Stage	Present altitude above sea level (feet)
Latest Duluth	1255
Sub-Duluth	1180
Higher Algonquin	1035
Lutsen	980
Marais	885
Tofte	815
Kodonce	725
Deronda	675
Nipissing	630

[a] After Sharp (1953, fig. 12).

is a strong temptation to attempt correlation between these shore features and the various Algonquin to Nipissing stages of the Huron basin, particularly because there are approximately the same number of distinct stages in the same time interval in the Superior and in the Huron basins. However, there is not sufficient information available from the vast area between the northwestern shore of Lake Superior and the Sault Ste. Marie area to support any correlations which might be suggested, and the subject is therefore left for future study.

It may be supposed that sometime shortly after the Algonquin stage (of the Huron and Michigan basins) the ice barrier in the eastern part of the Superior basin was removed and that then the Superior, Huron, and Michigan basins contained lakes which had common elevations through several stages. It appears quite reasonable to presume that at the time of the Chippewa and Stanley lowest-water stages of the Michigan and Huron basins, the Superior basin had a correlative low-water stage (Fig. 73) whose elevation was determined either by the outlet sill of the Superior basin, or by the elevation of the Stanley stage, whichever was higher.

Because the Korah beach of the Huron basin, as represented near Sault Ste. Marie, Ontario, is approximately 100 feet above the present sill elevation of the Superior basin, it may be assumed that the outlet river of Lake Superior was about 100 feet deep during the Korah stage; and that a further lowering of the lake in the Superior basin, which would amount to a 100-foot drop, could have occurred during the Stanley stage of the Huron basin. It is impossible to calculate the elevation of the Superior basin sill in Stanley-stage time, because the rates of uplift of the area are not known adequately. However, because the Algonquin beach in the vicinity of Sault Ste. Marie has been raised 410 feet, the Superior basin sill must have been 410 feet lower, or at an elevation of 190 feet above sea level, during Algonquin time; and because the Nipissing beach in the vicinity of Sault Ste. Marie has been raised 45 feet, the Superior basin sill must have been 45 feet lower, or at an elevation of 555 feet above sea level, in Nipissing time. Further, because Stanley (1937, p. 1681) indicated that the greater part of the post-Algonquin uplift occurred after Payette time, and possibly after the Stanley-stage time, it may be presumed that the Superior basin sill was at an elevation nearer its Algonquin time level during the Stanley stage. The present author ventures the guess that the elevation of the lowest stage of the Superior basin was approximately 300 feet above sea level, or 300 feet below the present surface of Lake Superior. It is hoped that future investigations in the field will yield information bearing on this subject.

A number of shore features occurring on Isle Royale, in northwestern Lake Superior, have been described in an unpublished thesis by Stanley (1932). Some of these were suggested as correlatives of Algonquin beaches, and others as correlatives of the Battlefield and Fort Brady beaches. In view of later work by Stanley (1936 and 1937), and particularly in view of his recommendation that use of the terms "Battlefield" and "Fort Brady" be suspended (Stanley, 1936, p. 1955), the correlations suggested in the unpublished work of 1932 are here ignored. The basic data of beach elevations are, however, in existence for possible future use.

A low-water stage of the Superior basin, occurring between the Algonquin and Nipissing, has been mentioned briefly by Stanley (1941); this is the Minong stage, represented by a beach on Isle Royale which has a present inclination of 2.7 feet per mile, rising northeastward, and represented by shore features occurring at intervals along the eastern shore of Lake Superior between Goulais and Montreal River. The Minong beach rises almost four feet per mile to the northward in this area. If the plane represented by this feature is projected southward from Goulais, where its elevation is 700 feet, it passes nearly 50 feet above the present outlet of Lake Superior. Because of this, it may be inferred that the Minong stage was not the lowest possible stage of the Superior basin. A further lowering of 50 feet, perhaps brief, appears to have been possible. No detailed account of the Minong stage has been published. Further reference to it may be found in a paper by Hubbs and Lagler, wherein personal communications from Stanley are quoted. These state that the Minong beach occurs on Isle Royale, where "it grades from about 165 feet above Lake Superior in the northeastern part of the island to only 80 feet near the southwestern end. By extrapolation of Stanley's estimates the Minong level near the Huron Mountains [Marquette County, Michigan] lies from 20 feet below to 20 above present Lake Superior" (Hubbs and Lagler, 1947, p. 89).

If the present outlet of Lake Superior was the outlet of that basin during the lowest low-water stage, the plane of the lowest-water stage will lie mainly above the present lake shores. This must be concluded from the understanding that almost the entire Superior basin lies north of a line of equal uplift running through the present outlet area (Leverett, 1929a, pl. 9). According to this line of reasoning, the plane of the lowest stage may lie below the present lake surface only in the extreme southwestern end of the lake and in an area in the vicinity of Marquette and Munising, Michigan (Fig. 73).

The possible existence of an alternate low-stage outlet of the Superior basin, to the James Bay area, perhaps via Long Lake north of Lake Superior, has been mentioned by Stanley (1941). This is possible only if glacial ice barriers to the northeast disappeared before the divides north and east of the Superior basin were raised to the level of the St. Marys River.

chapter fourteen

THE NIPISSING GREAT LAKES

INTRODUCTION

The Nipissing stage, shown in Fig. 74,[1] was the largest of all of the Great Lakes stages. It extended throughout the Huron, Michigan, and Superior basins at an altitude of 605 feet.

The first-described Nipissing beach is a feature near the shore of modern Lake Nipissing in the vicinity of the North Bay outlet, northeast of Georgian Bay. The feature occurs at a present elevation of 700 feet above sea level, along the base of a bluff one mile north of the Canadian Pacific Railway station in North Bay, Ontario (Leverett and Taylor, 1915, n. 4, p. 448). This beach has been correlated, with a high degree of assurance, with a strong strand line which occurs in the upper Huron basin, the upper Michigan basin, and the southeastern part of the Superior basin. It is this more extensive beach which has become the "type" Nipissing beach. Taylor (Leverett and Taylor, 1915) stated, "The name 'Nipissing beach' has from the first been applied to the shore line now known to have been formed during the two-outlet stage of the lakes, when the discharge was deserting North Bay and returning to Port Huron . . ." (p. 449).

The present author accepts this part of Taylor's concept of the Nipissing stage, but adds a third point of discharge (the Chicago outlet) for reasons to be stated later. Taylor, after defining the Nipissing beach as quoted in the foregoing paragraph, states further, "This shore line is therefore, in reality, the beach of a transition stage . . . in spite of the fact that this name properly

---
[1] Figures 53–75, lake stage maps, follow Chapter 16.

belongs to the slightly older beach made by the Nipissing Great Lakes when their whole discharge passed eastward to Ottawa River"; he then describes the concept of a low-water stage, draining through the North Bay outlet, and designates it as the "first" or "original" Nipissing (Leverett and Taylor, 1915, p. 449). The elevation of this first Nipissing beach is given as "530 ±" feet A.T. in his table on page 469 (1915). Taylor's first Nipissing beach dates from the first use of the North Bay outlet. This concept is rejected in the present compilation of Great Lakes history, because it would now have to coincide with the Chippewa-Stanley low-water stages.

THE STANLEY-NIPISSING TRANSITION

Stanley (1936 and 1937) has shown that the Algonquin and four lower beaches (Wyebridge, Penetang, Cedar Point, and Payette) are nearly parallel, and that the Nipissing beach crosses these as it is traced across country (Fig. 44). He points out that this relation of unconformity signifies the lapse of an amount of time not generally recognized before in the history of the Great Lakes, and that the greater part of the post-Algonquin uplift had occurred before the Nipissing beach was formed.

In present usage, therefore, the events occurring between the Chippewa-Stanley low stages and the formation of the generally accepted Nipissing beach are referred to as a Stanley-Nipissing transitional period. During this transitional period the North Bay outlet was rising, while carrying the entire discharge of the Huron, Michigan, and Superior basins. The water surface in those basins was rising relative to the land, everywhere south of an isobase extending through North Bay. It appears likely that no extensive well-developed beach features would be formed during the presumably continuous rise of the lake surface, and no such features have been found which can be assigned to this part of the history.

CHARACTERISTICS OF THE NIPISSING SHORELINE

The Nipissing beaches of the northern Huron, northern Michigan, and southeastern Superior basins are among the strongest

250   GEOLOGY OF THE GREAT LAKES

and most spectacular shoreline features of any age found in the Great Lakes region. They obviously record vigorous shore work of waves during a time of nearly static lake level. (The nearly static position of the lake surface is not considered to be a result of cessation of uplift of the North Bay outlet; it can be explained adequately as a result of discharge through the stable outlets to the south. This is discussed in subsequent paragraphs.)

The Nipissing beaches in the areas mentioned lie from 25 to 70 feet above the present surface of Lakes Huron and Michigan, because of upwarp of the land to the north since Nipissing time. So far as is known, the upwarping was continuous from the time of the Stanley stage, through Nipissing, and to the present, but its rate has been diminishing during the latter part of this period.

The strongly developed Nipissing beaches are remarkably fresh and undissected, but they are built on dissected topography in many places (Fig. 50). Above the Nipissing beach the land generally is creased by gullies or small valleys which terminate at the Nipissing beach, and the spurs of the divides between these valleys are truncated at the upper margin of the beach. The floors of the

Fig. 50. The Nipissing beach on the west side of Torch Lake, Antrim County, Michigan, one quarter-mile north of Persons Harbor. Well-dissected topography lies above the beach, and the beach truncates spurs of divides between valleys.

small valleys are graded to the Nipissing beach. The Nipissing beach crosses the mouths of larger valleys as a barrier ridge, enclosing lower ground, swamps, and lakes. These relationships are a result of the extreme low-water stage, which occurred during the long interval between the Algonquin and Nipissing stages, and during which the dissection of the land and the incision of the stream valleys occurred. Rise of water then flooded this topography, and wave work at the Nipissing level produced the erosional and depositional features associated with the beach.

NORTH-SOUTH CORRELATIONS OF THE NIPISSING BEACH

The strongly developed Nipissing beach in the northern parts of the Huron and Michigan basins has been raised by tilting or warping of the land. It descends to the southward, from an elevation of 700 feet above sea level at North Bay, Ontario, to an elevation of about 605 feet at the southern limits of the areas in which the beach can be recognized by direct tracing. Gilbert (1898, p. 605), as well as Spencer and Taylor in their early writings, stated that the Nipissing beach descended below present lake level in the southern parts of the Huron and Michigan basins. Taylor (Leverett and Taylor, 1915, p. 457) later stated that the Nipissing beach of the north descended to an elevation of about 15 feet above present lake level and extended throughout the southern parts of the Huron and Michigan basins at that elevation. A correlation thus was made between the strong Nipissing beach of the north and a weak beach in the south having an average elevation of 596 feet above sea level.

None of the Nipissing or older beaches in the north can be traced to a junction with any of the beaches in the southern parts of the basins. Shore erosion has removed all of the old beaches above present lake level in the critical areas where the tilted northern portions approach horizontality. This was recognized by Taylor, specifically in relation to the Nipissing: "On certain shores the Nipissing beach has been largely and in places almost wholly cut away at present lake level. This is notably the case along the east side of the 'thumb' north of Port Huron, on both sides of Lake Michigan, and on the east side of Lake Huron" (Leverett and Taylor, 1915, p. 450). In view of this, and in the absence of any other supporting data, Taylor's correlation of the 596-foot beach in the south with the Nipissing beach in the north is indefensible.

The author has published a revision (Hough, 1953c) in which the Nipissing beach becomes horizontal in the south at an altitude of 25 feet above present lake level, or 605 feet above sea level. It thus coincides with the earlier Algonquin level. The reasoning which led to this conclusion is as follows. During the main Algonquin stage the lakes of the Huron and Michigan basins (which were connected through the Straits of Mackinac area) discharged through both the Chicago outlet at the south end of the Michigan basin and the Port Huron outlet at the south end of the Huron basin. The main Algonquin stage had an average surface altitude of 605 feet above sea level. As explained in a foregoing chapter, the Algonquin stage was brought to a close by the opening of another outlet, to the east of Georgian Bay (probably by the reopening of the Kirkfield outlet), and the lake surface was lowered. The Chicago and Port Huron outlets thus were abandoned, and they carried no drainage except for extremely local runoff. The strongly developed Algonquin beaches of the region indicate that outlet channels remained stable for a relatively long time, and that the maximum discharge of the Algonquin-stage lakes was unable to cut the outlet channels deeper. The next beaches below the main Algonquin are well below the Algonquin level; this corroborates the interpretation that the Algonquin stage was ended by diversion of discharge to other, lower outlets.

In view of the foregoing, it seems that no downcutting of the southern outlet channels could occur between Algonquin and Nipissing times, while they were abandoned and dry. Because of this, the water rising to the Nipissing stage (as a result of uplift of the North Bay outlet) found the old southern outlets unchanged from the altitudes they had had during the Algonquin stage. It is concluded, therefore, that the Nipissing stage rose to the same altitude (605 feet) above sea level as the Algonquin stage, and that in the unwarped southern portions of the Huron and Michigan basins the Nipissing beach coincided with the old Algonquin beach.

The beach at 596 feet in the south, which Leverett and Taylor designated as Nipissing, is a much weaker feature than the Nipissing beach in the north, but it compares favorably with the Algoma beach of the northern areas, which lies in a position between the Nipissing and the present lake level. The Algoma beach is discussed in Chapter 15.

RADIOCARBON DATES AFFECTING CORRELATION OF NIPISSING FEATURES

When the first few radiocarbon dates of beach materials in the Great Lakes region became available, they did not fall into a reasonable sequence. These dates (Arnold and Libby, 1951, and Libby, 1951 and 1952), when listed according to the sequence of events they were assumed to represent, are as follows:

Age in Years

"Toleston level beach" (605 feet) in Chicago area .... 3469 ± 230
Nipissing beach, Lake Superior .... 3656 ± 640
"Nipissing level" (594 feet), Burley site, Lake Huron .... 2619 ± 220

If, however, it is assumed that the Nipissing-stage lake rose to the Algonquin level (605 feet), and the names of the beaches are adjusted accordingly, the radiocarbon dates fall into the correct sequence:

Age in Years

Nipissing beach, Lake Superior .... 3656 ± 640
Nipissing beach ("Toleston level beach," 605 feet) ...... 3469 ± 230
Algoma level (594 feet), Burley site, Lake Huron .... 2619 ± 220

From these data it may be concluded that the Nipissing-stage level coincided with the old Algonquin-stage level in the south, and occurred approximately 3500 years ago, and that the Algoma stage occurred nearly 1000 years later at a lower level.

Recent investigations by Dreimanis (1958, in press) have revealed evidence of the rise of the Nipissing-stage surface up to the Algonquin level, along the Lake Huron shore east of Sarnia,

Ontario. A section near Blackwell, Ontario, showed a deposit of beach sand and gravel ranging from 2 to 27 feet thick, with the crest of the beach bar lying at an elevation of 609 feet above sea level. The lower boundary of this beach deposit descended from 594 feet to 581 feet. This material was underlain by a deposit of reworked material consisting of sand, silt, lacustrine clay, and partly weathered till, with an admixture of fresh-water mollusks and plant remains, including several logs. The reworked material was underlain variously by oxidized till, by till with remnants of B and A soil horizons, and by unweathered lacustrine clay. The upper boundary of this lowest zone descended from 588 feet to 581 feet (present lake level). This section records the transgression of a beach zone across a previously subaerial terrane with mature soil development. The age of the transgression by a rising lake is indicated as 4610 ± 210 years to 4650 ± 200 years, by radiocarbon dates obtained from logs taken from the base of the uppermost beach deposit. These dates are much too young to be Algonquin; they must represent an immediately pre-Nipissing time, late in the rise of the lake waters toward the final Nipissing level.

EXTENT

From North Bay, Ontario, the type locality of the Nipissing beach, the Nipissing shoreline has been traced by correlation over short gaps or over longer gaps along lines of nearly equal uplift, with a fair degree of assurance. These portions of the Nipissing record show that the lake occupied the northern parts of the Huron and Michigan basins and extended from the Sault Ste. Marie area westward along the south side of the Superior basin. The correlation of these portions of the beach with shoreline features lying at an average altitude of 605 feet above sea level in the unwarped southern portions in the Huron and Michigan basins has been described in foregoing paragraphs. It is believed, therefore, that the Nipissing stage filled the Huron and Michigan basins and extended into the Superior basin as a body of water with a surface altitude of 605 feet.

The extent of the Nipissing lake in the Superior basin has not been determined by direct tracing or by correlation across small gaps in the record. It seems, however, that glacial ice must have disappeared from the Superior basin by Nipissing time, because the ice had disappeared from the North Bay area much earlier, and because the uplift of various older beaches throughout the northern Great Lakes region is presumed to have been in response to removal of the ice. In view of this, it may be assumed that the Nipissing lake filled the Superior basin, as is shown in Fig. 74.

Because of upwarp since Nipissing time, and because of large gaps in knowledge, it is not yet possible to identify the Nipissing shoreline in the northeastern and northern parts of the Superior basin with certainty. Recognition of Nipissing beaches in these areas has been based on projection of lines of equal uplift from the Huron basin into and across the entire Superior basin, and by selecting beaches with seemingly appropriate strength of development which occurred closest to the projection of the Nipissing water plane and adjusting the projected plane to the elevations of the beaches. The resulting adjustments of lines of equal uplift were shown by Leverett and Taylor (1915, pl. 9). A revision of their map is reproduced as Fig. 51 of the present book. The localities and elevations of beaches on the north shore of Lake Superior which have been identified by Leverett and Taylor as Nipissing are given in Table 16. Leverett and Taylor (1915) state, "Although these observations of the Nipissing beach on the northern shores are widely scattered, some of them being separated by intervals of more than 100 miles, yet this beach is so distinguished from all others by its great strength and individual characteristics that its identity seems to be beyond doubt at every locality where it has been reported" (p. 457).

Recent investigations have raised some question of the identity of the Nipissing beach in the northern part of the Superior basin.[2] Because of this, the lack of certainty of the correlations was emphasized in the foregoing discussion, and Table 16 was included to provide data for use in future work.

The unconformable relation of the Nipissing beach to earlier-formed topography, which resulted from dissection of a land

---

[2] Quimby, G. I., archeologist, Chicago Nat. History Mus., personal communication, Aug. 22, 1957.

256   GEOLOGY OF THE GREAT LAKES

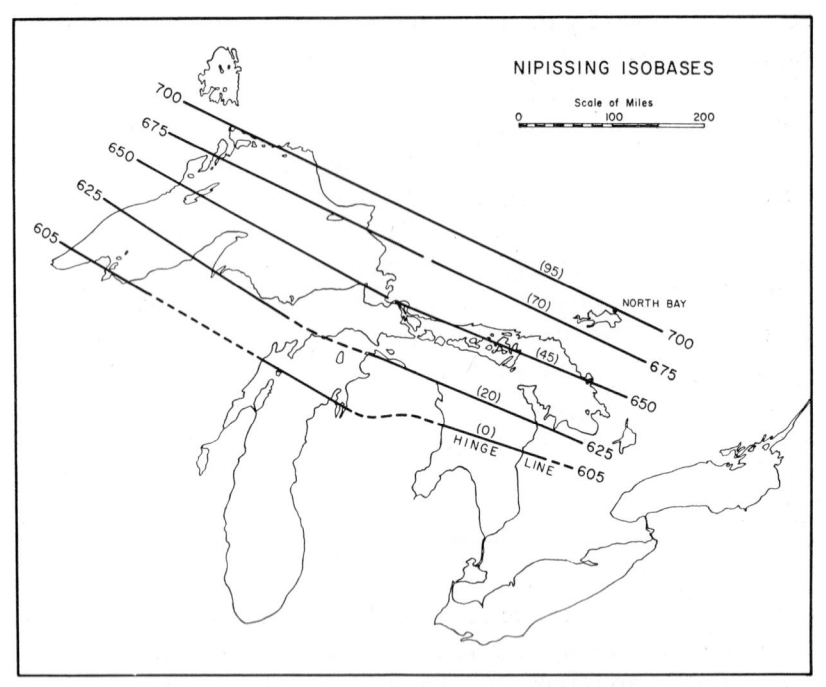

Fig. 51. Nipissing beach isobases. Lines of equal uplift of the beaches, with present altitudes above sea level in feet given by figures at ends of lines, and amount of uplift above original altitude in feet given by figures in parentheses. The line of zero uplift is the hinge line, and south of it the beaches are horizontal. (Modified from Leverett and Taylor, 1915, pl. 9.)

surface during low stages of the lakes followed by encroachment of the Nipissing stage, is a feature of some value in recognition of Nipissing beaches elsewhere in the region. Any locality lying on the isobase of the North Bay outlet would, however, have been raised the same amount as the outlet was raised; if the uplift began at the same time, and proceeded at the same rate, there would have been no encroachment by the lake at such localities. Along this line the lake shore would remain in a constant position from the Stanley low-water stage to the Nipissing stage. If the North Bay isobase lies south of the northeastern corner of Lake Superior, as is shown in Fig. 51, there would have been a recession of the shoreline in the area northeast of the isobase—again assuming that uplift began at the same time and proceeded at the same rate everywhere along the isobase.

The elevation of the Nipissing beach at Peninsula Harbor is

Table 16. *Altitudes of Nipissing beach on the north shore of Lake Superior.*[a]

Localities listed in order from southwest to northeast	Altitude (feet) above sea level
United States:	
Beaver Point, ordinary beach	606
Engles Cove, storm beach	612
Cove near Baptism River, storm beach	616
Kennedys Cove, ordinary beach	610
Little Marais, ordinary beach	612
Little Marais, storm beach	618
Pork Bay, ordinary beach	612
Thomasville, ordinary beach	615
Thomasville, storm beach	620
Cross River, ordinary beach	618
Cross River, storm beach	623
Lutsen by Poplar River, ordinary beach	624
Lutsen, storm beach	628
Carbou Point, ordinary beach	623
Carbou Point, storm beach	628
Good Harbor Bay, ordinary beach	627
Grand Marais, ordinary beach	630
Grand Marais, storm beach	638
Red Cliff Camp, ordinary beach	634
Horseshoe Bay	639
Chicago Bay, ordinary beach	638
Chicago Bay, storm beach	644
Grand Portage	638
Mount Josephine	643
Wauswaugoning Bay	644
McKellars Point	648
Carp River	652
Pigeon Point	657
Canada:	
Port Arthur	661
Mazokama	698
Simpson Island	693
Terrace Bay	696
Jackfish Bay	703
Peninsula Harbor	710

[a] Copied, with slight modification, from Leverett and Taylor (1915, p. 460).

given as 710 feet, in Table 16. With reference to this area, Quimby[3] has cited evidence reported by Bell (1928) as follows: "Around 1884, in the course of building the Canadian Pacific Railway, a copper hook or gaff and some charcoal were found in the bottom of a rock cavity under 18 feet of clay with a topping of gravel. The site is 600 feet east of Pic River along the railroad track. The elevation of the actual spot is about 100 feet above Lake Superior presumably to the locus of the copper, charcoal and wood" (p. 51). The copper tool and charcoal apparently were deposited on land, and the evidence cited indicates that water later covered the place to a depth of at least 25 feet. The most reasonable interpretation of this information is that the archeological material was emplaced sometime during a pre-Nipissing low-water stage and that the site was flooded by water rising to the Nipissing level. This interpretation does not seem, however, to be compatible with the conclusion reached if the position of the North Bay isobase as shown in Fig. 51 is correct. There are various possible explanations of the anomaly. One is as follows: even though the total post-Nipissing uplift is correctly represented by the North Bay isobase as projected across the Superior basin in Fig. 51, the uplift may have begun later along this line in the Superior basin than it did at North Bay. If this were true, water rising to the Nipissing level would have flooded the Pic River site and then would have receded after uplift began locally. The Nipissing shore at Pic River then would have been somewhat younger than the original Nipissing shore at North Bay.

An alternative explanation is that the North Bay isobase is incorrectly shown in Fig. 51, and that it should be drawn somewhere northeast of the northeastern shore of Lake Superior.

A further explanation is that the Superior basin may have had an outlet to the north before Nipissing time, and that there was a lake stage in the Superior basin lower than the Stanley low-water stage of the Huron basin; and that uplift of this unknown northern outlet raised the lake surface above the copper-and-charcoal site.

Additional detailed field work along the north side of the Su-

---

[3] Personal communication, Aug. 22, 1957.

perior basin is needed, and until more information is available the correlations of the Nipissing beaches in the area must be regarded as tentative.

OUTLETS

Before the Nipissing stage, as it is defined in the present book, came into existence, the entire discharge of the Huron, Michigan, and Superior basins went through the North Bay outlet and down the Mattawa and Ottawa rivers to the St. Lawrence.

When the North Bay outlet was raised sufficiently to bring the surface of the lake waters to the altitude of the lowest point in the sill of either of the old southern outlets, discharge to the southward began. In view of the proposed mechanism of development and adjustment of the Chicago and Port Huron outlets in Early Algonquin time, which is described in Chapter 8, it is assumed here that the sill of the Port Huron outlet may have been slightly higher than the sill of the Chicago outlet. If this were true, the first southward discharge of the Nipissing stage would have been through the Chicago outlet.

The difference between the sill elevations of the two old outlets must have been slight, however, and in a general summary of the Nipissing stage it seems satisfactory to assume that the initial southward discharge was practically simultaneously through the Port Huron and Chicago outlets. The Nipissing may, therefore, be characterized as a three-outlet stage essentially from its inception until the time when the North Bay outlet was eliminated.

Early in the Nipissing stage the North Bay outlet continued to carry a greater part of the discharge, and it was not until the sill of the North Bay outlet was raised above lake surface that discharge to the northeast ceased. During this part of the rise of land in the north the lake-surface elevation was controlled mainly by the discharge through the stable southern outlets. While a rise of the lake surface would result as more water was forced to flow through the southern outlets, these outlet channels could accommodate progressively greater quantities of flow as the water deepened in them. As a result of this, the rise of the lake surface would have been slow during the three-outlet stage.

The total rise of the lake surface during the three-outlet Nipissing stage probably amounted to about ten feet. This is deduced from the information that remnants of the Chicago outlet sill indicate a sill elevation there, in Nipissing time, of about 595 feet (Alden, 1918, p. 336). When the lake surface ultimately rose to 605 feet, the depth of water in the channel then was about ten feet.

If discharge at North Bay ceased while the lakes stood at the 605-foot Nipissing level, a two-outlet stage of the Nipissing ensued. If, however, the Port Huron outlet sill was cut down, thus lowering the lake surface and terminating the Nipissing, before discharge at North Bay ceased, the Nipissing remained a three-outlet stage throughout its history. This point is discussed in a subsequent paragraph.

CAUSE OF THE APPARENTLY STATIC LEVEL OF THE NIPISSING STAGE

Because the Nipissing stage is represented generally by a single strongly developed zone of shore features, it has been assumed that the lake surface stood at a nearly constant altitude for a long time. Such a static condition has been interpreted by some writers (Leverett and Taylor, 1915, p. 449, and Antevs, 1957, p. 139) as resulting from a halt in the uplift which brought the lake surface to the Nipissing level. Because later uplift raised the Nipissing beach in the northern areas, and because uplift of these areas is in progress at the present time, it is reasonable to question the suggested occurrence of a halt in the uplift during Nipissing time.

The rise of lake surface to initiate the Nipissing stage was caused by rise of the North Bay outlet; when discharge through the stable southern outlets began, the rise of the lake surface was slowed very considerably, and it may be assumed that the altitudes of the stable southern outlet channels would largely determine the level of the lake surface, whether uplift continued in the north or whether it ceased.

The question remains, however, as to whether a single strongly developed Nipissing beach could be formed in the northern areas if the land there continued to rise during the Nipissing stage. If it is assumed that rise of the land did continue, it may also be

THE NIPISSING GREAT LAKES

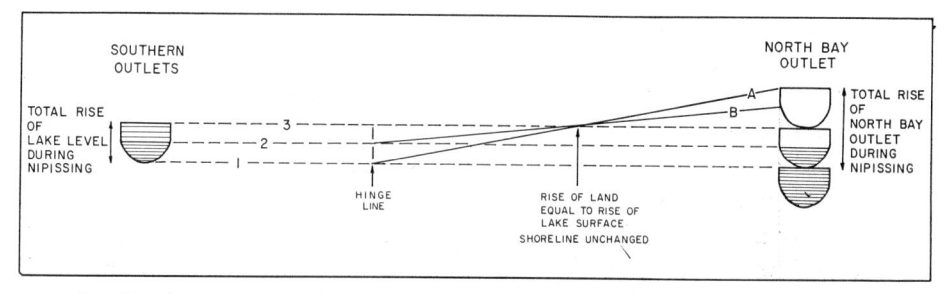

Fig. 52. Nipissing-stage variations in level of the lake surface and of the land. Line 1 represents lake level at the beginning of discharge to the south, when the surface of the outflowing stream at North Bay was raised to the level of the southern outlet sills. Line 2 represents lake level at a time of equal discharge through the southern oulets and through the North Bay outlet. Line 3 represents the Nipissing stage at its highest level, when the bottom of the North Bay outlet was raised to the level of the lake surface. The total rise from level 1 to level 3 was approximately ten feet. At a point between the hinge line and North Bay the rise of lake level was equal to the rise of land, and the shoreline was unchanged. Line A represents the tilted portion of line 1, at the time when lake surface stood at the level of line 3. Line B represents the tilted portion of line 2, at the time when lake surface stood at the level of line 3.

assumed that there was a rise of the lake surface during the Nipissing stage. As was pointed out in the discussion of outlets of the Nipissing stage, the entire discharge of the upper lakes was transferred from the North Bay outlet to the southern outlets gradually, and a rise of approximately ten feet must have occurred between the time of first discharge to the south, over sills with an altitude of about 595 feet, and the time of attainment of the highest (605-foot) level of the Nipissing stage. The rise of lake level was perhaps half as fast, on the average, than was the rise of the North Bay outlet. As a consequence of the rise of lake level, the Nipissing shore encroached on the land south of the hinge line, and it also encroached on the land north of the hinge for a considerable distance. Along a line approximately halfway between the hinge line and the North Bay outlet the amount of uplift during Nipissing time was equal to the amount of rise of the lake surface, and there the lake surface must have remained static with relation to the land. This relationship is illustrated in Fig. 52.

An isobase drawn halfway between the Nipissing hinge line (Fig. 51) and the North Bay outlet would lie fairly close to most of the type occurrences of the Nipissing beach in the area lying between Sault Ste. Marie and Mackinac Island and extending eastward through the Huron North Channel. In that area, therefore, only a single Nipissing shore zone would have occurred. South of that line of unchanging shoreline the Nipissing waters rose on the land. In the areas of transgressing shoreline, it is to be expected that the waves would obliterate the first-formed beaches, and by sweeping material up the beach would leave a record of static level only at the highest altitude reached.

The interpretation of the mechanism of maintenance of a nearly static level of the Nipissing stage which is presented in the foregoing paragraphs and in Fig. 52 is in accord with a more generally stated conclusion of Stanley: "I consider it far more logical that very gradual uplift proceeded uninterruptedly before, during, and after . . . the stage . . . represented by the Nipissing beach. The strength of . . . shorelines I attribute to slowly rising water level and to the time required to effect shift of outlet, considering depth of water in discharging rivers at thresholds." [4]

---

[4] Personal communication, Oct. 16, 1953.

chapter fifteen

THE TRANSITION FROM THE NIPISSING STAGE TO THE PRESENT GREAT LAKES

The Nipissing beach is the last extensive, strongly developed shoreline feature above the present Great Lakes. Several strand lines occur in the interval between the Nipissing and the present shores, in the northern parts of the Huron basin, but only one of these, the Algoma beach, is sufficiently extensive and well developed to have been named as a lake stage.

The history of the Nipissing to present transition in the three upper basins of the Great Lakes is here interpreted as one of continuous uplift of land in the north, producing a slow southward tilting of the region which hinged on a line lying within the lake basins. This tilting had begun long before the Nipissing stage came into existence, and it continues today. It is conceivable that there were halts in the tilting movement which allowed the Algoma and some of the other post-Nipissing beaches to form, but these shoreline evidences of static levels of the lake surface may also be explained by pauses in the downcutting of the Port Huron outlet.

CLOSE OF THE NIPISSING STAGE

The Nipissing stage was ended by downcutting of the Port Huron outlet. This is known from the facts that the Nipissing beach has been raised in the north but is horizontal in the southern parts of the Huron and Michigan basins, that it is traceable into the Port Huron and Chicago outlets, and that the Chicago outlet has been abandoned. The floor of the Chicago outlet is in part on bedrock, which could not be excavated by the outlet stream. The Port Huron outlet channel, represented by the present St. Clair River, Lake St. Clair, and the Detroit River, is still floored by unconsolidated material and it is therefore apparent that incision of this channel by the outflowing stream caused the lowering which brought the Nipissing stage to an end.

The downcutting of the Port Huron outlet must have been accomplished by a quantity of water which was greater than any previous discharge from the upper lakes. It was shown in Chapter 12 that the Algonquin stage, which had stood at the same altitude as the Nipissing and which had discharged through the same outlets, was restricted to the Huron and Michigan basins. During the Algonquin stage the Superior basin discharged through the St. Croix outlet to the upper Mississippi River. It was not until after the Algonquin stage had been ended by diversion of its discharge to a new, lower outlet to the east that the Superior basin was joined with the other two upper lakes.

The combined discharge of the Huron, Michigan, and Superior basins first flowed through one of the various outlets to the east of Georgian Bay, and the last of these outlets was the one at North Bay. When the North Bay outlet was raised sufficiently to bring the lake surface to the altitude of the floors of the old southern outlets at Chicago and at Port Huron, discharge to the southward began. For a long time after this the North Bay outlet carried a part of the discharge of the Nipissing stage. It was not until the floor of the North Bay outlet was raised to the level of the surface of the Nipissing lake that the entire discharge of that body of water could go through the southern outlets. In other words, the entire discharge of the upper three Great Lakes could not have gone through the two southern outlets until late in the Nipissing stage. It was a quantity of discharge in excess of the outflow of the previous Algonquin stage which initiated downcutting of the Port Huron outlet. A discharge exceeding that of the Algonquin would have been delivered through the southern outlets sometime before the entire discharge of the three upper lakes was diverted to the south. It seems possible, therefore, that downcutting of the Port Huron outlet may have begun before the discharge at North Bay had quite stopped. If the Nipissing stage was ended while there was still some discharge at North Bay, and this seems probable, the Nipissing was a three-outlet stage throughout its existence.

THE ALGOMA STAGE

The first shoreline features below the Nipissing beach which are distinctive enough to have been recognized and correlated over any considerable distance are those of the Algoma beach. This was named from an occurrence at an altitude of 630 feet at Algoma Mills, Ontario, on the North Channel of Lake Huron, by Taylor (Leverett and Taylor, 1915, p. 464). Correlative beaches have been reported for several localities in the northern Huron and Michigan basins, and from a few localities in the Superior basin (Leverett and Taylor, 1915, pp. 464–65). The author has observed a fairly extensive beach on Manitoulin Island southeast of Little Current and east of Sheguiandah at an average elevation of 628 feet which is correlated with the Algoma because it is the most distinctive feature between the Nipissing (at 650 feet) and the present beaches. In the area between Petoskey and Traverse City, Michigan, the Algoma beach is extensively developed as a terrace lying essentially parallel with, and a few feet below, the strongly developed Nipissing beach.

In the quest for a correlative of the Algoma beach in the southern parts of the Huron and Michigan basins, Leverett and Taylor (1915) were hindered by two factors. One of these was the absence of any older beaches above the present shores in the areas where the tilted older northern beaches approached horizontality (see Chaps. 12 and 14). The other was the fact that they had correlated the only well-developed post-Nipissing beach of the southern areas, a feature lying at an average altitude of 596 feet, with the Nippissing beach of the north. The author and Dreimanis have since shown that the Nipissing-stage lake stood at an altitude of about 605 feet (Chap. 14). It now seems reasonable to correlate the Algoma beach of the north with the beach lying at 596 feet in the south, particularly because the features so correlated represent the most strongly developed shoreline occurring between the Nipissing and the present shores in both areas.

The Algoma stage extended into the Superior basin, through a broad channel at Sault Ste. Marie, and stood at the same altitude there. The present sill of the St. Marys River channel holds Lake Superior at an altitude of 602 feet, but that sill had not been elevated sufficiently in Algoma time to raise the level of the water in the Superior basin above that in the lower lakes.

The Algoma beach is sufficiently well developed to require a period of static lake level for its formation. A pause in the post-Nipissing tilting of the northern areas could be postulated to account for the Algoma beach north of the hinge line; but this could not explain a static level of the lake south of the hinge line. The only plausible explanation seems to be a pause in the downcutting of the Port Huron outlet. Such a pause is indicated by the record of a comparatively recent shift of the Detroit River channel from a boulder-paved and bedrock-floored channel segment to a deeper channel in glacial drift, by lateral migration. This is described in detail by Leverett and Taylor (1915, pp. 494–96). Those writers proposed that the shift possibly was related to the first deepening of the Port Huron outlet which ended the Algonquin 605-foot stage of the upper lakes. The present author suggests that the downcutting which ended the Nipissing stage was brought to a halt when the Detroit River and the St. Clair River, above it, became graded to the temporary base level formed by the bedrock and boulder-paved channel segment; and that the Algoma stage resulted from this halt. The Algoma stage was then brought to a close when downcutting was resumed as a result of the lateral shift of the channel onto more easily eroded materials.

Another factor which may have had some bearing on the maintenance of a nearly static level in Algoma time was the existence of the two southern outlets. The sill altitude of the Chicago outlet has been cited (Alden, 1918, p. 336) as 595 feet. It appears, therefore, that a small amount of discharge continued to go through the Chicago outlet in Algoma time when the lake stood at an average elevation of 596 feet. While it is believed that the primary reason for the pause at the Algoma level was the existence of the rocky channel segment in the Detroit River channel, the tendency of the Port Huron outlet system to be graded to that temporary base level may have been strengthened by the fact that the Chicago outlet would discharge greater amounts of water whenever the lake surface rose. The peak flows through the Port Huron outlet thus were decreased, reducing the erosive power of the discharge. It was not until a further downcutting of the Port Huron outlet sill occurred, presumably in response to the lateral shift of the Detroit River channel onto more easily eroded material, that the lake was lowered sufficiently to prevent its high-water stages from spilling through the Chicago outlet.

If the sequence of events described in the foregoing paragraphs is correct, it may be concluded that the full discharge of the three upper Great Lakes did not pass through the Port Huron outlet and into Lake Erie until the close of the Algoma stage.

In the interval between the Algoma beach and the present shores of the upper Great Lakes there are several ridges of sand or gravel. None of these is of sufficient magnitude to be singled out as a representative of a distinct lake stage.

THE TRANSITION FROM THE ALGOMA STAGE TO THE PRESENT GREAT LAKES

The most reasonable view of this part of the lake history seems to be that uplift of the land in the north probably continued without halt, but possibly at a diminishing rate; and that downcutting of the Port Huron outlet probably continued without interruption, at least until the present level of the upper lakes was attained.

The several beach ridges which were formed during this interval of time constitute broad belts of distinctive topography in some places in the northern part of the region. In view of the statements made in the foregoing paragraph, the formation of the series of post-Algoma beaches must be explained by a mechanism which is not dependent on halts in either uplift of the land or of downcutting of the outlet. It is probable that the lake surface fluctuated in altitude both seasonally and through cycles of several years' duration, during all of post-Nipissing time just as it does today (Figs. 20A and 20B). A beach ridge formed by storm waves during a time of unusually high water would be left behind by fall of lake surface and rise of the land during several succeeding years. Storms occurring during the next high-water period then would construct a beach somewhat lower on the shore.

There is no evidence of a change in mean lake level in the Huron and Michigan basins in historic time, and it is generally assumed that the rate of downcutting of the Port Huron outlet at the present time is negligible. This condition has come about because the Port Huron to Lake Erie channel has been reduced to a low gradient, partly because of the downcutting which had occurred earlier and partly because the surface of Lake Erie has been raised by uplift of its outlet at the northeastern corner of the basin.

Lake Superior was raised above the Huron-Michigan level by uplift of its outlet, but the exact time at which this occurred is not known. It is possible that water level in the Superior basin was lowered below the Algoma level before the upwarp had raised the channel bottom sufficiently to produce a higher level in that basin. Since Lake Superior emerged as a separate lake it has been rising and flooding its southern-most shores because they lie south of the isobase of uplift which passes through the outlet.

chapter sixteen

RADIOCARBON CHRONOLOGY OF GREAT LAKES HISTORY

The events of lake history which have been reviewed in detail in Part 2 of this book were described without reference to an absolute time scale. In the present chapter an absolute time scale is given. This is based on the radiocarbon dates which are listed in Tables 17 through 21 and evaluated in the following discussion. Table 22 gives a correlation chart of Great Lakes events plotted according to the radiocarbon time scale. It must be understood that the scale is in "radiocarbon years" before the present. Any errors present in the dating method are reflected in the time scale. Further, only a few events of lake history have been dated and the positions of many of the boundaries shown in Table 22 are located by interpolation. The time scale is, therefore, only an approximation.

DATES OF THE TWO CREEKS INTERVAL AND THE VALDERS GLACIAL SUBSTAGE

The forest bed at Two Creeks, Wisconsin, marks one of the major events of Great Lakes history and it was one of the first geological horizons to be dated by the radiocarbon method. Because its age lies well within the range of the method and because abundant material was available, samples of Two Creeks wood have been run by various laboratories as interlaboratory checks on the dating method. The ten dates in Table 17 range from 10,877 ± 740 years to 12,168 ± 1500 years. Their simple average

Table 17. *Radiocarbon dates of Two Creeks–interval materials from northeastern Wisconsin.*

Lab. no.[a]	Description	Dates, C 14 yr.	Present assignment	References
C-308		10,877 ± 740		Arnold and Libby, 1951, p. 117.
C-365		11,437 ± 770		Arnold and Libby, 1951, p. 117.
C-366		11,097 ± 600		Arnold and Libby, 1951, p. 117.
C-536		12,168 ± 1500		Arnold and Libby, 1951, p. 117.
C-537	*Two Creeks, Wis.* Wood and peat from Two Creeks forest bed.	11,442 ± 640	Two Creeks interval	Arnold and Libby, 1951, p. 117.
M-342		10,700 ± 600		Crane, 1956, p. 669.
M-343		10,400 ± 600		Crane, 1956, p. 669.
W-42		11,350 ± 120		Suess, 1954, p. 471.
W-83		11,410 ± 180		Suess, 1954, p. 471.
Y-227		11,130 ± 350		Preston, Person, and Deevey, 1955, p. 958.
—	(average)	11,200		
C-630	*Kimberly, Wis.* Wood from depth of 10 feet in a 25-foot varved-clay section, apparently deposited in lake ponded in front of Valders ice.	10,676 ± 750	Two Creeks interval	Libby, 1952, p. 678.
C-800	*Appleton, Wis.* Wood from Valders till.	11,471 ± 500 10,241 ± 650	Two Creeks interval	Libby, 1954a, p. 139.
Y-237	*Menasha, Wis.* Wood from Valders till.	11,690 ± 370	Two Creeks interval	Preston, Person, and Deevey, 1955, p. 958.
Y-147X	*Green Bay, Wis.* Wood from Valders till.	11,940 ± 390	Two Creeks interval	Preston, Person, and Deevey, 1955, p. 958.

[a] C = Chicago; M = Michigan; W = Washington; Y = Yale.

is 11,200 years. Sixty per cent of the dates fall between 11,000 and 11,500 years, and the ranges of error of all of the dates lie within or extend into the 11,000- to 11,500-year bracket. The two dates with a small range of error average 11,380 years. It seems reasonable, therefore, to assign an age of about 11,400 radiocarbon years to the Two Creeks material.

Because the Two Creeks forest was flooded when advancing Valders ice closed the northeastern outlet of the Michigan basin (Chap. 5), the 11,400-year date represents an age a little greater than that of the maximum of the Valders glaciation. The Valders maximum is considered to have occurred about 11,000 radiocarbon years ago (Flint and Deevey, 1951, p. 263).

In accordance with the dates derived in the foregoing paragraphs, Table 22 shows the Valders maximum at 11,000 years and the close of the Two Creeks low-water stage at 11,375 years.

Dates from four other localities in northeastern Wisconsin are included in Table 17. These all represent the ages of wood imbedded in Valders till or in red clays apparently deposited in a lake ponded by Valders ice. While there is a considerable variation between the individual dates, they all are close enough to the age of the Two Creeks material to be correlated with it.

DATES OF LAKE STAGES

The radiocarbon dates given in Table 18 generally are single determinations and these involve a greater probable error than is present in the date of the Two Creeks material which was derived from ten determinations. Most of these other dates, however, are remarkably consistent with the ages which may be predicted for various events when they are considered in relation to the time of the Valders maximum.

The Algoma stage is dated by sample C-608 (Table 18) at 2619 ± 220 years, as described in Chapter 15. On the correlation chart (Table 22) it is given a range from 2250 to 3000 years, but the duration of the stage may have been considerably shorter.

The most reliable Nipissing date is that of sample C-504, 3656 ± 640 years, from Sand Island in Lake Superior, where the Nipissing beach is separate from the shore features of other stages.

Table 18. *Radiocarbon dates of various lake stages.*

Lab. no.[a]	Description	Dates, C 14 yr.	Present assignment	References
C-608	Burley site, near Port Franks, Ont. Charcoal from Indian dwelling site, "Occupational Horizon No. 1," altitude 594 feet.	2,619 ± 220	Algoma	Libby, 1952, p. 674; Dreimanis, 1952, p. 72; Hough, 1953c, p. 139; Dreimanis, 1958, in press.
C-364	Dalton, Ill. Wood from base of lake sand overlying till; from "Toleston level, Lake Chicago."	3,469 ± 230	Nipissing	Arnold and Libby, 1951, p. 115.
C-504	Sand Island, Bayfield County, Wis. Peat from Nipissing level.	3,656 ± 640	Nipissing	Arnold and Libby, 1951, p. 118.
S-25	Blackwell, Ont. Wood from base of Nipissing or late pre-Nipissing gravel and sand, altitude 582 feet.	4,610 ± 210	Late pre-Nipissing	Dreimanis, 1958, in press.
S-24	Blackwell, Ont. Wood from base of Nipissing or late pre-Nipissing gravel and sand, altitude 590 feet.	4,650 ± 200	Late pre-Nipissing	Dreimanis, 1958, in press.
Y-238	Marinette, Wis. Wood from a sewer trench, collected before 1943 and stored for several years. Stratigraphy of site not specified. According to Thwaites, the altitude of occurrence must lie between 580 and 600 feet, and the wood probably was emplaced during Nipissing stage, either in the lake or in the mouth of Menominee River.	4,880 ± 190	Late pre-Nipissing?	Preston, Person, and Deevey, 1955, p. 958.
L-312	Meaford, Ont. Wood from lake bed clay, at 597 feet, overlain by 8.2 feet of sand. Altitude indicates Nipis-	6,300 ± 150	Late pre-Nipissing?	Broecker and Kulp, 1957, p. 1325.

Table 18. *Radiocarbon dates of various lake stages (continued).*

Lab. no.[a]	Description	Dates, C 14 yr.	Present assignment	References
	sing or late pre-Nipissing age for base of sand.			
C-674	*Chicago, Ill.* Wood from Lake Chicago sands; at altitude of 578 feet, overlain by 14 feet of sand. "Bretz and Horberg . . . are of the opinion that this sand represents a Toleston and post-Toleston deposit and that at a depth of 14 feet the sand is probably Toleston."	8,200 ± 480	Algonquin	Libby, 1952, p. 674.
C-526	*Bellevue, Ohio.* Wood from altitude of 620 feet, Lundy level.	8,513 ± 500	Lundy?	Libby, 1951, p. 292.
W-199	*Marilla, N.Y.* Wood from lake clays located between sand bar and beach of Lake Warren, 32 inches deep in deposits believed to be of late Warren age.	9,640 ± 250	Lowest Warren?	Rubin and Suess, 1955, p. 483.
M-359	*Ontonagon County, Mich.* Wood buried under 80 feet of red lake clay, presumably from Glacial Lake Ontonagon. Should date the early stages of Lake Duluth in the Superior basin.	10,220 ± 500	Duluth	Crane, 1956, p. 669.
M-288A	*South Haven, Mich.* Wood from above silt layer and below bedded sands. "This sample should date the Two Creeks low-water stage of the Michigan basin."	11,200 ± 600	Two Creeks?	Crane, 1956, p. 668, and Zumberge and Potzer, 1956, p. 276.
W-167		10,860 ± 350		Rubin and Suess, 1955, p. 483.
Y-293B		10,550 ± 150		Preston, Person, and Deevey, 1955, p. 958.
—	(average)	10,870		

Table 18. *Radiocarbon dates of various lake stages (concluded).*

Lab. no.[a]	Description	Dates, C 14 yr.	Present assignment	References
C-801	Dyer, Ind. Wood from sand spit built westward over lagoon deposit of Lake Chicago, apparently of Glenwood stage.	10,661 ± 460 11,284 ± 600	Glenwood III	Libby, 1954a, p. 137.
W-161		12,200 ± 350		Rubin and Suess, 1955, p. 483.
—	(average)	11,383		
W-140	Dyer, Ind. Wood from peat of lagoon deposit of Lake Chicago, apparently of Glenwood stage, beneath the sand spit represented by C-801 and W-161.	12,650 ± 350	Glenwood II	Rubin and Suess, 1955, p. 483.
Y-240	Bellevue, Ohio. Wood fragments imbedded in beach sediments of Lake Whittlesey, 4.5 miles southeast of Bellevue.	12,800 ± 250	Whittlesey	Barendsen, Deevey, and Gralenski, 1957, p. 912.
W-33	Cleveland, Ohio. Twigs, roots, and leaves from a deposit representing a Lake Arkona lagoon, at an altitude of 690 feet, and overlain by 10 to 12 feet of sand and silt deposited as the lake rose to the Whittlesey stage.	13,600 ± 500	Lowest Arkona	Suess, 1954, p. 469, and Flint and Rubin, 1955, p. 649.

[a] C = Chicago; L = Lamont; M = Michigan; S = Saskatchewan; W = Washington; Y = Yale.

Sample C-364 (Dalton, Illinois) is from a location and an elevation which might have contained material from any of the high-water stages from the Glenwood to the latest Nipissing stage. Its date, 3469 ± 230 years, serves as a minimum for the Nipissing.

The next four samples of Table 18, S-25, S-24, Y-238, and L-312, all are from altitudes below the level of the Nipissing water plane. Because of this, they may have been desposited late in the transition from the Chippewa-Stanley low-water stage to the Nipissing level. The fact that all of them have ages in excess of 4600 years strongly suggests, however, that the Nipissing may be of somewhat greater age than is shown on the correlation chart (Table 22). The principal objection to placing the beginning of the Nipissing stage at a date earlier than 4000 years ago is that the end of the Algonquin has been placed at 8000 years ago, and the intervening 4000-year period seems to be required to accommodate the several post-Algonquin low stages and the upwarp of the land which returned the lake surface to the 605-foot level.

The Algonquin-Nipissing interval is represented in a section at South Haven, Michigan, by a 30-inch peat layer (Zumberge and Potzer, 1956). Samples from various levels in the peat are listed in Table 19. Samples C-849 and M-291, from the top of the peat, give ages of 4816 ± 290, 4000 ± 300, and 4000 ± 350 years, the average of which is 4272 years. Wood from the middle of the peat layer, represented by samples C-848, L-214, M-290, and Y-169B, has been dated at from 6659 ± 350 to 5000 ± 400 years by various laboratories, and the average of its dates is 5644 years. One sample of peat from a position seven inches above the base of the peat layer (M-289) was dated at 6330 ± 400 years. Three samples from the base of the peat yielded dates ranging from 6744 ± 530 to 10,790 ± 200 years, the average of which is 8346 years. If the two extreme dates are discarded, the remaining two determinations (of sample M-288) average 7925 years. The data of Table 19 thus may be interpreted to signify that the Algonquin-Nipissing interval existed between approximately 8000 and 4000 years ago. This is in accord with the conclusions of Zumberge and Potzer (1956, p. 271).

Returning to the data of Table 18, it may be seen that sample C-674 (Chicago) gives an additional date which may be interpreted as late Algonquin in age. The sample is from a location which might have contained material dating from any of the high-water stages from the Glenwood to the Algoma stage. The age of the sample, 8200 ± 480 years, apparently is late Algonquin.

Sample C-526, from Bellevue, Ohio, is wood from a fossil forest overlain by fine-grained sediment, at the altitude of the Lundy beach, and it presumably should yield a date for the Lundy stage. Its date, 8513 ± 500 years, is too young according to the correla-

Table 19. *Radiocarbon dates of the Algonquin to Nipissing section, South Haven, Michigan.*[a]

Lab. no.[b]	Description	Dates, C 14 yr.	Present assignment	References
C-849	South Haven, Mich. Peat from top of 30-inch peat layer underlying dune sand and overlying lacustrine sand and gravel, which overlies organic layer represented by M-288A (Table 18).	4,816 ± 290	Early Nipissing	Libby, 1954b, p. 735, and Zumberge and Potzer, 1956, p. 276.
M-291		4,000 ± 300 4,000 ± 350		Crane, 1956, p. 668.
—	(average)	4,272		
C-848		6,232 ± 310 6,659 ± 350		Libby, 1954b, p. 735, and Zumberge and Potzer, 1956, p. 276.
L-214	South Haven, Mich. Wood from middle of 30-inch peat layer underlying dune sand and overlying lacustrine sand and gravel.	5,130 ± 110	Nipissing-Algonquin interval	Broecker, Kulp, and Tucek, 1956, p. 161.
M-290		5,000 ± 400 5,185 ± 400		Crane, 1956, p. 668.
Y-169B		5,660 ± 100		Preston, Person, and Deevey, 1955, p. 958.
—	(average)	5,644		
M-289	South Haven, Mich. Peat from 7 inches above base of 30-inch peat layer underlying dune sand and overlying lacustrine sand and gravel.	6,330 ± 400	Nipissing-Algonquin interval	Crane, 1956, p. 668, and Zumberge and Potzer, 1956, p. 276.
C-846		6,744 ± 530		Libby, 1954b, p. 735, and Zumberge and Potzer, 1956, p. 276.
M-288	South Haven, Mich. Peat from base of 30-inch peat layer underlying dune sand and overlying lacustrine sand and gravel.	8,350 ± 500 7,500 ± 500	Latest Algonquin	Crane, 1956, p. 668.
Y-293A		10,790 ± 200		Preston, Person, and Deevey, 1955, p. 958; corrected by Flint, 1956, p. 270.
—	(average)	8,346		

[a] Additional dates from South Haven are given in Table 18.
[b] C = Chicago; L = Lamont; M = Michigan; Y = Yale.

tion chart of Table 22, where the Lundy stage is shown as terminating 9750 years ago.

Sample W-199, from Marilla, New York, is from lake clays believed to be of late Warren age. Its date, 9640 ± 250 years, is obviously too young in view of the generally accepted idea that the latest Warren stage was formed by the advance of the Valders ice, which reached its maximum about 11,000 years ago.

Glacial Lake Ontonagon, an ice-margin lake in the Superior basin formed early in the retreat of the Valders ice front, is dated by sample M-359 at 10,220 ± 500 years. This is in good agreement with the correlations shown in Table 22.

The Two Creeks low-water stage of the Michigan basin is recorded at South Haven, Michigan, by a zone of organic matter overlying a blue-gray silt, presumably deposited in deep water in an earlier lake stage, and underlying bedded gravels and sands containing streaks of organic matter, presumably deposited in a later lake stage (Zumberge and Potzer, 1956, p. 276). The zone of organic matter yielded dates ranging from 10,550 ± 150 to 11,200 ± 600 years. While the average of these dates, 10,870 years, seems slightly low, the zone is considered to be of Two Creeks age.

Dating of samples from sand spit and lagoon deposits at Dyer, Indiana, has had an interesting history. The first two dates were obtained for sample C-801, from the younger sand spit which obviously was built at the Glenwood (640-foot) level. The dates obtained, 10,661 ± 460 and 11,284 ± 600 years (Table 18), were much too young according to the interpretations of lake history given by Bretz (1951b, p. 421). Additional samples from the locality were dated, and C-871 from the sand spit yielded a date of 18,500 ± 500 years, and C-872 from the underlying lagoon deposit yielded a date in excess of 21,000 years (Libby, 1954b, p. 735). Further dating by the U.S. Geological Survey laboratory gave a date of 12,200 ± 350 years (sample W-161) for the younger sand spit and a date of 12,650 ± 350 years (sample W-140) for the underlying lagoon deposit. The extreme values of samples C-871 and C-872 are considered erroneous and they are not included in Table 18. The Dyer samples listed in Table 18 indicate that Lake Chicago stood at the Glenwood level as late as 12,650 ±

350 years ago and possibly as late as 10,661 ± 460 years ago. The average of the dates for C-801 and W-161, representing the younger sand spit, is 11,383 years. This suggests that the sand spit was formed in the third Glenwood stage shown in Table 22. The date of 12,650 ± 350 years for W-140, representing the older lagoon deposit, suggests that that deposit was formed during the second Glenwood stage shown in Table 22.

The Whittlesey stage in the Erie basin is dated by sample Y-240, from near Bellevue, Ohio, as 12,800 ± 250 years old. This confirms the correlation of the Whittlesey with a Glenwood stage of Lake Chicago, as shown in Table 22. Because the Whittlesey stage was formed by the Port Huron glacial ice advance, the Port Huron (Mankato) glacial substage maximum is assigned an age of 13,000 years.

A late Arkona stage in the Erie basin is dated by sample W-33, from Cleveland, Ohio, as 13,600 ± 500 years old. The Arkona date is from organic debris of a lagoon deposit which is overlain by sand and silt deposited as the lake rose to the subsequent Whittlesey stage. The Arkona and Whittlesey dates are in the correct relationship to each other.

The Arkona date, 13,600 ± 500 years, is the oldest date available which can be related directly to events of lake history. Because of this the time scale of Table 22 is not extended beyond 14,000 years. All of the Maumee stages, the earlier Arkona stages, and possibly a Cary–Port Huron low-water stage occurred prior to the latest Arkona. Three major retreats and two major advances of the Cary glacial ice front appear to be correlated with these earlier lake stages, as shown in Table 22. It therefore appears that 2000 to 3000 years is a reasonable estimate of the duration of this earlier part of the Great Lakes history.

THE ST. LAWRENCE VALLEY

The middle portion of the St. Lawrence valley, from Montreal to Quebec, was covered with glacial ice from early Wisconsin time until after the Valders substage maximum. This has been shown by detailed studies of the Pleistocene deposits, with radiocarbon dating of significant horizons as listed in Table 20.

Because the St. Lawrence valley was blocked by ice from the time of the earliest known Great Lakes until after the Valders maximum, all of the eastward discharge, including that occurring during the Cary–Port Huron interval and the Two Creeks interval, must have passed down the Hudson River valley.

The dates of the post-Valders St. Lawrence Sea deposits, ranging from 10,630 ± 330 to 11,370 ± 360 years, are all greater than was expected. According to the correlation chart (Table 22) the marine embayment reached the upper St. Lawrence valley not earlier than about 8000 years ago. There is thus an apparent discrepancy of about 2500 to 3000 years in the dates. Flint (1956, p. 278) has rejected the three dates, suggesting that because they were all obtained from carbonate shell material it is possible that old carbonate was incorporated in the shells during growth.

THE NORTH BAY OUTLET AND THE COCHRANE PROBLEM

The North Bay outlet was first used by discharge from the Great Lakes sometime after the Algonquin stage and probably not much earlier than the Stanley low-water stage. No radiocarbon dates are available which apply to specific events in the Algonquin to Nipissing sequence. The Chippewa and Stanley stages have been placed arbitrarily as beginning at about 6500 years ago in Table 22.

Radiocarbon dates from the Cochrane district, Ontario, listed in Table 21, indicate minimum ages of the last glaciation there to range from 5300 ± 300 to 6730 ± 200 years. These dates apparently do not allow sufficient time between the disappearance of ice from North Bay, if it occurred 6500 years ago as shown in Table 22, and the disappearance of ice from the Cochrane district, 180 miles to the north, though the discrepancy is not great. Flint (1956, p. 279) has suggested that glacial ice may have persisted longer in the North Bay area than it did at Cochrane.

LAKE BARLOW-OJIBWAY

The Cochrane ice advances, described in the foregoing paragraph, were over the sediments of Lake Barlow-Ojibway. That lake was ponded between the Great Lakes–Hudson Bay divide and the James Bay lobe of the waning ice sheet, and it may have

Table 20. *Radiocarbon dates from the St. Lawrence valley.*

Lab. no.[a]	Description	Dates, C 14 yr.	Present assignment	References
Y-215	Hull, Que. Shells from Champlain Sea deposit.	10,630 ± 330	St. Lawrence Sea,[b] post-Valders	Preston, Person, and Deevey, 1955, p. 956.
Y-216	Uplands, Ont. Shells from Champlain Sea deposit.	10,850 ± 330		
Y-233	Notre Dame des Neiges, Que. Shells from 3 feet or less below surface of Champlain Sea deposit.	11,370 ± 360		
Y-254	Les Vieilles Forges, Que. Wood from below 204 feet of stratified deposits including till and silt. Believed to be correlative with Y-242.	>29,630	Pre-Wisconsin or early Wisconsin	Preston, Person, and Deevey, 1955, p. 957.
Y-255	Les Vieilles Forges, Que. Peat from same layer as the wood of Y-254.	>30,840		
Y-256	Pierreville, Que. Wood and compressed peat from below 38.5 feet of thin-bedded silt and sand, overlying thick section including varved sediments believed to be correlative with varved material overlying Y-242 and Y-255.	>29,630		
Y-242	St. Pierre–Les Becquets, Que. Wood underlying varved silt, which is overlain elsewhere by gray till.	>29,630 >30,840		
W-189		>40,000		Rubin and Suess, 1955, p. 485.

[a] W = Washington; Y = Yale.
[b] "St. Lawrence Sea" is used in preference to "Champlain Sea" in this book.

Table 21. *Radiocarbon dates from the Cochrane, Ontario, district.*

Lab. no.[a]	Description	Dates, C 14 yr.	Present assignment	References
W-176	Cochrane, Ont. Wood and peat overlying latest till. Serves as a minimum date for latest glaciation of area.	5,300 ± 300		Rubin and Suess, 1955, p. 485.
W-136		6,380 ± 350		Rubin and Suess, 1955, p. 485.
Y-222	*Dugwal, Ont.* Peat from a bog; underlain by gray homogeneous clay. The bog overlies till laid down by the latest glaciation in the Cochrane district.	6,730 ± 200	Post-glacial	Preston, Person, and Deevey, 1955, p. 957.

[a] W = Washington; Y = Yale.

had an east-west length of 400 miles. The discharge apparently was southeastward via the Ottawa River to the St. Lawrence River (Flint, 1957, p. 350). The life of Lake Barlow-Ojibway apparently was short, because it must have existed between the time of disappearance of ice from the North Bay–Ottawa district and the time of the Cochrane ice advances.

LAKE AGASSIZ

Lake Agassiz, which existed northwest of the Great Lakes region in Ontario, Manitoba, Minnesota, and North Dakota, has not been discussed in the foregoing chapters because that lake had a history which was mainly separate from that of the Great Lakes. However, Lake Agassiz discharged into the Superior basin during certain brief periods. Recent work reported by Elson (1957) has shown the routes of discharge and gives radiocarbon dates which permit correlations between some of the events occurring in the Agassiz basin and in the Great Lakes region.

Two principal stages of the northwestern lake are recognized: Lake Agassiz I and Lake Agassiz II. The following correlations are possible. The Agassiz I stage coincided in part with the waning of the Port Huron (Mankato) ice and with an early part of Lake Keweenaw. It discharged to the western end of the Superior basin, probably via Brule Creek. An Agassiz I–Agassiz II interval, during which the lake was drained, coincided with the Two Creeks interval and Lake Keweenaw. Lake Agassiz II was formed by the Valders-substage ice, and at the time of the Valders maximum it discharged southward to the Mississippi drainage system. During retreat of the Valders ice front Lake Agassiz II discharged to the Superior basin, first down the Black Sturgeon spillway and later down the Pikitigushi and other northern spillways to Lake Nipigon (Elson, 1957, p. 1001).

Table 22. *Correlation chart of events of Great Lakes history plotted on a radiocarbon-dated time scale.*[a]

DATES C14 yr. B.P. (approx.)	GLACIAL EVENT	SUPERIOR BASIN	MICHIGAN BASIN	HURON BASIN	ERIE BASIN	ONTARIO BASIN
1,000		SUPERIOR (602)	MICHIGAN (580)	HURON (580)	ERIE (573)	ONTARIO (246)
2,000		(St. Marys R.) →	(Straits of Mackinac) →	(St. Clair - Detroit R.) →		
3,000		ALGOMA (596) ← (Des Plaines R.)		(St. Clair - Detroit R.) →		
4,000		NIPISSING (605) ← (Des Plaines R.)	(North Bay: Ottawa - St. Lawrence R.)	(St. Clair - Detroit R.) →		
5,000		TRANSITION (St. Marys R.) →	TRANSITION (Straits of Mackinac) →	TRANSITION (North Bay: Ottawa R.) →		EARLY ONTARIO (O) (St. Lawrence R.) →
6,000		SUB-MINONG ? (?) MINONG DERONDA KODONCE (St. Marys TOFTE valley to MARAIS Huron) → LUTSEN SUB-DULUTH (?) → (Au Train - Whitefish - Green Bay) →	CHIPPEWA (230) ← (Submerged Mackinac R.) KORAH SHEGUIANDAH PAYETTE (465) CEDAR POINT (493) PENETANG (510) WYEBRIDGE (540) "UPPER GROUP"	STANLEY (200-?) (North Bay: Ottawa R.) → (?) → (Fossmill: Ottawa R.) → (Unknown outlets) → (Trent valley) →		ST. LAWRENCE MARINE EMBAYMENT (St. Lawrence valley) →
7,000						
8,000		DULUTH	ALGONQUOIN (605) ← (Des Plaines R.) KIRKFIELD (565?) TOLESTON - EARLY ALGONQUIN (605) ← (Des Plaines R.)	(St. Clair - Detroit R.) → (Trent valley: L. Iroquois) →	EARLY ERIE (540?) (Niagara R.) →	FRONTENAC (Covey: to east) → IROQUOIS (330?) (Rome: Mohawk & Hudson R.) → DAWSON→ Ice → EARLY IROQUOIS
9,000		← (St. Croix R.)				
10,000		VARIOUS ICE-MARGIN LAKES ← (St. Croix R.)	CALUMET (620) ← (Des Plaines R.) GLENWOOD III (640) ← (Des Plaines R.)	← (to L. Calumet) ← (to L. Glenwood) ← (Grand R.)	LUNDY (620) GRASSMERE (640) LOWEST WARREN (675) WAYNE (655)	DANA (590?) (to Mohawk & Hudson R.) → Glacial ice (to Mohawk & Hudson R.) →
11,000	VALDERS	Glacial ice ← (?)				
12,000		KEWEENAW (to Michigan & Huron) → ← (?)	TWO CREEKS LOW-WATER STAGE (Trent valley?) → (St. David filled gorge) → (to Hudson R.?) →			
13,000	PORT HURON (MANKATO)	Glacial ice	GLENWOOD II (640) ← (Des Plaines R.)	WARREN (690-682) ← (Grand R.) SAGINAW in Saginaw Bay ← (Grand R.) ← (ice margin) LOWEST ARKONA II (695) ← (Grand R.)	WHITTLESEY (738) ← (Ubly channel)	VANUXEM II and HALL II ? (to Mohawk & Hudson R.) →
14,000		← (?) ? ← (?)	CARY - PORT HURON LOW-WATER STAGE			(Unknown eastern outlets) →
	CARY / LAKE BORDER	Glacial ice	GLENWOOD I (640) ← (Des Plaines R.)	ARKONA (710-700-695) ← (Grand R.) EARLY SAGINAW in Saginaw Bay ← (Grand R.)	Glacial ice in Huron MIDDLE MAUMEE (780) (Fort Wayne: to Wabash R.) LOWEST MAUMEE (760) ← (to Saginaw Bay via ice margin)	VANUXEM I (Syracuse: Mohawk & Hudson R.) → ← (to Erie) HALL NEWBERRY (to Susquehanna R.)
	TINLEY-DEFIANCE		Glacial ice EARLY LAKE CHICAGO ← (Des Plaines R.)	Glacial ice	HIGHEST MAUMEE (800) (Fort Wayne: to Wabash R.)	
	VALPARAISO		Glacial ice		Glacial ice	Glacial ice

[a] Lake stage names are shown in capital letters and are followed by their original altitudes in feet above sea level (if known). Outlets of the lakes are shown in lower-case letters in parentheses. Glacial ice is shown in italics.

LAKE STAGE MAPS

Figures 53-75

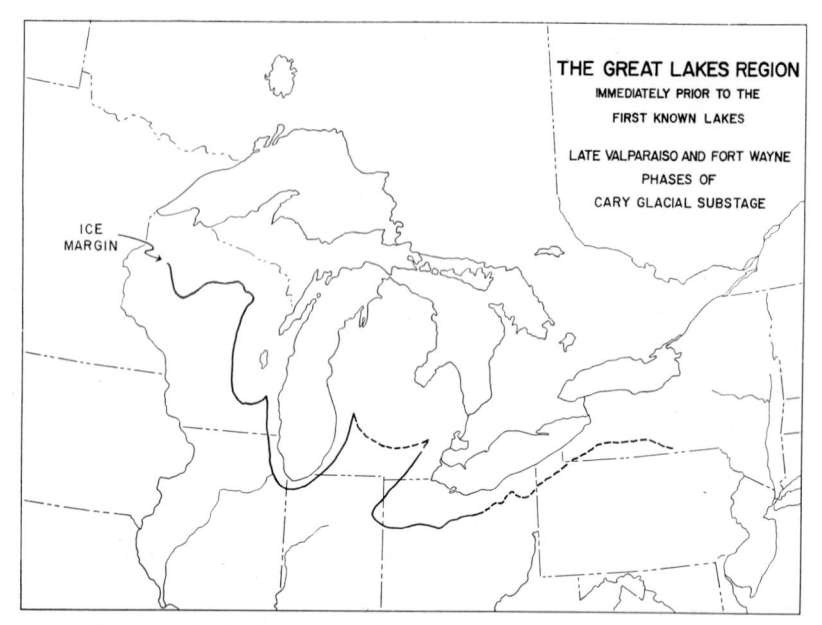

Fig. 53. The Great Lakes region immediately prior to the first known lakes; the late Valparaiso and Fort Wayne phases of the Cary glacial substage.

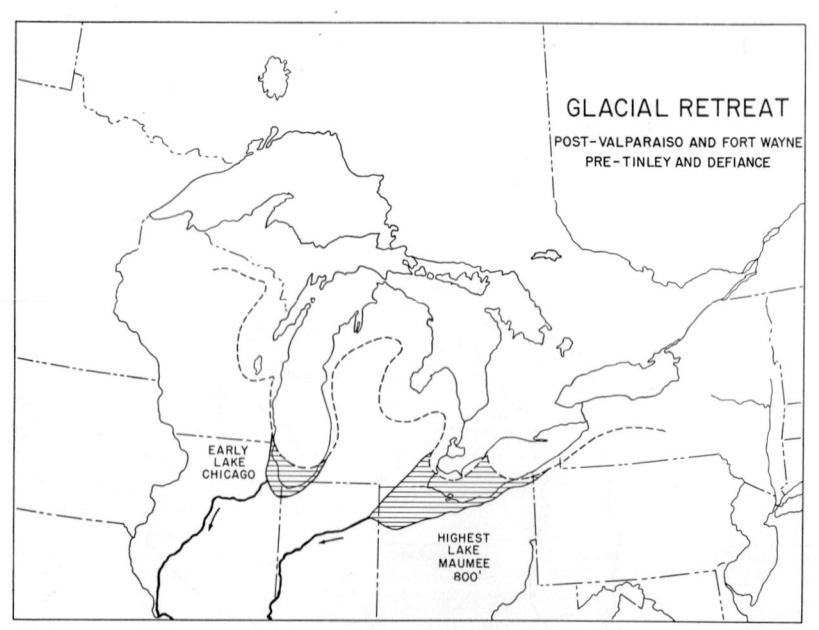

Fig. 54. Early Lake Chicago and Highest Lake Maumee, occurring during the glacial retreat between the Valparaiso–Fort Wayne advances and the Tinley-Defiance advances. Position of ice front conjectural.

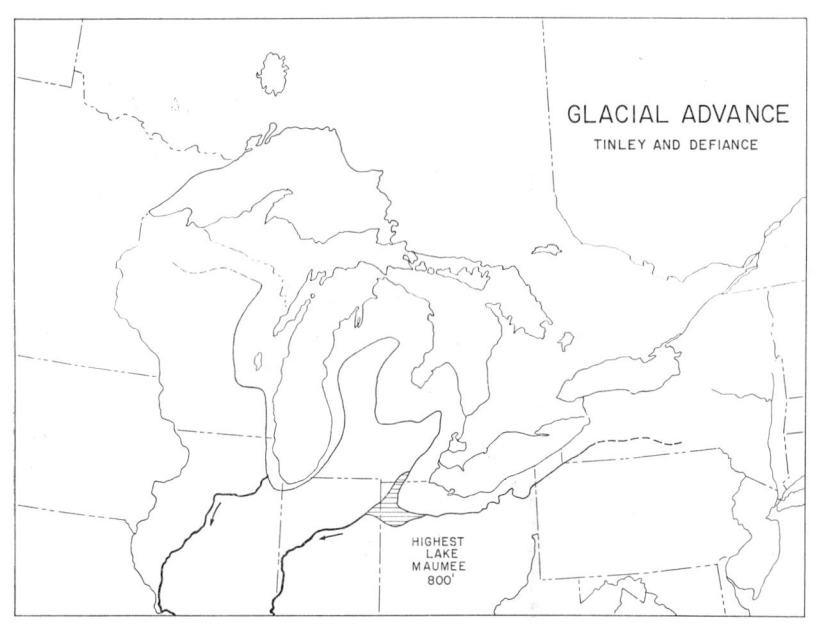

Fig. 55. Highest Lake Maumee (continued). Advance of ice to the Defiance moraine constricted Lake Maumee, and advance of ice to the Tinley moraine filled the Michigan basin.

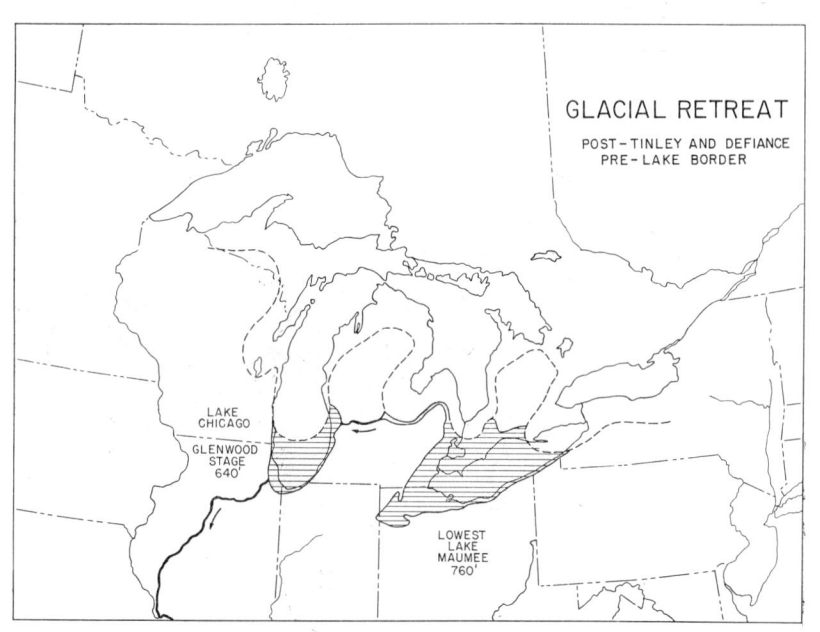

Fig. 56. Glenwood I stage of Lake Chicago and Lowest Lake Maumee of the Erie basin, which discharged down Grand River to Lake Chicago; occurring during the glacial retreat between the Tinley and Defiance advance and the Lake Border advances. Position of ice front conjectural.

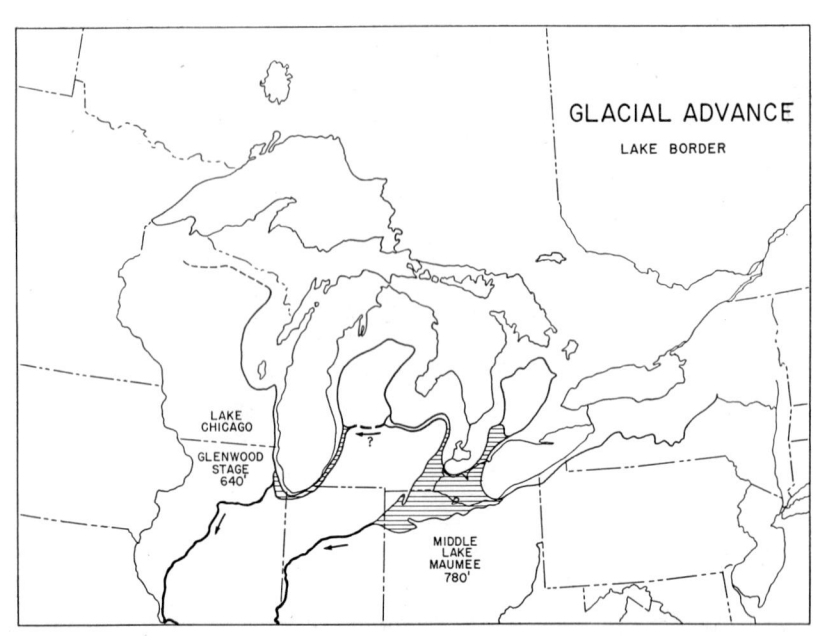

Fig. 57. Glenwood I stage of Lake Chicago (continued) and Middle Lake Maumee of the Erie basin. Advance of ice to Lake Border moraines constricted Lake Chicago, returned discharge of the Erie basin to the Fort Wayne outlet, and probably eliminated discharge down the Grand River.

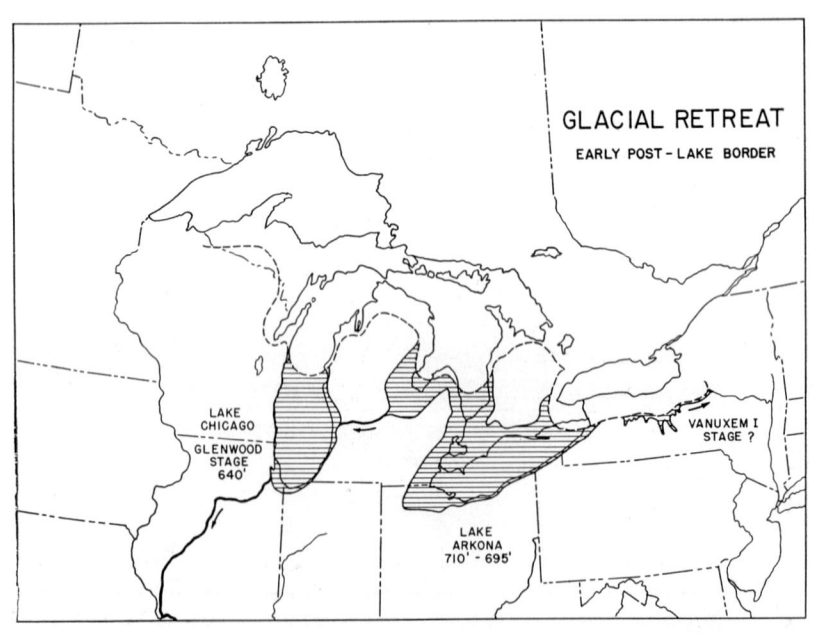

Fig. 58. Glenwood I stage of Lake Chicago (continued) and Highest Lake Arkona of the Erie and Huron basins, which discharged down Grand River to Lake Chicago; occurring during the glacial retreat from the latest Lake Border moraine. The Vanuxem I stage of the Ontario basin probably occurred at this time.

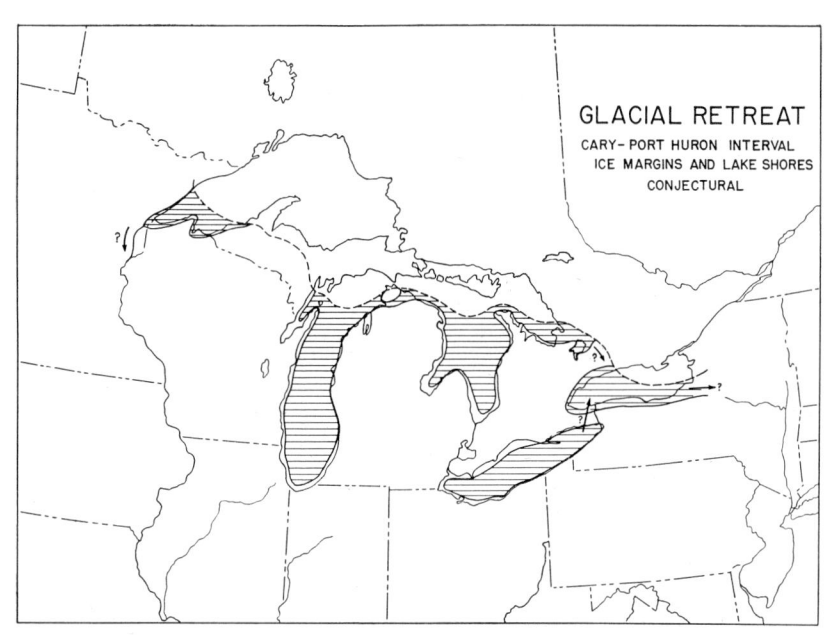

Fig. 59. Low-water stages of the Cary–Port Huron interval. Extent of glacial retreat unknown; positions of lake shores conjectural.

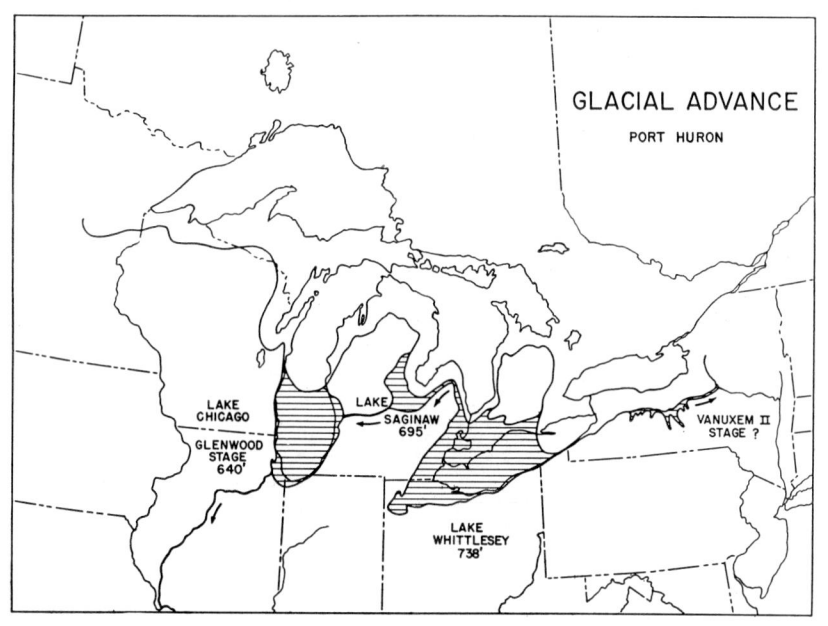

Fig. 60. Glenwood II stage of Lake Chicago, Lake Whittlesey of the Erie basin, and Lake Saginaw of Saginaw Bay, occurring at the time of maximum extent of the Port Huron (Mankato) glacial substage. The Vanuxem II stage of the Ontario basin probably occurred at this time.

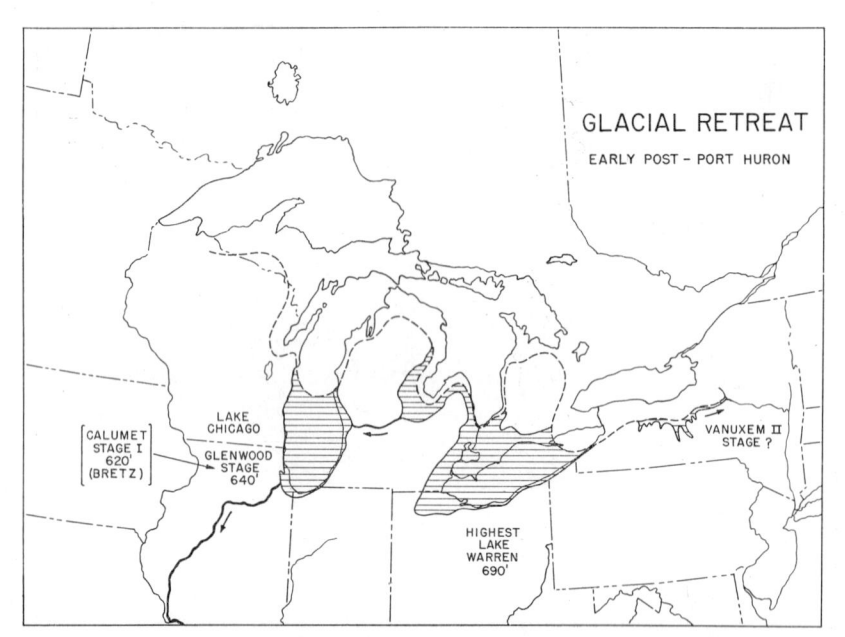

Fig. 61. Glenwood II stage of Lake Chicago (continued) and Highest Lake Warren of the Erie and Huron basins, occurring during the glacial retreat from the Port Huron maximum position. The Vanuxem II stage probably continued in the Ontario basin.

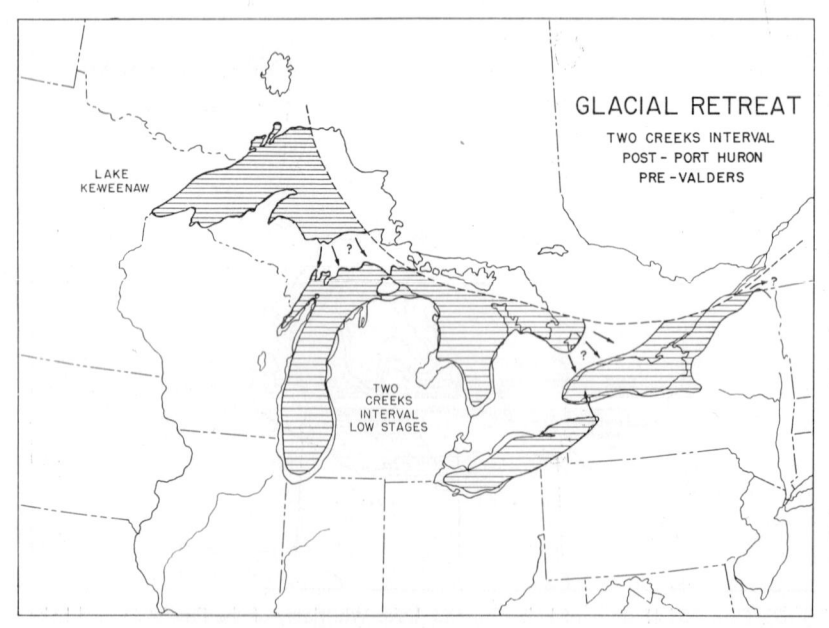

Fig. 62. Two Creeks low-water stages of the four lower lake basins and Lake Keweenaw of the Superior basin, occurring during the Two Creeks interval between the Port Huron (Mankato) glacial substage and the Valders glacial substage.

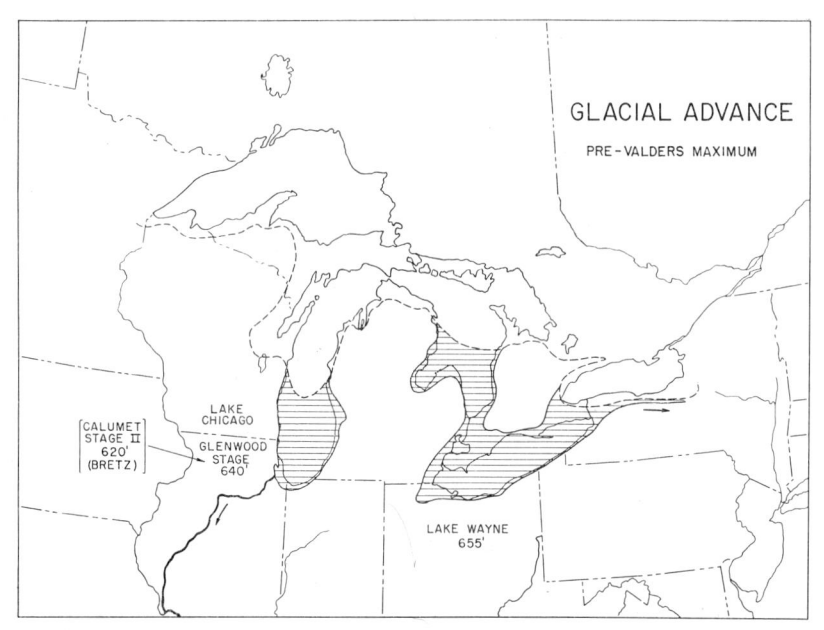

Fig. 63. Glenwood III stage of Lake Chicago, and Lake Wayne of the Huron, Erie, and Ontario basins which discharged eastward; occurring during the advance of Valders glacial substage ice.

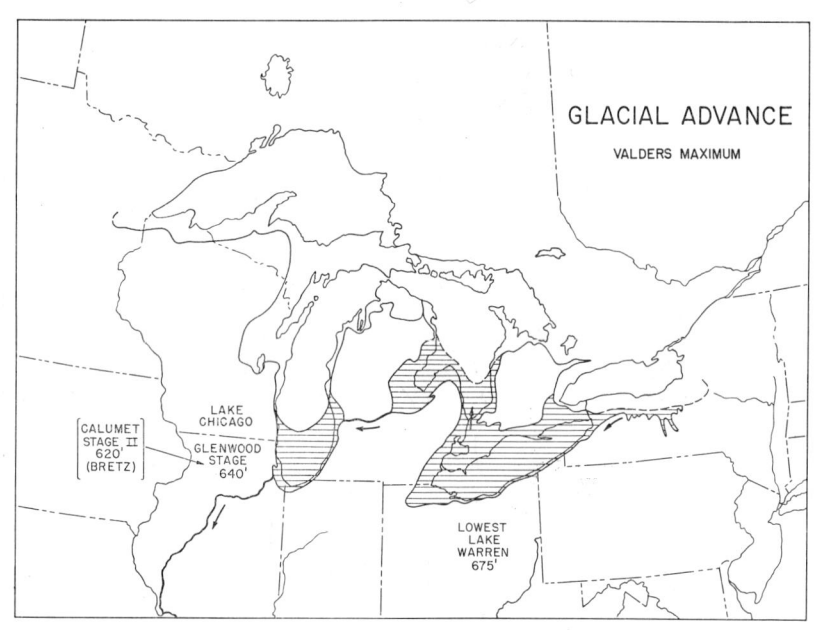

Fig. 64. Glenwood III stage of Lake Chicago (continued), and Lowest Lake Warren of the Huron, Erie, and Ontario basins which discharged down Grand River to Lake Chicago; occurring at the time of the Valders glacial substage maximum.

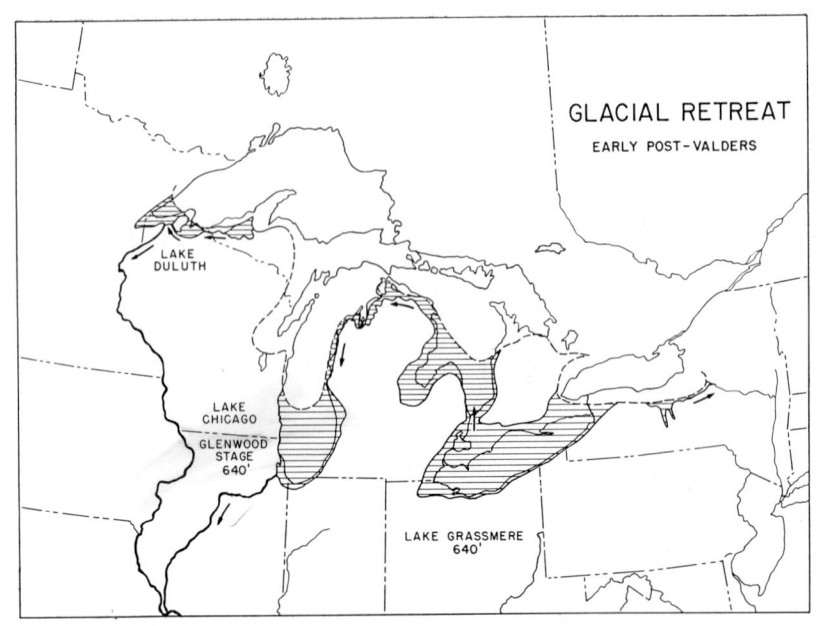

Fig. 65A. Glenwood III stage of Lake Chicago (continued), Lake Grassmere of the Erie and Huron basins which discharged northwestward to Lake Chicago, and early Lake Duluth of the Superior basin; occurring early in the retreat of the Valders ice. (See Fig. 65B for an alternative interpretation.)

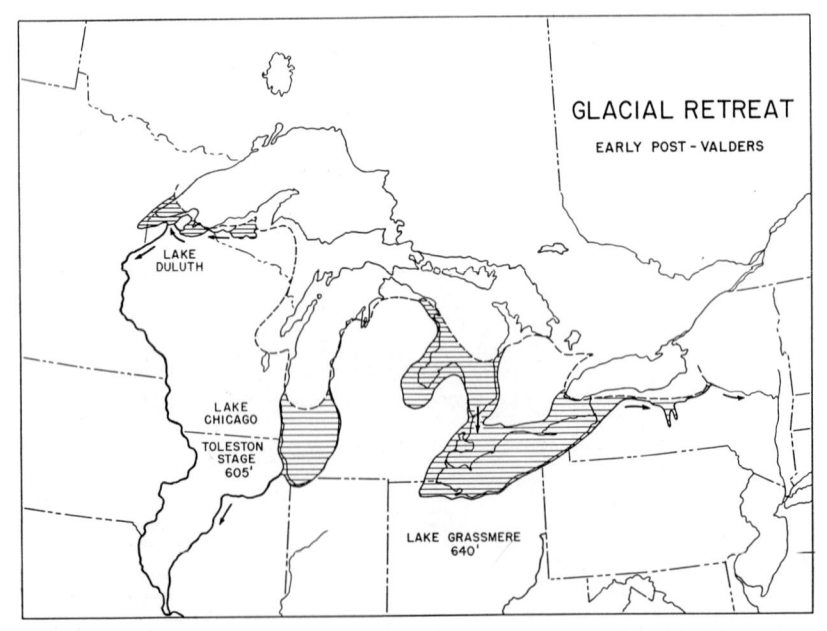

Fig. 65B. Alternative interpretation (of Bretz); Toleston stage of Lake Chicago, and Lake Grassmere of the Erie and Huron basins which discharged eastward early in the retreat of the Valders-substage ice.

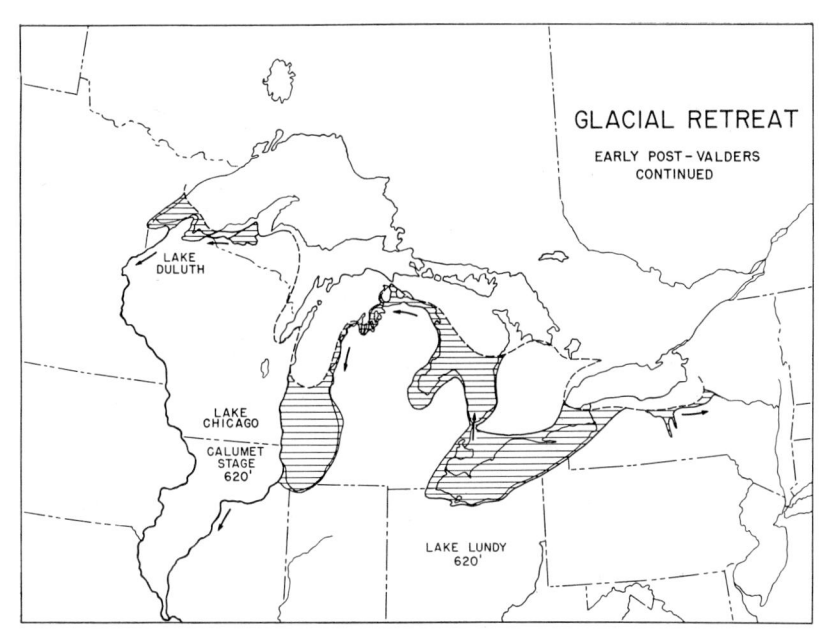

Fig. 66A. Calumet stage of Lake Chicago, Lundy stage of the Erie and Huron basins which discharged northwestward to Lake Chicago, and early Lake Duluth of the Superior basin; occurring during continued retreat of the Valders ice. (See Fig. 66B for an alternative interpretation.)

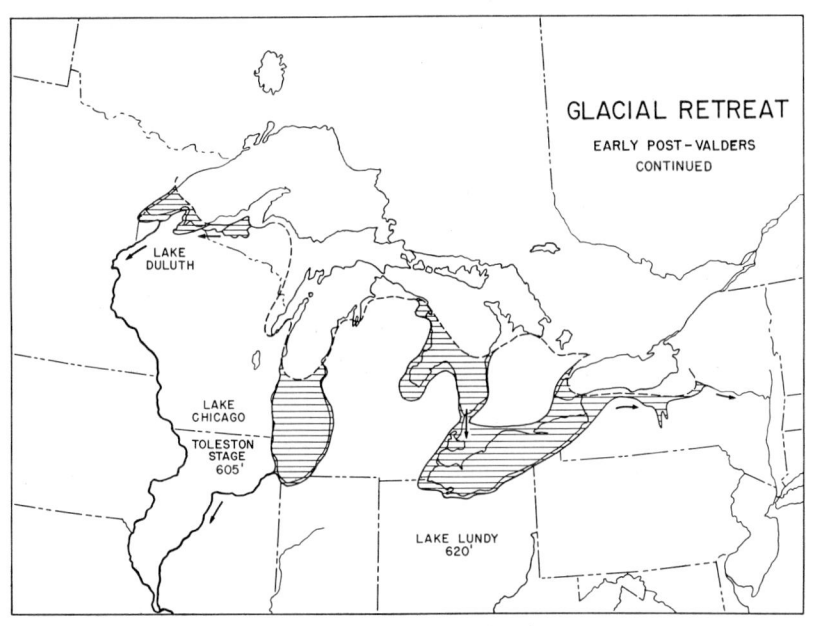

Fig. 66B. Alternative interpretation (of Bretz); Toleston stage of Lake Chicago, and Lundy stage of the Erie and Huron basins which discharged eastward.

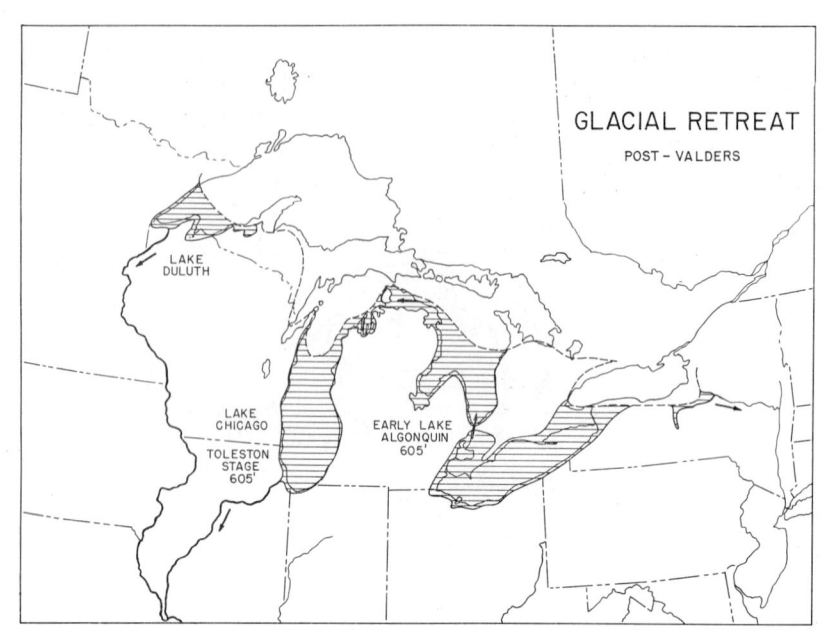

Fig. 67A. Toleston stage of Lake Chicago, Early Lake Algonquin of the Erie and Huron basins which discharged northwestward to Lake Chicago, and Lake Duluth of the Superior basin. (See Fig. 67B for an alternative interpretation.)

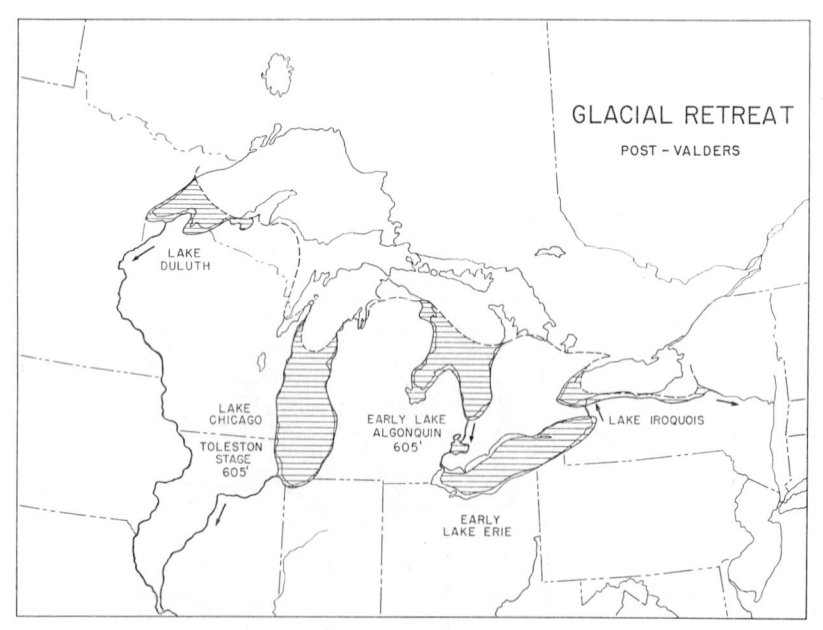

Fig. 67B. Alternative interpretation (of Leverett and Taylor); Toleston stage of Lake Chicago, Early Lake Algonquin of the Huron basin discharging southward, and early Lake Erie discharging to Lake Iroquois.

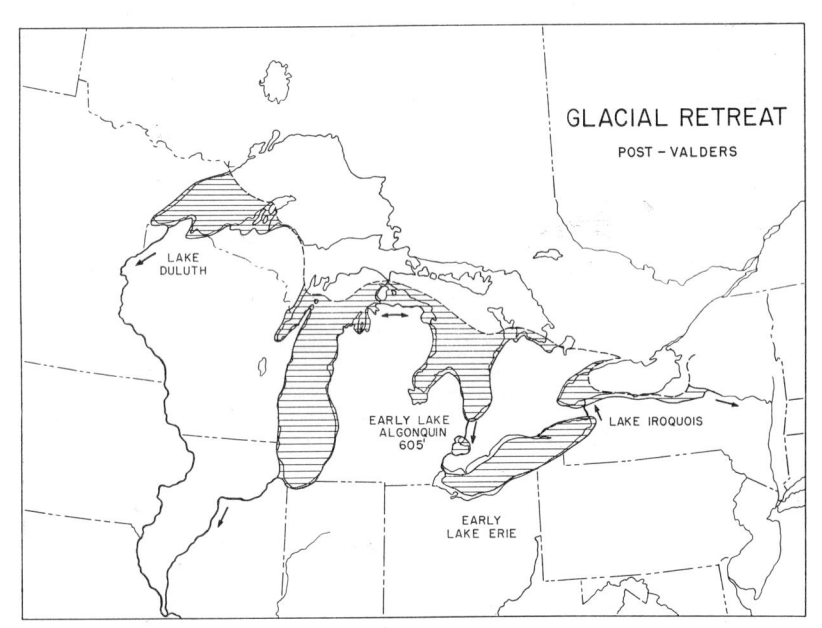

Fig. 68. Early Lake Algonquin of the Michigan and Huron basins, early Lake Erie, Lake Iroquois of the Ontario basin, and Lake Duluth.

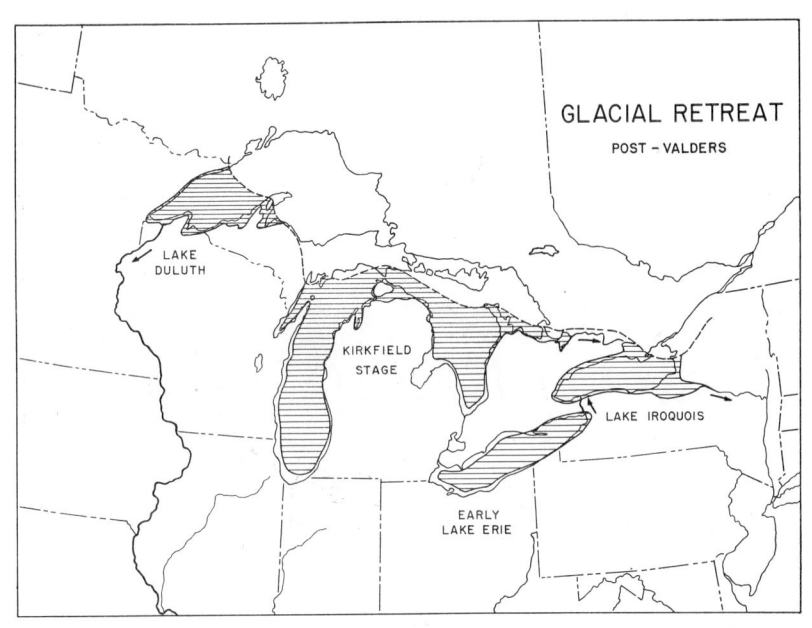

Fig. 69. Kirkfield stage of the Michigan and Huron basins discharging down Trent valley to Lake Iroquois of the Ontario basin; early Lake Erie, and Lake Duluth.

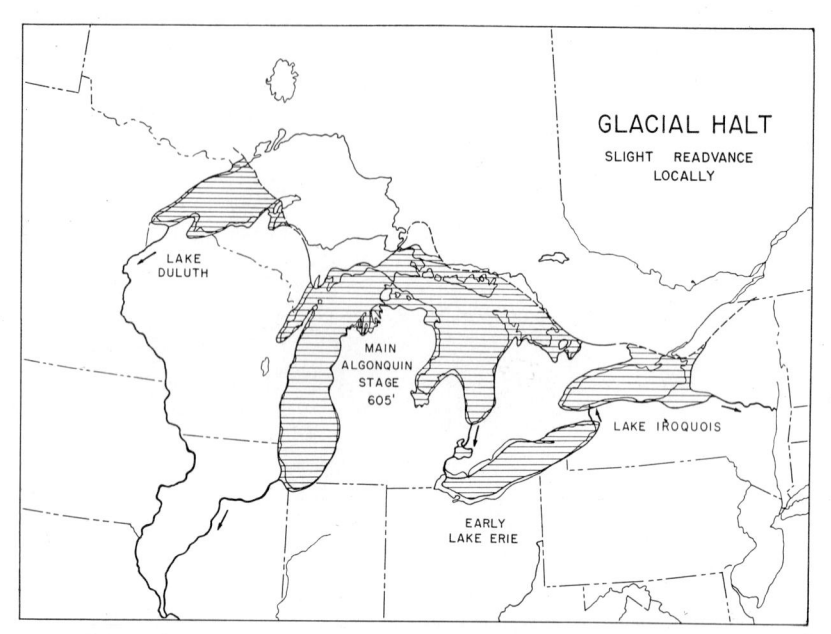

Fig. 70. Main Algonquin stage of the Michigan and Huron basins, early Lake Erie, Lake Iroquois, and Lake Duluth.

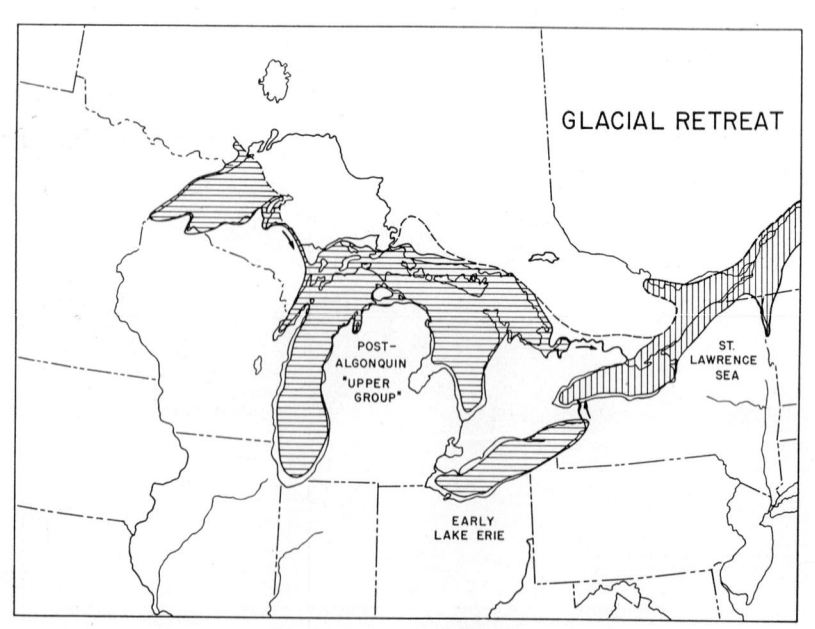

Fig. 71. Post-Algonquin "upper group" lake stages discharging down Trent valley to a marine embayment of the Ontario basin; early Lake Erie, and a sub-Duluth stage of the Superior basin discharging to Green Bay of the Michigan basin.

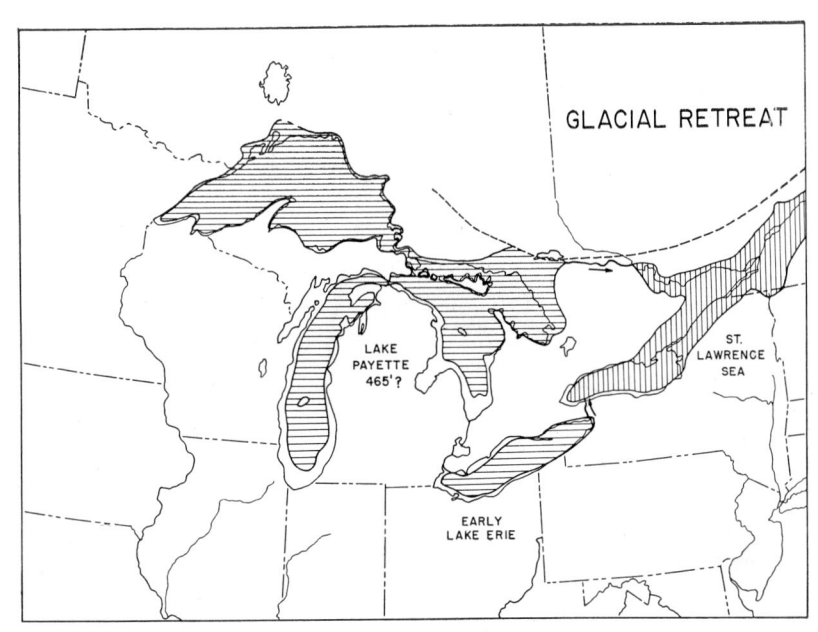

Fig. 72. Lake Payette, one of several post-Algonquin low-water stages of the Michigan and Huron basins. One of the post-Duluth low-water stages of the Superior basin is correlative.

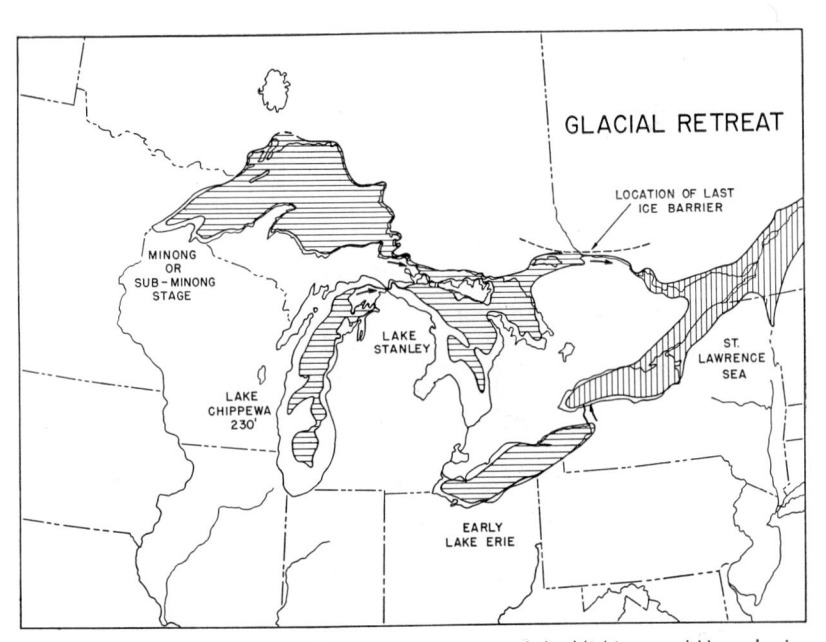

Fig. 73. Lakes Chippewa and Stanley, the lowest stages of the Michigan and Huron basins, discharging at North Bay, Ontario, to the Ottawa River and St. Lawrence Sea. The lowest stage of the Superior basin (Minong or a sub-Minong stage) is correlative.

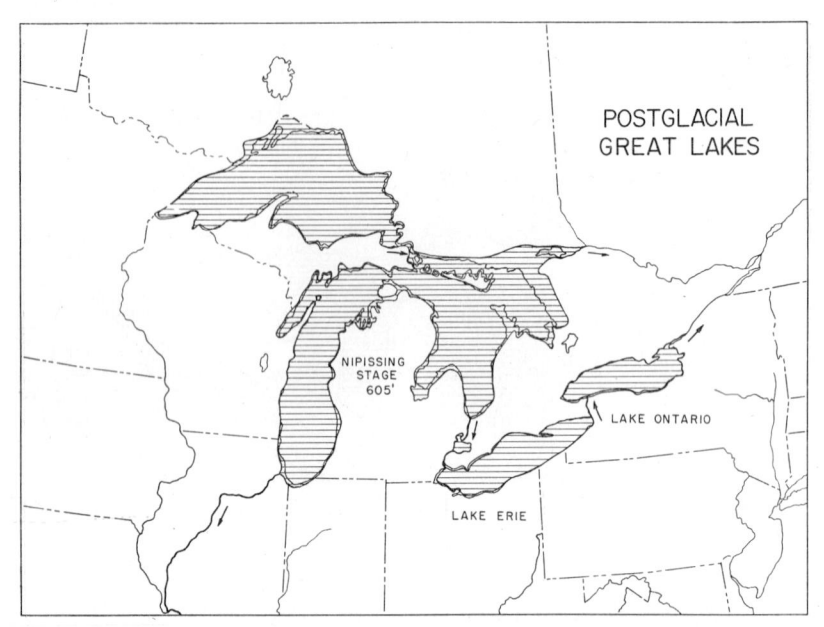

Fig. 74. Lake Nipissing, confluent in the three upper Great Lakes basins and discharging through three outlets (North Bay, Chicago, and Port Huron); Lake Erie, and early Lake Ontario.

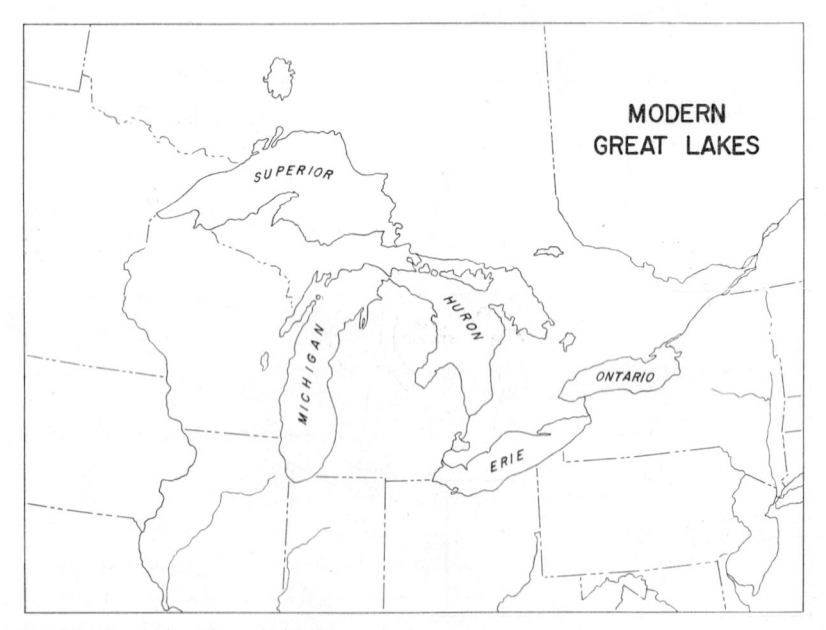

Fig. 75. The modern Great Lakes.

BIBLIOGRAPHY

Ahrens, L. H., 1955, Oldest rocks exposed: Crust of the earth: Geol. Soc. of America Spec. Paper 62, pp. 155–68.

Alden, W. C., 1918, The Quaternary geology of southeastern Wisconsin: U.S. Geol. Survey Prof. Paper 106.

Anderson, E. C., and others, 1947, Radiocarbon from cosmic radiation: Science, n.s., v. 105, pp. 576–77.

Andrews, E., 1870, The North American lakes considered as chronometers of postglacial time: Chicago Acad. Sci. Trans., v. 2, art. 1, pp. 1–24.

Antevs, E., 1928, The last glaciation: Am. Geog. Soc. Res. Ser. 17, p. 292.

———, 1931, Late-glacial correlations and ice recession in Manitoba: Canada Geol. Survey Mem. 168.

———, 1953, Geochronology of the Deglacial and Neothermal ages: Jour. Geology, v. 61, pp. 195–230.

———, 1955, Varve and radiocarbon chronologies appraised by pollen data: Jour. Geology, v. 63, pp. 495–99.

———, 1957, Geological tests of the varve and radiocarbon chronologies: Jour. Geology, v. 65, pp. 129–48.

Arnold, J. R., and Libby, W. F., 1949, Age determinations by radiocarbon content: Checks with samples of known age: Science, n.s., v. 110, pp. 678–80.

——— and Libby, W. F., 1951, Radiocarbon dates: Science, n.s., v. 113, pp. 115–18.

Ayers, J. C., Anderson, D. V., Chandler, D. C., and Lauff, G. H., 1956, Currents and water masses of Lake Huron: Univ. of Michigan Great Lakes Res. Inst. Tech. Paper 1, and Ontario Dept. of Lands and Forests, Div. of Research, Res. Rept. 35.

Baker, F. C., 1920, The life of the Pleistocene or Glacial Period: Univ. of Illinois Bull., v. 17, pp. 1–476.

Barendsen, G. W., Deevey, E. S., and Gralenski, L. J., 1957, Yale natural radiocarbon measurements III: Science, n.s., v. 126, pp. 908–19.

Bell, C. N., 1928, An implement of prehistoric man: 36th Ann. Archeological Rept., pp. 51–54: *in* Rept. Minister of Education, Ontario.

Bergquist, S. G., 1936, The Pleistocene history of the Tahquamenon and Manistique drainage region of the Northern Peninsula of Michigan: Michigan Geol. Survey Pub. 40, Geol. Ser. 34.

Bramlette, M. N., and Bradley, W. H., 1940, Geology and biology of the North Atlantic deep-sea cores: U.S. Geol. Survey Prof. Paper 196-A.

Bretz, J H., 1951a, Causes of the glacial lake stages in Saginaw basin, Michigan: Jour. Geology, v. 59, pp. 244–58.

———, 1951b, The stages of Lake Chicago: Their causes and correlations: Am. Jour. Sci., v. 249, pp. 401–29.

———, 1953, Glacial Grand River, Michigan: Michigan Acad. Sci. Papers, v. 38, pp. 359–82.

———, 1955, Geology of the Chicago region, Part II: The Pleistocene: Illinois Geol. Survey Bull. 65.

Broecker, W. S., and Kulp, J. L., 1957, Lamont natural radiocarbon measurements IV: Science, n.s., v. 126, pp. 1324–34.

———, Kulp, J. L., and Tucek, C. S., 1956, Lamont natural radiocarbon measurements III: Science, n.s., v. 124, pp. 154–65.

Bryan, K., and Ray, L. L., 1940, Geologic antiquity of the Lindermeier site in Colorado: Smithsonian Misc. Coll., v. 99, no. 2.

Burkholder, P. R., 1929, Biological significance of the chemical analyses, pp. 65–72: *in* Fish, C. J., 1929, Preliminary report on the cooperative survey of Lake Erie: Buffalo Soc. Nat. Sci. Bull., v. 14, pp. 7–220.

Butler, B. S., and Burbank, W. S., 1929, The copper deposits of Michigan: U.S. Geol. Survey Prof. Paper 144.

Carman, J. E., 1946, The geologic interpretation of scenic features in Ohio: Ohio Jour. Sci., v. 46, pp. 241–83.

Chamberlin, T. C., 1895, Glacial phenomena of North America, pp. 724–75: *in* Geikie, J., The great ice age: New York, D. Appleton & Co.

Chandler, D. C., and Weeks, O. B., 1945, Limnological studies of western Lake Erie: Ecological Monog., v. 15, pp. 435–57.

Chapman, D. H., 1937, Late-glacial and postglacial history of the Champlain valley: Am. Jour. Sci., 5th ser., v. 34, pp. 89–124.

Chapman, L. J., 1954, An outlet of Lake Algonquin at Fossmill, Ontario: Geol. Assoc. of Canada Proc., v. 6, pp. 61–68.

────── and Putnam, D. F., 1951, The physiography of southern Ontario: Toronto, Univ. of Toronto Press.

Church, P. E., 1942, The annual temperature cycle of Lake Michigan, I: Cooling from late autumn to the terminal point, 1941–42: Univ. of Chicago Inst. of Meteorology Misc. Repts. 4.

──────, 1945, The annual temperature cycle of Lake Michigan, II: Spring warming and summer stationary periods, 1942: Univ. of Chicago Inst. of Meteorology Misc. Repts. 18.

Clarke, F. W., 1924, The data of geochemistry: U.S. Geol. Survey Bull. 770.

Coleman, A. P., 1900, Marine and fresh-water beaches of Ontario: Geol. Soc. of America Bull., v. 12, pp. 138–43.

──────, 1904, The Iroquois beach in Ontario: Ontario Bur. of Mines Rept., v. 13, pp. 225–44.

──────, 1936, Lake Iroquois: Ontario Dept. of Mines, 45th Ann. Rept., v. 45, pt. 7, pp. 1–36.

Cornish, V., 1910, Waves of the sea and other water waves: London, T. Fisher Unwin.

Corps of Engineers, U.S. Army, 1956, Great Lakes pilot: U.S. Lake Survey, Detroit.

──────, 1958, Great Lakes pilot: U.S. Lake Survey, Detroit.

Crane, H. R., 1956, University of Michigan radiocarbon dates I: Science, n.s., v. 124, pp. 664–72.

Dawson, G. M., 1886, Notes to accompany a geological map of the northern portion of the Dominion of Canada: Canada Geol. Survey Ann. Rept., v. 2, pp. 1R–62R.

──────, 1890, On the glaciation of the northern part of the Cordillera, with an attempt to correlate the events of the Glacial Period in the Cordillera and Great Plains: Am. Geologist, v. 6, pp. 153–62.

Deane, R. E., 1950, Pleistocene geology of the Lake Simcoe district, Ontario: Canada Geol. Survey Mem. 256.

Deason, H. J., 1932, A study of the surface currents in Lake Michigan: The Fisherman, v. 1, p. 12.

Dietz, R. S., and Menard, H. W., 1951, Origin of the abrupt change in slope at Continental Shelf margin: Am. Assoc. of Petroleum Geologists Bull., v. 35, pp. 1994–2016.

Division of Waterways, State of Illinois, 1952, Interim report for erosion control Illinois shore of Lake Michigan: Div. of Waterways, Dept. of Public Works, State of Illinois.

Dreimanis, A., 1952, Age determination of the Burley site at Port Franks, Ontario, by geological methods: Univ. of Western Ontario Mus. of Indian Archeology and Pioneer Life Bull. 9, pp. 72–75.

———, 1957, Depths of leaching in glacial deposits: Science, n.s., v. 126, pp. 403–4.

———, 1958, Beginning of the Nipissing phase of Lake Huron: Jour. Geology, v. 66, p. 592.

——— and Reavely, G. H., 1953, Differentiation of the lower and the upper till along the north shore of Lake Erie: Jour. Sedimentary Petrology, v. 23, pp. 238–59.

Elson, J. A., 1957, Lake Agassiz and the Mankato-Valders problem: Science, n.s., v. 126, pp. 999–1002.

Emery, K. O., 1951, Bathymetric chart of Lake Michigan: Univ. of Minnesota Inst. of Technology, Eng. Exper. Sta. Tech. Paper 77.

Ewing, M., Press, F., and Donn, W. L., 1954, An explanation of the Lake Michigan wave of 26 June 1954: Science, n.s., v. 120, pp. 684–86.

Fairchild, H. L., 1905, Ice erosion theory a fallacy: Geol. Soc. of America Bull., v. 16, pp. 13–74.

———, 1907, Gilbert Gulf (marine waters in Ontario basin): Geol. Soc. of America Bull., v. 17, p. 112.

———, 1909, Glacial waters in central New York: New York State Mus. Bull. 127, pp. 5–66.

Fish, C. J., 1929, Preliminary report on the cooperative survey of Lake Erie: Buffalo Soc. Nat. Sci. Bull., v. 14, pp. 7–220.

Flint, R. F., 1943, Growth of the North American ice sheet during the Wisconsin age: Geol. Soc. of America Bull., v. 54, pp. 325–62.

———, 1947, Glacial geology and the Pleistocene epoch: New York, John Wiley and Sons, Inc.

———, 1953, Probable Wisconsin substages and late-Wisconsin events in northeastern United States and southeastern Canada: Geol. Soc. of America Bull., v. 64, pp. 897–919.

———, 1956, New radiocarbon dates and late-Pleistocene stratigraphy: Am. Jour. Sci., v. 254, pp. 265–87.

———, 1957, Glacial and Pleistocene geology: New York, John Wiley and Sons, Inc.

——— and Deevey, E. S., Jr., 1951, Radiocarbon dating of late-Pleistocene events: Am. Jour. Sci., v. 249, pp. 257–300.

——— and Rubin, M., 1955, Radiocarbon dates of pre-Mankato events in eastern and central North America: Science, n.s., v. 121, pp. 649–58.

——— and others, 1945, Glacial map of North America: Geol. Soc. of America Spec. Paper 60.

Folk, R. L., 1951, Stages of textural maturity in sedimentary rocks: Jour. Sedimentary Petrology, v. 21, pp. 127–30.

Gadd, N. R., 1955, Pleistocene geology of the Becancour map-area, Quebec: unpublished Ph.D. thesis, Univ. of Illinois, Urbana.

Gilbert, G. K., 1898, Recent earth movements in the Great Lakes region: U.S. Geol. Survey, 18th Ann. Rept., pt. 2, pp. 601–47.

Goldthwait, J. W., 1907, The abandoned shore-lines of eastern Wisconsin: Wisconsin Geol. and Nat. History Survey Bull. 17.

———, 1908a, Physical geography of the Evanston-Waukegan region: Illinois Geol. Survey Bull. 7.

———, 1908b, A reconstruction of water-planes of the extinct glacial lakes in the Michigan basin: Jour. Geology, v. 16, pp. 459–76.

———, 1910, An instrumental survey of the shore-lines of the extinct lakes Algonquin and Nipissing in southwestern Ontario: Canada Geol. Survey Mem. 10, pp. 1–57.

Grabau, A. W., 1901, Guide to the geology and paleontology of Niagara Falls and vicinity: New York State Mus. Bull. 45.

Greenman, E. F., and Stanley, G. M., 1943, The archeology and geology of two early sites near Killarney, Ontario: Michigan Acad. Sci. Papers, v. 28, pp. 505–30.

Grogan, R. M., 1945, Shape variation of some Lake Superior beach pebbles: Jour. Sedimentary Petrology, v. 15, pp. 3–10.

Gutenberg, B., 1933, Tilting due to glacial melting: Jour. Geology, v. 41, pp. 449–67.

Harrington, M. W., 1895, The surface currents of the Great Lakes: U.S. Weather Bur. Bull. B.

Harris, R. A., 1907, Tides in lakes and wells: Manual of tides: Rept. Supt. Coast and Geodetic Survey, Append. 6, pp. 463–86.

Hayford, J. F., 1922, Effects of winds and of barometric pressures on the Great Lakes: Carnegie Inst. of Washington Pub. 317.

Holmes, C. D., 1952, Drift dispersion in west-central New York: Geol. Soc. of America Bull., v. 63, pp. 993–1010.

Horberg, L., 1950, Bedrock topography of Illinois: Illinois Geol. Survey Bull. 73.

——— and Anderson, R. C., 1956, Bedrock topography and Pleistocene glacial lobes in central United States: Jour. Geology, v. 64, pp. 101–16.

Hotchkiss, W. O., 1923, The Lake Superior geosyncline: Geol. Soc. of America Bull., v. 34, pp. 669–78.

Hough, J. L., 1932, Suggestion regarding the origin of rock bottom areas in Massachusetts Bay: Jour. Sedimentary Petrology, v. 2, pp. 131–32.

———, 1934, Redeposition of microscopic Devonian plant fossils: Jour. Geology, v. 42, pp. 646–48.

———, 1935, The bottom deposits of southern Lake Michigan: Jour. Sedimentary Petrology, v. 5, pp. 57–80.

———, 1950, Pleistocene lithology of Antarctic ocean-bottom sediments: Jour. Geology, v. 58, pp. 254–60.

———, 1952, Fathogram indications of bottom materials in Lake Michigan: Jour. Sedimentary Petrology, v. 22, pp. 162–72.

———, 1953a, Final report on the project, Pleistocene chronology of the Great Lakes region: Office of Naval Research contract No. N 6 ori-07133, Project NR-018-122. Univ. of Illinois, Urbana (mimeographed report).

———, 1953b, Pleistocene climatic record in a Pacific Ocean core sample: Jour. Geology, v. 61, pp. 252–62.

———, 1953c, Revision of the Nipissing stage of the Great Lakes: Illinois State Acad. Sci. Trans., v. 46, pp. 133–41.

———, 1954, Geologic history of Great Lakes beaches: Proc. Fourth Conf. on Coastal Eng., Council on Wave Research, Univ. of California, Berkeley, pp. 79–100.

———, 1955, Lake Chippewa, a low stage of Lake Michigan indicated by bottom sediments: Geol. Soc. of America Bull., v. 66, pp. 957–68.

Hubbard, B., 1840, Report on Lenawee, Hillsdale, Branch, St. Joseph, Cass, Berrien, Washtenaw, Oakland, and Livingston counties, with notes on the lake ridges and Great Lakes: Michigan State Geol. Survey Ann. Rept. 3, pp. 77–111.

Hubbs, C. L., and Lagler, K. F., 1947, Fishes of Isle Royale, Lake Superior, Michigan: Michigan Acad. Sci. Papers, v. 33, pp. 73–133.

Johnson, F., and others, 1951, Radiocarbon dating: Soc. Am. Archeology Mem. 8.

Johnston, W. A., 1916a, The Trent valley outlet of Lake Algonquin and the deformation of the Algonquin waterplane in Lake Simcoe district, Ontario: Canada Geol. Survey Mus. Bull. 23.

———, 1916b, Late Pleistocene oscillations of sea-level in the Ottawa valley: Canada Geol. Survey Mus. Bull. 24.

———, 1928, The age of the upper great gorge of Niagara River: Royal Soc. of Canada Trans., v. 22, pp. 13–29.

———, 1933, Quaternary geology of North America in relation to the migration of man, pp. 11–45: *in* Jenness, D. (ed.), The American aborigines: Toronto, Univ. of Toronto Press.

Karlstrom, T. N. V., 1956, The problem of the Cochrane in late Pleistocene chronology: U.S. Geol. Survey Bull. 1021, pp. 303–31.

Kay, G. F., 1931, Classification and duration of the Pleistocene period: Geol. Soc. of America Bull., v. 42, pp. 425–66.

Kindle, E. M., 1925, The bottom deposits of Lake Ontario: Royal Soc. of Canada Trans., 3rd ser., v. 19, pp. 17–72.

——— and Taylor, F. B., 1913, Niagara folio, New York: U.S. Geol. Survey Geol. Atlas, Folio 190.

Krumbein, W. C., and Griffith, J. S., 1938, Beach environment at Little Sister Bay, Wisconsin: Geol. Soc. of America Bull., v. 49, pp. 629–52.

Ladd, H. S. (ed.), 1957, Treatise on marine ecology and paleoecology: Geol. Soc. of America Mem. 67.

Landes, K. K., Ehlers, G. M., and Stanley, G. M., 1945, Geology of the Mackinac Straits region: Michigan Geol. and Biol. Survey Pub. 44.

Lane, F. C., 1948, The world's great lakes: New York, Doubleday and Co., Inc.

Lee, T. E., 1954, The first Sheguiandah expedition, Manitoulin Island, Ontario: Am. Antiquity, v. 20, pp. 101–11.

Leighton, M. M., 1933, The naming of the subdivisions of the Wisconsin glacial age: Science, n.s., v. 77, p. 168.

———, 1957, The Cary-Mankato-Valders problem: Jour. Geology, v. 65, pp. 108–11.

——— and Willman, H. B., 1950, Loess formations of the Mississippi valley: Jour. Geology, v. 58, pp. 599–623.

Leith, C. K., Lund, R. J., and Leith, A., 1935, Pre-Cambrian rocks of the Lake Superior region: U.S. Geol. Survey Prof. Paper 184.

Leverett, F., 1899, The Illinois glacial lobe: U.S. Geol. Survey Monog. 38.

———, 1902, Glacial formations and drainage features of the Erie and Ohio basins: U.S. Geol. Survey Monog. 41.

———, 1929a, Moraines and shore lines of the Lake Superior region: U.S. Geol. Survey Prof. Paper 154-A.

Leverett, F., 1929b, Pleistocene glaciations of the Northern Hemisphere: Geol. Soc. of America Bull., v. 40, pp. 745-60.
———, 1930, Relative length of Pleistocene glacial and interglacial stages: Science, n.s., v. 72, pp. 193-95.
———, 1932, Quaternary geology of Minnesota and parts of adjacent states: U.S. Geol. Survey Prof. Paper 161.
———, 1935, Patrician ice movements: Pan-Am. Geologist, v. 63, pp. 5-8.
——— and Taylor, F. B., 1915, The Pleistocene of Indiana and Michigan and the history of the Great Lakes: U.S. Geol. Survey Monog. 53.
Libby, W. F., 1951, Radiocarbon dates II: Science, n.s., v. 114, pp. 291-96.
———, 1952, Chicago radiocarbon dates III: Science, n.s., v. 116, pp. 673-81.
———, 1954a, Chicago radiocarbon dates IV: Science, n.s., v. 119, pp. 135-40.
———, 1954b, Chicago radiocarbon dates V: Science, n.s., v. 120, pp. 733-42.
Logan, W. E., and others, 1863, Geology of Canada: Canada Geol. Survey Progress Rept. to 1863.
Lyell, C., 1845, Travels in North America; with geological observations on the United States, Canada, and Nova Scotia, v. 2: London, J. Murray.
MacClintock, P., and Apfel, E. T., 1944, Correlation of the drifts of the Salamanca re-entrant, New York: Geol. Soc. of America Bull., v. 55, pp. 1143-64.
Martens, J. H. C., 1935, Beach sands between Charleston, S. C., and Miami, Fla.: Geol. Soc. of America Bull., v. 46, pp. 1563-96.
Martin, H. M., 1955, Map of the surface formations of the Southern Peninsula of Michigan: Michigan Geol. Survey Pub. 49.
Mather, K. F., 1917, The Champlain Sea in the Lake Ontario basin: Jour. Geology, v. 25, pp. 542-54.
McDonald, W. E., 1954, Variation in Great Lakes levels in relation to engineering problems: Proc. Fourth Conf. on Coastal Eng., Council on Wave Research, Univ. of California, Berkeley, pp. 249-57.
Melhorn, W. H., 1954, Valders glaciation of the Southern Peninsula of Michigan: unpublished Ph.D. thesis, Univ. of Michigan, Ann Arbor.
Moore, E. S., 1927, Batchawana area, District of Algoma: Ontario Dept. of Mines Ann. Rept., v. 35, pt. 2, pp. 53-85.

Moore, R. C., 1949, Introduction to historical geology: New York, McGraw-Hill Book Co., Inc.

Moore, S., 1948, Crustal movement in the Great Lakes area: Geol. Soc. of America Bull., v. 59, pp. 697–710.

Murray, R. C., 1953, The petrology of the Cary and Valders tills of northeastern Wisconsin: Am. Jour. Sci., v. 251, pp. 140–55.

Newberry, J. S., 1874a, On the structure and origin of the Great Lakes: New York Lyceum Nat. History Proc., v. 2, pp. 136–38.

———, 1874b, Geology of Ohio: Ohio Geol. Survey Rept., v. 2, pt. 1, pp. 1–80.

Parmenter, R., 1929, Hydrography of Lake Erie, pp. 25–50: *in* Fish, C. J., 1929, Preliminary report on the cooperative survey of Lake Erie: Buffalo Soc. Nat. Sci. Bull., v. 14, pp. 7–220.

Pegrum, R. H., 1929, Topography of the Lake Erie basin, pp. 17–24: *in* Fish, C. J., 1929, Preliminary report on the cooperative survey of Lake Erie: Buffalo Soc. Nat. Sci. Bull., v. 14, pp. 7–220.

Pettijohn, F. J., 1931, Petrography of the beach sands of southern Lake Michigan: Jour. Geology, v. 39, pp. 432–55.

———, 1949, Sedimentary rocks: New York, Harper and Bros.

———, and Lundahl, A. C., 1943, Shape and roundness of Lake Erie beach sands: Jour. Sedimentary Petrology, v. 13, pp. 69–78.

——— and Ridge, J. D., 1932, A textural variation series of beach sands from Cedar Point, Ohio: Jour. Sedimentary Petrology, v. 2, pp. 76–88.

——— and Ridge, J. D., 1933, A mineral variation series of beach sands from Cedar Point, Ohio: Jour. Sedimentary Petrology, v. 3, pp. 92–94.

Piggot, C. S., and Urry, W. D., 1941, Time relations in ocean sediments: Geol. Soc. of America Bull., v. 53, pp. 1187–1210.

Pincus, H. J., Metter, R. E., Humphris, C. C., Kleinhampl, F. J., and Bowman, R. S., 1953, 1951 investigations of Lake Erie shore erosion: Ohio Geol. Survey Rept. Inv. 18.

———, Roseboom, M. L., and Humphris, C. C., 1951, 1950 investigation of Lake Erie sediments, vicinity of Sandusky, Ohio: Ohio Geol. Survey Rept. Inv. 9.

Pohlman, J., 1884, Life History of Niagara River (abs.): Am. Assoc. Adv. Sci. Proc., v. 32, p. 202.

Preston, R. S., Person, E., and Deevey, E. S., 1955, Yale natural radiocarbon measurements II: Science, n.s., v. 122, pp. 954–60.

Ridgeway, R., 1912, Color standards and color nomenclature: Washington, D.C., published by the author.

Rubin, M., and Suess, H. E., 1955, U.S. Geological Survey radiocarbon dates II: Science, n.s., v. 121, pp. 481–88.

Ruhe, R. V., 1952, Classification of the Wisconsin glacial stage: Jour. Geology, v. 60, pp. 398–401.

Ruttner, F., 1953, Fundamentals of limnology (transl. by Frey, D. G., and Fry, F. E. J.): Toronto, Univ. of Toronto Press.

Schalk, M., 1938, A textural study of the outer beach of Cape Cod, Massachusetts: Jour. Sedimentary Petrology, v. 8, pp. 41–54.

Schermerhorn, L. Y., 1887, Physical characters of the northern and northwestern lakes: Am. Jour. Sci., 3rd ser., v. 33, pp. 278–84.

Schuchert, C., 1943, Stratigraphy of the eastern and central United States: New York, John Wiley and Sons, Inc.

Schwartz, G. M., 1949, The geology of the Duluth metropolitan area: Minnesota Geol. Survey Bull. 33.

Shaffer, P. R., 1954a, Extension of Tazewell glacial substage of western Illinois and eastern Iowa: Geol. Soc. of America Bull., v. 65, pp. 443–56.

———, 1954b, Farmdale drift: Science, n.s., v. 119, pp. 693–94.

———, 1956, Farmdale drift in northwestern Illinois: Illinois Geol. Survey Rept. Inv. 198.

Sharp, R. P., 1953, Shorelines of the glacial Great Lakes in Cook County, Minnesota: Am. Jour. Sci., v. 251, pp. 109–39.

Shelford, V. E., 1937, Animal communities in temperate America (2nd ed.): Chicago, Univ. of Chicago Press.

Shepard, F. P., 1937, Origin of the Great Lakes basins: Jour. Geology, v. 45, pp. 76–88.

———, 1948, Submarine geology: New York, Harper and Bros.

——— and Suess, H. E., 1956, Rate of postglacial rise of sea level: Science, n.s., v. 123, pp. 1082–83.

Shepps, V. C., 1953, Correlation of the tills of northeastern Ohio by size analysis: Jour. Sedimentary Petrology, v. 23, pp. 34–48.

Smith, S. I., 1871, Dredging in Lake Superior under the direction of the U.S. Lake Survey: Am. Jour. Sci., 3rd ser., v. 2, pp. 373–74.

———, 1874, Sketch of the invertebrate fauna of Lake Superior: U.S. Fish. Comm. Rept. (1872–73), pp. 690–707.

Snodgrass, D. B., 1952, A study of Lake Michigan bottom sediments: unpublished Master's thesis, Univ. of Illinois, Urbana.

Spencer, J. W., 1890, The deformation of Iroquois beach and birth of Lake Ontario: Am. Jour. Sci., 3rd ser., v. 40, pp. 443–51.

———, 1891a, Origin of the basins of the Great Lakes of America: Am. Geologist, v. 7, pp. 86–97.

———, 1891b, Deformation of the Algonquin beach and birth of Lake Huron: Am. Jour. Sci., 3rd ser., v. 41, pp. 12–21.

———, 1894, The duration of Niagara Falls: Am. Jour. Sci., 3rd ser., v. 48, pp. 455–72.

———, 1907, The Falls of Niagara: Canada Geol. Survey Pub. 970, p. 277.

Spurr, S. H., and Zumberge, J. H., 1956, Late Pleistocene features of Cheboygan and Emmet counties, Michigan: Am. Jour. Sci., v. 254, pp. 96–109.

Stanley, G. M., 1932, Abandoned strands of Isle Royale and northeastern Lake Superior: unpublished Ph.D. thesis, Univ. of Michigan, Ann Arbor.

———, 1934, Pleistocene potholes in the Cloche Mountains of Ontario: Michigan Acad. Sci. Papers, v. 19, pp. 401–12.

———, 1936, Lower Algonquin beaches of Penetanguishene Peninsula: Geol. Soc. of America Bull., v. 47, pp. 1933–60.

———, 1937, Lower Algonquin beaches of Cape Rich, Georgian Bay: Geol. Soc. of America Bull., v. 48, pp. 1665–86.

———, 1938a, Impounded Early Algonquin beaches at Sucker Creek, Grey County, Ontario: Michigan Acad. Sci. Papers, v. 23, pp. 477–95.

———, 1938b, The submerged valley through Mackinac Straits: Jour. Geology, v. 46, pp. 966–74.

———, 1941, Minong beaches and waterplane in Lake Superior basin (abs.): Geol. Soc. of America Bull., v. 52, p. 1935.

Stetson, H. C., 1938, The sediments of the Continental Shelf off the eastern coast of the United States: Papers in Physical Oceanography and Meteorology: Massachusetts Inst. of Technology and Woods Hole Oceanographic Inst., v. 5.

Stimpson, W., 1870, On the deep-water fauna of Lake Michigan: Am. Naturalist, v. 4, pp. 403–5.

Suess, H. E., 1954, U.S. Geological Survey radiocarbon dates I: Science, n.s., v. 120, pp. 467–73.

Sverdrup, H. U., Johnson, M. W., and Fleming, R. H., 1942, The oceans: New York, Prentice-Hall, Inc.

Swain, F. M., and Prokopovich, N., 1957, Stratigraphy of upper part of sediments of Silver Bay area, Lake Superior: Geol. Soc. of America Bull., v. 68, pp. 527–42.

Tanton, T. L., 1920, Shore of Lake Superior between Port Arthur and Nipigon: Canada Geol. Survey Summary Rept. for 1919, pt. E, p. 3e.

Tanton, T. L., 1931, Fort William and Port Arthur and Thunder Cape map areas, Thunder Bay district, Ontario: Canada Geol. Survey Mem. 167, pp. 64, 86–87.

Taylor, F. B., 1895a, Changes of level in the region of the Great Lakes in recent geological time: Am. Jour. Sci., 3rd ser., v. 49, pp. 69–71.

———, 1895b, Niagara and the Great Lakes: Am. Jour. Sci., 3rd ser., v. 49, pp. 253–58.

———, 1929, New facts on the Niagara gorge: Michigan Acad. Sci. Papers, v. 12, pp. 251–65.

Thomson, J. E., 1954, Geology of the Mamainse Point copper area: Ontario Dept. of Mines Ann. Rept., v. 62.

Thornbury, W. D., 1940, Weathered zones and glacial chronology in southern Indiana: Jour. Geology, v. 48, pp. 449–75.

Thwaites, F. T., 1934, Outline of glacial geology: Ann Arbor, Michigan, Edwards Bros., Inc.

———, 1935, Sublacustrine topographic map of the bottom of Lake Superior: Kansas Geol. Soc., Guidebook, Ninth Ann. Field Conf., Wichita, Kansas, pp. 226–28.

———, 1943, Pleistocene of part of northeastern Wisconsin: Geol. Soc. of America Bull., v. 54, pp. 87–144.

———, 1946, Outline of glacial geology: Ann Arbor, Michigan, Edwards Bros., Inc.

———, 1949, Geomorphology of the basin of Lake Michigan: Michigan Acad. Sci. Papers, v. 33, pp. 243–51.

Townsend, C. McD., 1916, The currents of Lake Michigan and their influence on the climate of the neighboring states: Jour. Western Soc. Eng., v. 21, pp. 293–309.

Tyrell, J. B., 1894, Notes on the Pleistocene of the Northwest Territories of Canada: Geol. Mag., 4th ser., v. 1, pp. 394–99.

———, 1895, A second expedition through the Barren Grounds of northern Canada: Geog. Jour., v. 6, pp. 438–48.

———, 1896, The genesis of Lake Agassiz: Jour. Geology, v. 4, pp. 811–15.

———, 1898, The glaciation of north central Canada: Jour. Geology, v. 6, pp. 147–60.

———, 1913, Hudson Bay exploring expedition 1912: Ontario Bur. of Mines, 22nd Ann. Rept., v. 22, pt. 1, pp. 161–209.

———, 1914, The Patrician glacier south of Hudson Bay: Twelfth Intern. Geol. Cong. (1913), Comptes Rendus, Ottawa, Ontario, pp. 523–34.

―――, 1935, Patrician center of glaciation: Pan-Am. Geologist, v. 63, pp. 1–5.

United States Lake Survey: *see* Corps of Engineers, U.S. Army.

Upham, W., 1896, Origin and age of the Laurentian lakes and of Niagara Falls: Am. Geologist, v. 18, pp. 169–77.

―――, 1914, Fields of outflow of the North American ice-sheet: Twelfth Intern. Geol. Cong. (1913), Comptes Rendus, Ottawa, Ontario, pp. 515–22.

Urry, W. D., 1942, The radio-elements in non-equilibrium systems: Am. Jour. Sci., n.s., v. 240, pp. 426–36.

―――, 1949, Concentrations of the radio-elements in marine sediments of the Southern Hemisphere: Am. Jour. Sci., n.s., v. 247, pp. 257–75.

Van Hise, C. R., and Leith, C. K., 1911, The Geology of the Lake Superior region: U.S. Geol. Survey Monog. 52.

Volney, C. F., 1803, Tableau de climat et du sol des Etats-Unis d'Amérique: Paris, Imprimeur-Libraire.

von Engeln, O. D., and Caster, K. E., 1952, Geology: New York, McGraw-Hill Book Co., Inc.

Weinberg, E. L., 1948, Deep water sediments of western Lake Huron: unpublished Master's thesis, Univ. of Illinois, Urbana.

Welch, P. S., 1935, Limnology: New York, McGraw-Hill Book Co., Inc.

White, G. W., 1951, Illinoian and Wisconsin drift of the southern part of the Grand River lobe in eastern Ohio: Geol. Soc. of America Bull., v. 62, pp. 967–77.

―――, 1953, Geology and water-bearing characteristics of the unconsolidated deposits of Cuyahoga County, pp. 36–42: *in* Winslow, J. D., White, G. W. and Webber, E. E., 1953, The water resources of Cuyahoga County, Ohio: Ohio Dept. Nat. Res., Div. of Water, Bull. 26.

Whittlesey, C., 1838, Report on the geology and topography of a portion of Ohio: Ohio Geol. Survey Ann. Rept. 2, pp. 41–71.

Wilson, L. R., 1932, The Two Creeks forest bed, Manitowoc County, Wisconsin: Wisconsin Acad. Sci. Trans., v. 27, pp. 31–46.

―――, 1936, Further studies of the Two Creeks forest bed, Manitowoc County, Wisconsin: Torrey Bot. Club Bull., v. 63, pp. 317–25.

Workman, L. E., 1925, A Pleistocene section in the vicinity of the Thornton reef: unpublished Master's thesis, Univ. of Chicago.

Wright, H. E., 1955, Valders drift in Minnesota: Jour. Geology, v. 63, pp. 403–11.

Wright, S., 1931, Bottom temperatures in deep lakes: Science, n.s., v. 74, p. 413.

Youngquist, C. V., and others, 1953, Lake Erie pollution survey: Ohio Dept. Nat. Res., Final report, pp. 1–201, Supplement, pp. 1–125.

Zumberge, J. H., 1956, Late Pleistocene history of the Lake Michigan basin: Guidebook, Friends of the Pleistocene Annual Meeting, Traverse City, Michigan, p. 6.

―――― and Potzer, J. E., 1956, Late Wisconsin chronology of the Lake Michigan basin correlated with pollen studies: Geol. Soc. of America Bull., v. 67, pp. 271–88.

―――― and Wilson, J. T., 1954, Effect of ice on shore development: Proc. Fourth Conf. on Coastal Eng., Council on Wave Research, Univ. of California, Berkeley, pp. 201–5.

INDEX

Agassiz, Lake, 281-82
Algoma, Lake, 263-67
Algonquin, Early Lake, 152-62
Algonquin, Lake, 207-24
Algonquin lakes, lower, 225-29
Arkona, Lake, 146-48
Ashland, Lake, 189

Barlow, Lake, 279
Battlefield beach, 231
Bedrock formations, 13-30
Bowmanville lake stage, 172-74
Brule, Lake, 188

Calumet, Lake, 181-82
Canadian shield, 13, 16, 25, 81-83, 113
Cary glacial substage, 94, 96-99, 114, 139, 163-69, 185-87, 194-95, 278
Cedar Point, Lake, 229-33
Champlain Sea. See St. Lawrence marine embayment
Chemistry, water, 59-64
Chicago, Early Lake, 164-65
Chicago, Lake, 164-83
Chippewa, Lake, 236-42
Circulation, water, 50-57
Coal, 84
Currents, surface, 35-44

Dana, Lake, 198, 200
Dawson, Lake, 198, 201
Depths, 7-13
Deronda, Lake, 244-45
Drainage systems, preglacial, 86-89
Duluth, Lake, 186, 189-92

Erie, Lake
  depths of, 7, 11
  description of, general, 3-7
  geologic setting of, 13-16, 27
  sediments in, 65-75 passim
  stages of, 139-62, 209-11, 268
  topography of, 11, 13
  water characteristics of, 31-64 passim
Erosion, preglacial, 86-89
Evaporation, 58-59

Fort Brady, Lake, 236
Frontenac, Lake, 198, 202

Glacial lakes. See individual listings
Glaciation
  centers of, 109-10
  Pleistocene, 90-115
Glenwood, Lake, 165-81
Grand Haven River, 241-42
Grand River, 134, 144-45, 149-51, 177-78, 181-82, 212
Grassmere, Lake, 152-61
Great Lakes
  chemistry of water of, 59-64
  circulation of water of, 50-57
  currents of, surface, 35-44
  depths of, 7-13
  description of, general, 3
  evaporation in, 58-59
  geologic setting of, 13
  history of, dating events of, 116-32
  history of stages of (table), 282
  ice in, 49-50
  levels of, 46-48
  precipitation in, 58-59
  sediments in, 65-75
  seiches in, 44-45
  thermal stratification in, 50-57
  tides in, 35
  topography of, bottom, 13
  water characteristics of, 31-64
  wave action of, 31-35

Hall, Lake, 195-99
History, lake, dating events of, 116-32
History of region
  glacial, 90-115
  preglacial, 76-89

INDEX

Huron, Lake
  depths of, 7, 10
  description of, general, 3-7
  geologic setting of, 13-16, 25-27
  sediments in, 65-67 *passim*
  stages of, 139-62, 207-16, 223-38, 242-44, 248-56, 259-68
  topography of, 10, 13
  water characteristics of, 31-64 *passim*

Ice, 49-50
Illinoian age, 93
Interglacial lakes, 113-15
Iron formations, 79
Iroquois, Lake, 198, 200-202

Kansan age, 93
Keweenaw, Lake, 186-88
Kirkfield, Lake, 211-12
Kodonce, Lake, 244-45
Korah, Lake, 234-36

Levels, 46-48
Lundy, Lake, 152-61
Lutsen, Lake, 244-45

Mackinac, Straits of, 3, 20, 23-26, 163-64, 237-38, 240-42, 252
Mackinac breccia, 24-25, 27
Mackinac River, 237-38
Mankato glacial substage, 94, 99-108, 123, 186, 187, 278. *See also* Port Huron glacial substage
Marais, Lake, 244-45
Maumee, Lake, 140-45
Michigan, Lake
  depths of, 7, 9
  description of, general, 3-7
  geologic setting of, 13-16, 20-25
  sediments in, 65-75 *passim*
  stages of, 163-83, 207-12, 214-30, 237-42, 248-54, 259-68
  topography of, 9, 13
  water characteristics of, 31-64 *passim*
Minong, Lake, 246-47

Nebraskan age, 91
Nemadji, Lake, 188
Newberry, Lake, 195-98
Niagara Falls, 6, 16, 30, 156
Niagara gorge, 6, 107, 119-22, 151, 153, 155-56, 161
Niagara River, 6, 107, 155-56
Niagaran cuesta (escarpment), 20, 26, 119
Niagaran Dolomite, 15, 20-21, 23-26, 29-30
Nipissing, Lake, 248-62

Ojibway, Lake, 279, 281
Ontario, Lake
  depths of, 7, 12
  description of, general, 3-7
  geologic setting of, 13-16, 30
  sediments in, 65-75 *passim*
  stages of, 194-206
  topography of, 12-13
  water characteristics of, 31-64 *passim*
Ontonagon, Lake, 189

Paleozoic
  era, 81-85
  history, 81-85
  rocks, 14-16, 20-30, 81-85
Payette, Lake, 229-33
Penetang, Lake, 229-33
Pleistocene glaciation, résumé of, 90-115
Port Huron glacial substage, 94, 99-108, 148-50, 160-61, 170, 175-77, 185-88, 194-99, 278. *See also* Mankato glacial substage
Post-Paleozoic time, 85-89
Precambrian
  era, 77-81
  history, 77-81
  rocks, 14-15, 77-81
Precipitation, 58-59

Radiocarbon
  dates, 269-81
  method, 127, 129
  time scale, 269, 282(table)

INDEX

Saginaw, Lake, 145-46, 148
St. Clair, Lake, 6, 11, 264
St. Clair River, 6, 155-56, 158-59, 264, 266
St. Lawrence marine embayment, 198, 203-4
St. Lawrence River, 7, 202-3, 259
St. Lawrence valley, 278-79
St. Marys River, 3, 185, 220-21, 247, 266
Salt formations, 23-24, 27, 84
Scour of basins, glacial, 111-12
Sediments, 65-75
Seiches, 44-45
Sheguiandah, Lake, 234-36
Stages. *See* individual listings
Stanley, Lake, 236-37, 242
Superior, Lake
  depths of, 7-8
  description of, general, 3-7
  geologic setting of, 13-19
  sediments in, 65-75 *passim*
  stages of, 184-93, 208, 217-23, 243-49, 253-59, 264-66, 268
  topography of, 8, 13
  water characteristics of, 31-64 *passim*

Thermal stratification, 50-57
Tides, 35
Time scale
  geologic, 76, 78
  radiocarbon, 269, 282(table)
Tofte, Lake, 244-45
Toleston, Lake, 182-83
Topography, bottom, 7-13
Two Creeks interval, 102-7, 150, 170-74, 187, 199, 269-71, 277, 282

Valders glacial substage, 94, 99-108, 121, 123, 125, 151, 160-61, 175-83, 187-88, 199-200, 203, 271, 277-79, 282
Vanuxem, Lake, 195-99
Varve correlation, 122-24

Warren, Lake, 149-52, 198-200
Water
  characteristics of, 31-64
  chemistry of, 59-64
  circulation of, 50-57
Wave action, 31-35
Whittlesey, Lake, 148-49
Wisconsin age, 93-108
Wyebridge, Lake, 229-33